WATER SUPPLY AND WASTE WATER ENGINEERING

For Polytechnic Students

AF656440

D.Lal
A.K. Upadhyay

WATER SUPPLY AND WASTE WATER ENGINEERING

A.K. Upadhyay

B.E. (Civil), M.E. (Soil and Foundation Engg.)

D.Lal

S.K. KATARIA & SONS®

Publisher of Engineering Books

4885/109, Prakash Mahal, Dr. Subhash Bhargav Lane,
Opposite Delhi Medical Association, Daryaganj, New Delhi-110002 (INDIA)
Phones: +91-11-23243489, 43551243; Mobile: +91-9871775858
e-mail: katariabook@gmail.com; katariabooks@yahoo.com
Website : www.skkatariaandsons.com

Published by:
S.K. Kataria & Sons®
4885/109, Prakash Mahal, Dr. Subhash Bhargav Lane,
Opposite Delhi Medical Association, Daryaganj, New Delhi-110002
Phones: +91-11-23243489, 43551243; Mobile: +91-9871775858
e-mail: katariabook@gmail.com; katariabooks@yahoo.com
Website: www.skkatariaandsons.com

First Edition: 2001; Second Edition: 2002, Third Edition: April, 2016 Reprint : 2017
Reprint : 2018 Reprint: 2019 Reprint: 2021 Reprint: 2023

Price: ₹ 395/-

Laser Typesetting: Plus Computers, Delhi *Printed at :* AL AZIZ BOOK BINDING HOUS

Preface to the Third Edition

This book has been thoroughly revised as per the suggestions received from the readers. Numerous additions have been made in various chapters to cover the syllabus of "Boards of Technical Education" of different states. Model questions, MCQ's, fill in the blanks questions given at the end of each chapter will be very helpful for the students to test themselves.

Any errors and suggestions for the improvement of this book, brought to our notice, will be highly appreciated.

Authors

Dedicated
to
Our Parents

Contents

PART-B : Waste Water Engineering 203–392

PART-A

WATER SUPPLY ENGINEERING

1. Introduction
2. Quantity of Water
3. Sources of Water
4. Quality of Water
5. Water Treatment
6. Conveyance of Water
7. Laying Out Pipes
8. Building Water Supply
9. Water Pollution Control and Management

Objective Type Questions with Answers

CHAPTER 1

INTRODUCTION

Learning Objectives

- General
- Necessity
- Water Supply Engineering
- Importance of Water Supply Engineering
- Water Supply System
- Requirements of Water Supply System
- Hydrological Cycle
- Quality of Water for Public Water Supply
- Duties of Public Health Engineer

1.1. GENERAL

Water is the most essential commodity for the sustenance of life. Water makes everything grow but, there is a saying "water water every where but not a drop to drink." It emphasises the importance of portable water. Water for human consumption ought to be pure and wholesome. Water plays an important role in making human life not only survive but comfortable and luxurious. Water is used for drinking, cooking, bathing, air-conditioning (deser coolers etc.), flushing of sewers, fire fighting, entertainment (water sports), industrial processes, power generation etc.

Water is and has been used for human consumption, since ages. Even in ancient cities there used to be a network of water supply and its proper disposal. Water then was carried through earthen and stone pipes and was usually untreated. But now-a-days the importance of treated water is well understood and it is treated (usually) before supplying it to community. With the advancement in technology new methods of treatment devices have been evolved and are being adopted. Water now-a-days is conveyed through plastic/PVC pipes which are more durable and do not corrode. Different types of pumps have been invented to lift water for supply.

This chapter deals with the necessity and importance of water supply. It also deals with the requirements of water supply system and water cycle.

1.2. NECESSITY

After air the water is the most essential commodity for the continuation of life. Water is also a cleaning agent. An adequate and clean water supply is the basic requirement for domestic use. Water is also required for various types of industries.

Water in good quantity is required for public use, like flushing of sewer, steel sprinkling, parks and grounds, fire fighting etc.

1.3. WATER SUPPLY ENGINEERING

The branch of civil engineering which deals with the supply of water for various purposes e.g. domestic, industrial, commercial, and public is called water supply engineering.

1.4. IMPORTANCE OF WATER SUPPLY ENGINEERING

Water supply is used for the following purposes:

(*i*) Drinking and cooking

(*ii*) Bathing and washing

(*iii*) Street washing

(*iv*) Watering of lawns and gardens

(*v*) Fire extinguishing

(*vi*) Swimming pools and display of fountains

(*vii*) Air-conditioning and heating system

(*viii*) Industries and irrigation

(*ix*) Disposal of household sewage

(*x*) Recreational purposes

1.5. WATER SUPPLY SYSTEM

The complete layout of getting water from the source, treating it for removing impurities and then, distributing it to the public for various uses, is called water supply system.

To ensure adequate supply of potable water a well organised water supply system is essential. The engineer has to work out the demand for domestic, commercial and public use and has the knowledge of the source of supply which can provide adequate supply to meet the demand of the city.

Rain is important source of supply. Water storage, streams, lakes etc., are the surface sources of supply whereas wells, springs, infiltration galleries are the goundwater supplies. The quantity of water depends upon the sources of supply. The goundwater source may not contain much impurities but the surfaces normally contain impurities and water requires treatment before use.

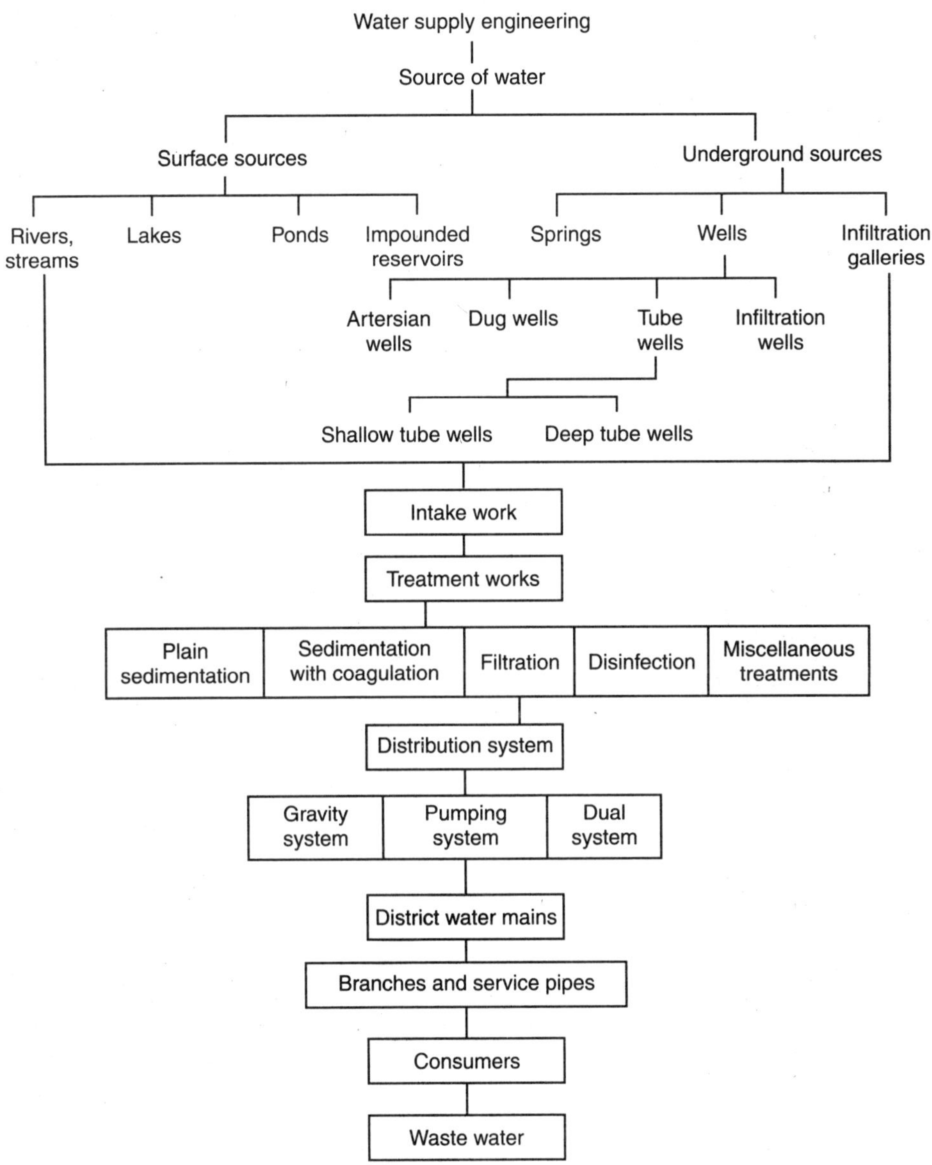

Fig. 1.1. Flow diagram of water supply system.

The transportation of water from source to treatment plants and then to distribution system requires skill and management technique to make it economical. The arrangement from the source of supply to distribution and to the consumers is known as the water supply system and a public health engineer must have full knowledge of all the operation to make a success of water supply system.

1.6. REQUIREMENTS OF WATER SUPPLY SYSTEM

It should satisfy the following requirements:

(*i*) Water should be available in adequate quantity.

(*ii*) Water to be supplied should be free from impurities and bacteria.

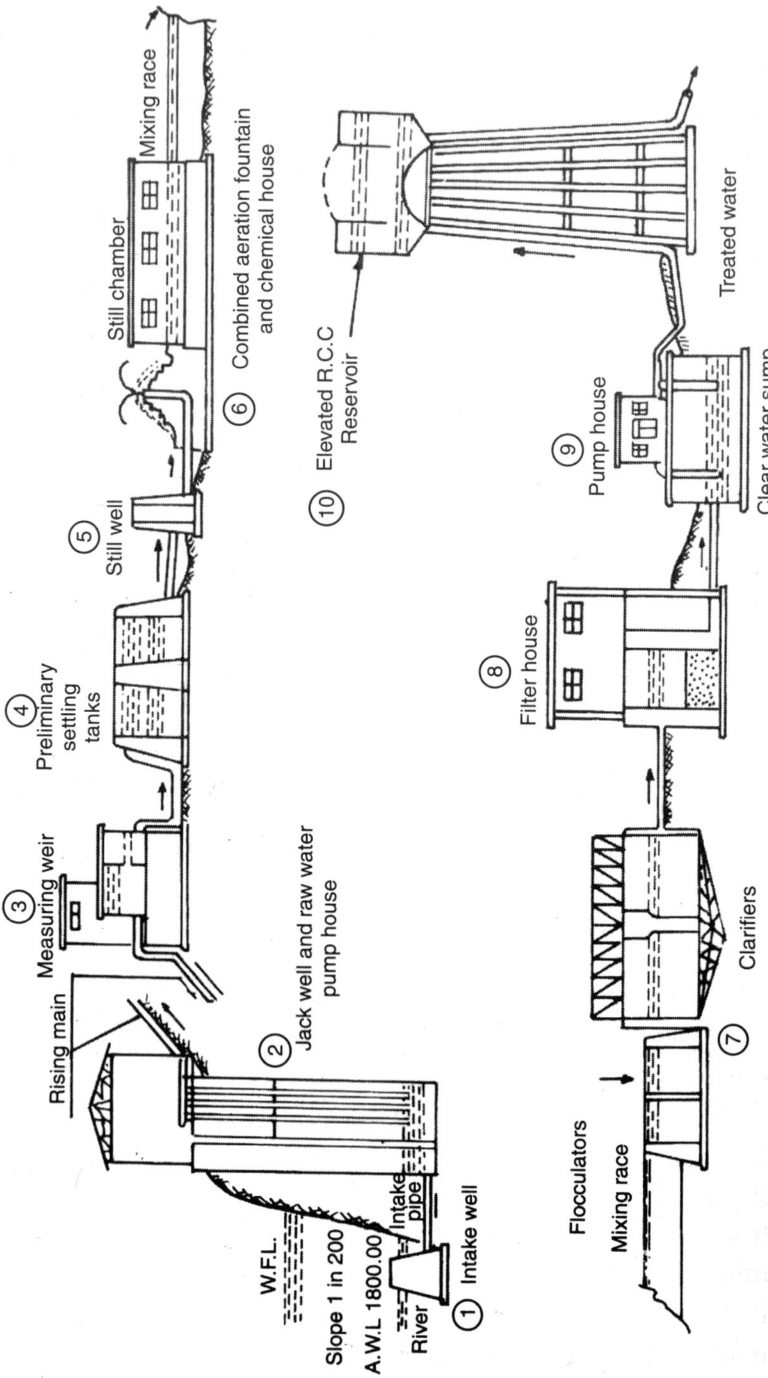

Fig. 1.2. General layout of water works

(*iii*) It should be cheap.
(*iv*) Water should be readily available to the consumers.
(*v*) It must be readily disposed off after it has served its purpose.

1.7. HYDROLOGICAL CYCLE

Fig. 1.3 shows the hydrological cycle of water. Water from earth's surface rivers, lakes etc., evapourates due to sun's heat and rises upward. Due to reduction in the atmospheric pressure at high altitude, these water vapours expand by absorbing energy from the surrounding air. The air cools down and when the temperature falls below the dew point, the excessive moisture starts falling in the form of rain, dew, frost, hails or precipitation. This processes of evaporation and precipitation continue for ever and such a cycle is known as hydrological cycle or water cycle.

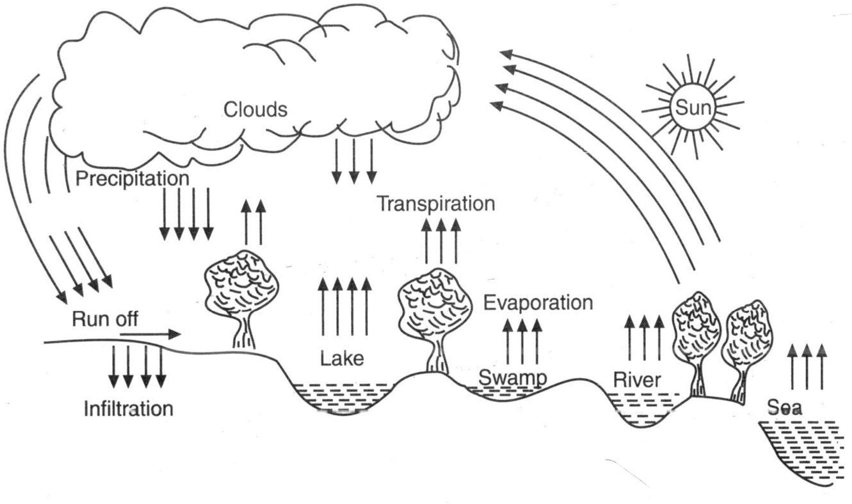

Fig. 1.3. Hydrological cycle

1.8. QUALITY OF WATER FOR PUBLIC WATER SUPPLY

Absolutely-pure water is never found in nature. Water served to the consumers must have the following qualities:

(*i*) It should be cool and colourless.
(*ii*) It should be odourless and tasteless.
(*iii*) It should be free from disease carrying bacteria.
(*iv*) It should be fresh.
(*v*) It should be free from excessive amounts of organic matter and mincral.
(*vi*) It should not corrode the pipe.
(*vii*) It should be free from poisonous materials.

1.9. DUTIES OF PUBLIC HEALTH ENGINEER

The duties of public health engineers are:

1. He should be well conversant with the planning, designing, construction and maintenance of water works.
2. He should be capable to design the water supply schemes in the best possible way with maximum economy and efficiency.
3. He should be able to protect the source of water as well as treated water from contamination.
4. He should be capable to alter the method of purification depending on the types of impurities and bacteria present in the water.
5. He should be capable to operate the water works without fail.
6. He should supply potable water to the public at required pressure at various points in the locality.
7. He should be capable of performing laboratory tests of water samples to check its quality and presence of any disease bacteria.
8. Public health engineer should be aware of latest techniques of water supply engineering.
9. He must be aware of latest purification and distribution methods of water.

MODEL QUESTIONS

1. **Fill in the blanks:**
   (*i*) The demand of water will be ______ if the population increases.
   (*ii*) During fire, the demand rate of water is ______.
   (*iii*) Water served to the consumers should be ______ from disease carrying bacteria.
   (*iv*) In case of water supply system water should be available in ______ quantity.
   (*v*) Water served to the consumers should be ______ from poisonous materials.
2. What do you understand by water-supply engineering?
3. Why water supply system is needed in the city?
4. Briefly describe the water supply system with the help of flow diagram (only water cycle).
5. Give a flow line diagram of a canal based water supply scheme.
6. Briefly describe, using a flow diagram, the various stages involved in the purification of public water supply for a town from a near by river.

□□□

CHAPTER 2

QUANTITY OF WATER

Learning Objectives

- Water Demand
- The per Capital Demand
- Factors Affecting Per Capita Demand
- Variation in Rate of Demand (or Consumption)
- Population: Present and Future
- Methods of Forecasting Population
- Total Quantity of Water for a Town

2.1. WATER DEMAND

During planning a water supply scheme, it is the duty of the engineer to carefully examine the various types of water demand a city may have. The demand of water may be broken into the following classes:

(*i*) Domestic water demand

(*ii*) Industrial water demand

(*iii*) Demand for public uses

(*iv*) Fire demand

(*v*) Water required to compensate the losses in wastes, theft etc.

For correct estimate of total water demand all the above demands should be considered.

(*i*) Domestic Water Demand: This includes the water required in private building for cooking, washing, bathing, gardening, sanitary purposes etc. The domestic water demand depends upon the living conditions of the consumer. As per IS : 1172-1963 water required for domestic purposes for average Indian conditions per head per day may be taken as 135 litres. Table 2.1 shows the details of water requirement for domestic purposes. The total domestic water consumption may amount to 50 to 60% of the total water consumption.

Table 2.1. Detail of water requirement for domestic purposes

S.No.	Description	Consumption of water per head per day in litres
1.	Bathing	55
2.	Washing of clothes	20
3.	Drinking	5
4.	Cooking	5
5.	Washing of utensils	10
6.	Washing of house	10
7.	Flushing of W.C.	30
	G. Total	135 Litres

Table 2.2. (a) Water demand for factories

S.No.	Factory products	Litres/kg. of products
1.	Coal mining	14
2.	Sugar	10
3.	Oil refining	12
4.	Butter	13
5.	Cheese	20
6.	Steel	5
7.	Glass	75
8.	Paper	170
9.	Synthetic fibres	225
10.	Ice	1.85

Table 2.2. (b) Water requirement for public use

Purpose	Water requirements
Public parks	1.5 litres/m^2/day
Street washing	1.0–1.5 litres/m^2/day
Sewer cleaning	5.0 litres/head/day

Table 2.2. (c) Water requirements for live stock

S.No.	Live stock	Water requirements in litre/animal/day
1.	Cows	70
2.	Dogs	16
3.	Chickens	0.1
4.	Goats	12
5.	Horses	40
6.	Sheep	12

Table 2.2. (d) Water requirements for irrigation in towns/cities

S.No.	Purpose	Water requirements
1.	Public parks	17000 litres/hectare/day
2.	Private gardens	17000 litres/hectare/day
3.	Roadside trees	28000 litres/km/day

(ii) **Industrial Water Demand:** This consumption includes water used in factories, offices, hotels, hospitals etc. This demand depends upon the nature of the city, number and types of industries. On an average, 20 to 25% of the total water demand may be allowed for this type of demand in the design. The approximate quantity of water required for some of the manufacturing units are shown in Table 2.2.

(iii) **Demand for Public Use:** Public demand includes the quantity of water required for public utility purposes, such as watering of public parks, gardening, sprinkling on roads, use in public fountains etc. In many water supply schemes these demands are not believed as essential and a nominal among not exceeding 5% of the total demand is kept on an arbitrary basis.

(iv) **Fire Demand:** It is the quantity of water required for fighting a fire outbreak. For high value cities, water requirement for this purpose is particularly essential. The quantity of water required for this purpose can be found out by applying certain empirical formulae. The most common are:

(a) National Board of Fire Underwriters Formula:

$$Q = 1020\sqrt{P}(1 - 0.1\sqrt{P})$$

When converted into metric units then it is expressed as

$$Q = 64\sqrt{P}(1 - 0.01\sqrt{P}) \text{ in litres per sec.}$$

(b) Free-Man's Formula:

$$Q = 1136.5\left(\frac{P}{10} + 10\right)$$

where Q = Fire demand in US gallons per minute

P = Population in thousands

The above formulae usually give quite high results which is not suitable for Indian conditions. Therefore, following formula of Kuichling gives satisfactory results.

$$Q = 3182\sqrt{P} \text{ litres per minute.}$$

The above high quantity of water for fire fighting is required few times in a year and that too short time of 4 to 5 hours only. If this quantity is calculated on yearly basis, it will be very small amount which will not much affect the total requirement. The minimum pressure at fire hydrant should be 1–1.5 kg/sq. cm. which should be available even after 4 to 5 hours of constantly using fire hydrant.

For Indian conditions a moderate allowance of one litre-per head per day for fire demand may be quite sufficient.

Let us assume that in a city whose population is about 40 lakhs, four fires break out in a day. Let the duration of each fire is four hours. Each hydrant has three streams, one stream for the property which is burning and one each for adjacent-property on either side of the burning property. The discharge of each stream is 1100 litres/minute.

The Total Quantity of water required = 3 × 1100 × 4 × 60 × 4

= 31,68,000 litres

Which is very nominal. See table 2.3 for details of required fire flow, as per National Board of Fire Underwriters:

Table 2.3. Details of require fire flow
(As per National Board of Fire Underwriters)

Population	Required fire flow in average city in litres per minute
1,000	4,500
2,000	6,750
4,000	9,000
6,000	11,000
10,000	13,500
17,000	18,000
22,000	20,250
40,000	27,000
60,000	31,500
80,000	36,000
1,00,000	40,500
1,25,000	45,000
1,50,000	49,500
2,00,000	54,000

Note: (*i*) The fire flow requirement is the controlling factor in the design of a water distribution on system for a small city.

(*ii*) National Board of Fire Underwriters requires provision of 5 hours fire-flow in places of less then 2500 population and provision for 10 hours flow in largest places.

EXAMPLE 2.1. Determine fire-flow and storage required for fire protection in a city of 40,000 population according to the requirements of the National Board of Fire Underwrites?

Solution: Refer Table 2.3

Required fire flow = 27000 litres/minute for 10 hours.

$\therefore$ Fire flow to be provided by storage or by

$$\text{other means} = 27000 \times 60 \times 10 = 1{,}62{,}00{,}000 \text{ litres}$$

This must be in addition to a reserve for normal used during the fire.

(*v*) Water required to compensate losses in wastes, thefts etc.: This includes the quantity of water lost in leakage at the joint and fittings, stolen water due to unauthorised water connections, etc. These losses can be reduced by providing good fittings, proper inspection and investigation against water thefts. This quantity is generally taken as 15% of total water consumption for a well managed water works.

2.2. THE PER CAPITA DEMAND

It is the annual average amount of daily water required by one person and includes the domestic, industrial and pubic use and the water lost in thefts and wastes.

If Q = total quantity of water required by a city per year (in litres)

P = Population of the city

then per capita demand (in litre per day) = $\dfrac{Q}{P \times 365}$

2.3. FACTORS AFFECTING PER CAPITA DEMAND

The various factors which affect the per capita demand are:

1. **Climatic conditions:** Water requirements during summer are more than winter. During summer everybody takes bath twice or thrice and washes clothes, more water is used for drinking and also more water is consumed in running coolers etc. Hence, water consumption is much more in summer than that in winter.
2. **Size of city:** Generally the demand of water per head will be more in big cities than that in small cities. In big cities lot of water is required for maintaining clean and healthy environments while in small towns it is not required.
3. **Habits of people:** High class community uses more water due to their better standard of living and higher economic status. Middle class people use water at average rate. And for poor people a single water tap may be sufficient for several families.
4. **Industries:** More water will be required in highly industrialised city.
5. **Cost of water:** More costly is the water less will be rate of demand. Hence the cost at which water is supplied to the consumer may also affect the rate of demand.

6. **Quality of water:** A water works system having a protected and good quality of water supply would always be more popular with consumers. Hence more quantity of water will be consumed if the quality is good.
7. **Pressures in the Distribution System:** These would be of great importance in the case of localities having a number of two or three storeyed buildings. Adequate pressure would mean an uninterrupted and constant supply of water.
8. **Sewage Facilities:** Town having water carriage system will consumer more water.
9. **System of Supply:** The system of water supply may be continuous or intermittent. In continuous system water is supplied all the twenty four hours while in the case of intermittent system, water is supplied for certain fixed hours of the day only, result in some reduction in the consumption. This may be due to decrease in losses and other wasteful use.
10. **Method of charging:** In the town where metering is done, less quantity of water will be used than the city without metering system. A metered supply ensures minimum of waste as the consumer then known that he has to pay for the water used by him and consequently is more careful in use.

Table 2.4. Effect of population on rate of consumption

Population	Consumption in Litres per capita per day
Upto 5,000	90
5,000–20,000	110
20,000–50,000	135
50,000–2,00,000	180
Over 2,00,000	200

2.4. VARIATION IN RATE OF DEMAND (OR CONSUMPTION)

The per capita demand, which we have studied in the previous articles, is the average consumption of the year. In practice it has been seen that this demand does not remain uniform throughout the year but it varies from season to season, even from hour to hour. Variation in rate of demand may be termed as:

(*i*) Seasonal or monthly variations
(*ii*) Daily variation
(*iii*) hourly variation

(*i*) Seasonal Variation: The water demand varies from season to season. In summer the water demand is maximum, because people will use more water in bathing, cooling, lawn watering, street sprinkling, etc. This demand goes on reducing and in winter it becomes minimum, because less water will be used in bathing and there will be no lawn watering. This fluctuation may be upto 150% of the average annual consumption.

***(ii)* Daily Variation:** The rate of demand for water may vary from day to day also. This is due to habits of the consumers, climatic conditions, holidays etc. Water demand on Sundays is generally more than other days of the week. On Sunday everybody takes bath lisurery, washes his clothes etc. Moreover, on the day of mass marriages or some fair, the rate of water demand will be more. On hot and dry day, water requirements will be more as compared to a rainy day. The maximum daily consumption may be much as 180 percent of the average annual consumption.

***(iii)* Hourly Variation:** The rate of demand for water during 24 hours does not remain uniform and it varies according to hours of the day. A typical graph showing hourly rate of consumption is shown in Fig. 2.1.

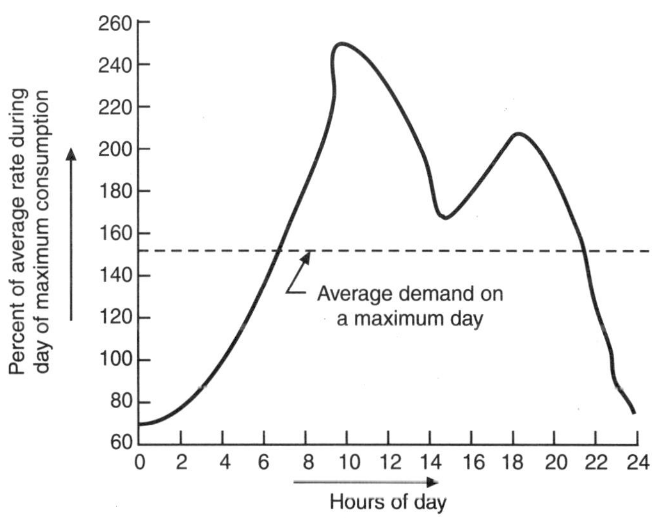

Fig. 2.1. Hourly variation

On Sundays and other holidays the peak hours may be about 8 a.m. due to late awakening whereas it may be 6 a.m. on the other workings days. The peak flow hours may be between 6 a.m. – 10 a.m., and 4 p.m. – 8 p.m. and minimum flow may be between 12 p.m. – 4 a.m. Certain industries may be working in day and night shifts and consuming more water. It is, therefore, necessary to study the characteristics of the district before estimating upon rate of consumption. The maximum hourly consumption may rise upto 200% that of average daily demand.

2.5. POPULATION: PRESENT AND FUTURE

After deciding the quantity of water required by an individual, the next step in the design of water supply scheme will be to find the total quantity required by a community. Knowing the present population from the most recent census, it is possible to determine the present demand, but for the design purpose future demand is required. Generally population will be growing every year and hence there will be growing demand for water also. The design period taken is generally 20 to 30 years.

2.6. METHODS OF FORECASTING POPULATION

The following are the methods used for population forecast:

***(i)* Arithmetical Increase Method:** In this method, the increase in population is assumed to be constant and an average increase of the last 4 to 5 decades is calculated and added in the present population to determine population of the next future decade. The population can be found out at the end of *n* year or *n* decades.

$$P_n = P + n \times i$$

where P = Present population

i = Yearly or per decade increase of population.

***(ii)* Geometrical Increase Method:** In this method the average percentage of growth of last few decades is determined, the forecasting is done on the basis that percentage increase per decade will be the same. Thus the population at the end of *n* years or decades is given as:

$$P_n = P\left(1+\frac{i}{100}\right)^n$$

where i = Yearly or per decade percentage rate of increase.

***(iii)* Incremental Increase Method:** This method is an improvement over the arithmetical increase method. The average increase in the population is calcated by the arithmetical method and to this is added the average of the net incremental increase once for each future decade.

EXAMPLE 2.2. Field Problem: The following data have been obtained from the census department regarding population of HARYANA Town:

Year	1960	1970	1980	1990
Population	12,000	17,000	22,500	28,500

Calculate the probable population in the year 2000, 2010 and 2020.

Solution: The average increase in population after every decade

$$= \frac{1}{3}(5000 + 5500 + 6000)$$

$$= 5500 \text{ persons.}$$

The probable population in various decades will be calculated by the formula.

$$P_n = P + n \times i$$

Year	Probable	Population
2000	28500 + 1 × 5500 =	34,000
2020	28500 + 3 × 5500 =	45,000

EXAMPLE 2.3. A growing Town has population as given by the census Deptt. as under:

Census Year	1941	1951	1961	1971	1981	1991	2001
Population	24,831	25,293	25,423	27,263	38,284	49,909	67,105

Estimate the population of the Town as in Apring 2021.

Solution: According to arithmetic increase method we known

$$P_n = P + n \times i$$

where n = Number of decades

$$i = \frac{1}{6}(462 + 130 + 1840 + 11021 + 11625 + 17196)$$

$$= \frac{1}{6}(42274) = 7046 \text{ persons}$$

Year	Probable	Population
2011	67105 + 1 × 7046 =	74,151
2021	67105 + 2 × 7046 =	81,197

Hence the population of the Town in 2021 will be 81,197 persons.

***(iv)* Changing Rate of Increase:** It is similar to the Geometrical Increase method except that a changing rate rather than a constant rate of increase is assumed. The changing rate for large and grown up cities is usually considered to be a decreasing rate. This method gives quite rational results.

***(v)* Graphical Method:** In this method the population of last few decades (as per census) is correctly plotted to the scale on the graph-paper. The curve is smoothly extended to find the future population. Sometimes population trend lines are plotted and similar expected population trend lines are drawn.

***(vi)* Comparative Method:** Cities having similar characteristics to the city whose population is to be estimated, are selected. Then on the basis of those cities, the population of the required city is forecasted. Graph may be plotted if necessary.

EXAMPLE 2.4. Estimate the population in the year 2020 from the following census data for 'Nilokheri' Town by using arithmetical increase method, geometrical increase method and incremental increase method.

Year	1970	1980	1990	2000
Population in Thousands	10	14	19	25

Solution:

Year	Population	Increase	Incremental Increase	% age Increase
1970	10,000	–	–	–
1980	14,000	4,000	–	40.00
1990	19,000	5,000	1,000	35.70
2000	25,000	6,000	1,000	31.60
Total	...	15,000	2,000	107.30
Average	...	5,000	1,000	35.76

(*i*) Arithmetic Increase Method

Population in 2020 = 25000 + 2 × 5000 = 35,000

(*ii*) Geometrical Increase Method

$$\text{Population in 2020} = 25000 + \left(1 + \frac{35.76}{100}\right)^2 = 46{,}077$$

(*iii*) Incremental Increase Method

Population in 2020 = 35000 + 2 × 1000 = 37,000

The results of different methods are given below:

Method Used	Probable Population (YEAR)	
	2010	2020
(*i*) Arithmetic Increase	30,000	35,000
(*ii*) Geometrical Increase	33,940	46,077
(*iii*) Incremental Increase	31,000	37,000

EXAMPLE 2.5. Assuming the geometric rate of growth of population of town, calculate the population of the town in the year 2021 and 2031 with the help of the following census records of the population.

Year	1971	1981	1991	2001	2011
Population in Thousands	258	495	735	1250	2148

Solution:

Year	Population	Increase	Percentage Increase
1971	2,58,000	–	–
1981	4,95,000	2,37,000	91.86
1991	7,35,000	2,40,000	48.48
2001	12,50,000	5,15,000	70.07
2011	21,48,000	8,98,000	71.84
Total		18,90,000	282.85
Average		4,72,500	56.45

Using Geometrical Increase Method:

Year **Population**

$$2021 = 2148000\left(1+\frac{56.45}{100}\right)^1 = 3360546$$

$$2031 = 2148000\left(1+\frac{56.45}{100}\right)^2 = 5257574$$

2.7. TOTAL QUANTITY OF WATER FOR A TOWN

As we have discussed above that the water supply project is required for domestic, industrial, trade purposes etc. For estimating the total quantity of water required by a town, the water used for various purposes should be added and 30 to 40 percent wastage allowance is also given. But this total requirement will vary from town to town which depend on many factors e.g., climatic conditions, habits of people, industry, cost of water etc. For an average Indian town, the water requirement may be as follows:

(*i*)	Domestic uses (As per BIS 1172–1963)	= 135 litres/head/day
(*ii*)	Industrial use	= 50 litres/head/day
(*iii*)	Public use	= 15 litres/head/day
(*iv*)	Fire Fighting	= 15 litres/head/day
(*v*)	Thefts, Wastages and losses	= 55 litres/head/day
	Total	= 270 litres/head/day

This 270 litres per capita demand will be multiplied with the population of the town to obtain the total quantity of water required.

Table 2.5. Consumption rate of water (App.) in some Indian towns

Town	Consumption, lit./ capita per day	Town	Consumption, lit./ capita per day
Agra	120	Hydrabad	215
Allahabad	135	Kanpur	210
Bangalore	100	Lucknow	110
Bhopal	110	Chennai	170
Mumbai	320	Mathura	125
Kolkata	200	Patna	120
Delhi	175	Poona	280
Gaya	205	Varanasi	120

MODEL QUESTIONS

1. **Fill in the blanks:**
   (*i*) Is consumption ______ throughout the day? (Constant/variable)
   (*ii*) Per capita consumption of water supply in Urban area is ______ and in Rural area is ______.
   (*iii*) Future population is considered for ______ years.
   (*iv*) Design period for water supply system in India is ______ years.
   (*v*) Rate of consumption in litres per capita per day for a town having 40,000 population ______. (90/110/135)
   (*vi*) Domestic requirements amounts to about ______ of the total consumption. (50/15/10%)
2. Why water supply system is needed in the city?
3. What do you understand by the term 'per capita' demand? In a town or city for what purposes generally water is required?
4. What allowances would you make for the following:
   (*i*) Domestic use
   (*ii*) Industrial use
   (*iii*) Fire protection
   (*iv*) Public use like gardens, parks, streets etc.
5. Mention different methods of forecasting the future population of a given town. Describe in detail any one of the methods.
6. Give the difference between:
   (*i*) Domestic and Industrial use of water.
   (*ii*) Seasonal and Daily Variation.
   (*iii*) Arithmetical Increase and Geometrical Increase method of population forecast.
   (*iv*) Water requirements for schools and hospitals.
7. Write advantages and disadvantages of metering policy of water supply.
8. What is fire demand? Give a suitable allowance in a water supply scheme.
9. Mention the factors that influence per capita demand.
10. (*a*) What do you mean by Water Allowance?
    (*b*) What water allowance would you propose for:
      (*i*) A village in Hissar District.
      (*ii*) A town like Nilokheri.
      (*iii*) A modern residential colony (Officers Colony at Medical College Campus Rohtak). Please give detailed justifications for your proposals.
11. Assuming the geometric rate of growth of population of town, calculate the population of the town in the year 2021 with the help of the following census records of the population:

Year	Population
1971	12,000
1981	17,000
1991	22,500

12. (*a*) How do you estimate the quantity of water required while arranging a water supply scheme for a Town.

(*b*) How does requirement of water vary from season to season, day to day, hour to hour and in what way does it effect different components of a water supply scheme?

13. (*a*) What are the various factors which govern the "per capita" demand of water supply?

(*b*) Discuss briefly how the quantity of water required in a Water Supply Project is determined.

14. The population of a city in various years was as follows:

1961	1971	1981	1991
66,350	1,32,070	1,94,150	3,00,610

Calculate its population in the year 2021 by:

(*i*) Geometric Increase Method.

(*ii*) Arithmetic Increase Method.

(*iii*) Incremental Increase Method.

15. What factors influence the value of demand rate?

16. (*a*) What is meant by per capita demand. Give the per capita consumption of water for Domestic, Industrial, Public and Fire Fighting as per BIS.

(*b*) Discuss the variation of water demand.

❑❑❑

CHAPTER 3

SOURCE OF WATER

Learning Objectives

- General
- Sources of Water
- Rainfall
- Classification
- Surface Sources
- Quality and Quantity of Surface Water
- Occurrence
- Ground Water/Sub Surface Sources
- Quality and Quantity of Groundwater
- Sub-surface Sources of Water
- Choice of Source for Water Supply Scheme
- Well
- Classification of Wells
- Yield Measurement of a Well
- Water Table Well or Gravity Well
- Confined Water Well or Artesian Well
- Specific Yield
- Numerical Example
- Well Tests
- Comparison between Surface and Sub-surface/Groundwater
- Intakes
- Location of Intakes
- Types of Intake

PART-A : SURFACE WATER

3.1. GENERAL

As discussed earlier for efficient water supply scheme, we require permanent source of water. The capacity of the source should be such as to meet all the water requirements of the people. The best available source should be selected for metting the demand.

3.2. SOURCES OF WATER

Precipitation in the form of rain, snow, dew etc., which falls on the ground is the primary source of water. Most important part of precipitation is rainfall which is carried off in different ways as maintained below:

(*i*) **Run-off:** Run-off is that portion of rainfall which flows on the ground and reaches rivers and streams.

(*ii*) **Percolation:** It is that portion of rainfall which penetrates into the ground and is the part of ground-water.

(*iii*) **Transpiration:** Transpiration is that part of rainfall which is taken up and disposed off by the leaves of trees.

(*iv*) **Evaporation:** The portion of rainfall which is lost to the atmosphere due to heat of sun.

3.3 RAINFALL

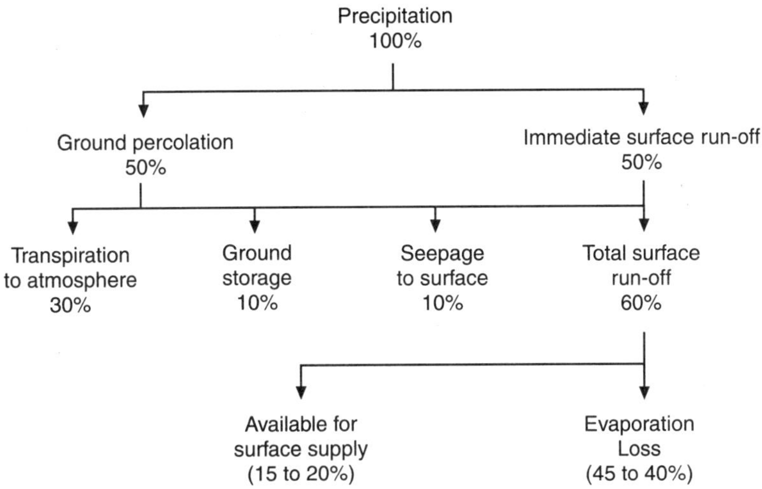

Fig. 3.1. Flow diagram showing rainfall distribution

3.4. CLASSIFICATION

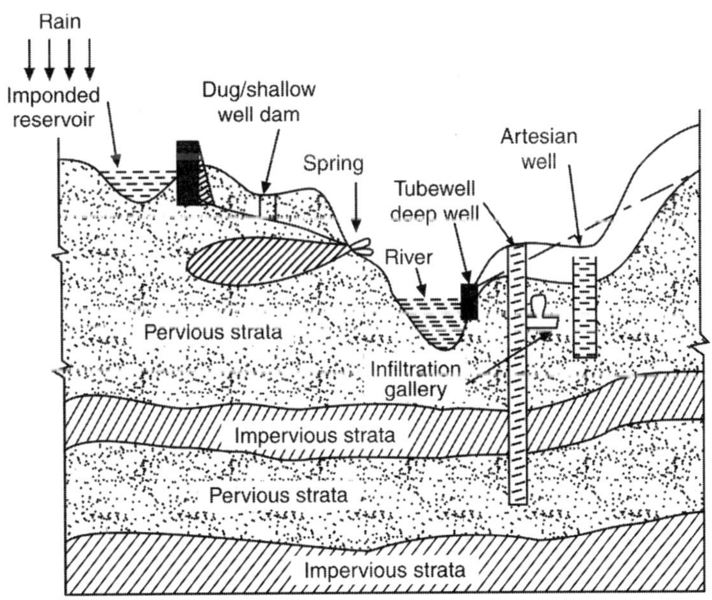

Fig. 3.2. Various source of water

The sources of water supply may be generally classified as:

1. **Surface Sources**

   (*a*) Rivers (*b*) Lakes

   (*c*) Streams (*d*) Ponds

   (*e*) Impounding reservoirs

2. Ground Sources

(*a*) Spring

(*b*) Infiltration galleries

(*c*) Wells

We will study only about surface sources of water in this chapter.

3.5. SURFACE SOURCES

The sources of water in which the water flows over the earth's surface are called surface sources.

The sources are mainly classified as:

(*i*) Rivers, Streams and Lakes: As discussed earlier they are formed by rainfall runoff i.e., rainwater flowing along the ground into these natural drainage depressions. Quantity varies depending on the catchment. Hilly rivers have good quantity of water while that of rivers in the plains is very bad. River water should always be used after necessary treatment.

(*ii*) Ponds: Ponds are depressions in plains, in which water is collected during rainy season. Generally the quantity of water in pond is very small and contains large amount of impurities, hence it is never recommended for water-supply purpose.

(*iii*) Impounding Reservoir:

(*a*) **Definition:** An impounding reservoir may be defined as an artificial lake created by the construction of a dam across a valley containing a water course. The object is to impound or store a portion of the stream flow so that it may be used for water supply. The reservoir consists of three parts:

(*i*) a dam to hold back water

(*ii*) a spillway through which excess stream flow may discharge

(*iii*) a gate chamber containing the necessary valves for regulating the flow of water from the reservoir.

(*b*) **Location:** The following points affect the location of the impounding reservoirs:

(*i*) The quantity of water available. It should be sufficient to meet all the demands throughout the year. This will depend on the intensity of rainfall, run off and catchment area.

(*ii*) The quality of water available. It should be seen that at the site bed soil should not contain soluble salts, minerals which may affect the quality of water.

(*iii*) Existence of suitable dam site. The shortest dam to impound the required volume of storage is the best. This would be possible if the river flows through a narrow gauge and the valley rapidly widens upstream from the site.

(*iv*) Distance and elevation of the reservoir with reference to the point of distribution. A longer distance means greater cost of the conduits while proper elevation of the reservoir ensures adequate supplies through gravity flow.

(*v*) Density and distribution of population over the catchment area. To avoid stream pollution there should be small density of population per sq. km. on the catchment area above the reservoir.

(*vi*) Existence of towns, highways, cultivated land and railways. This should be excluded from the submerged area of the reservoir.

(*vii*) Freedom of water shed from swamp areas. The latter form extensive growth of micro-organisms inside water, rendering it more difficult to be treated, hence objectionable.

(*viii*) Geological conditions of the storage basin. If the rocks are porous and deeply fissured, there will be considerable loss of water due to percolation.

3.6. QUALITY AND QUANTITY OF SURFACE WATER

Rivers, streams etc., give large supplies of water but real quantities vary according to seasons. Discharges of rivers etc., are observed to give an idea of the quantity of water available from them.

Surface water is generally of bad quality, being highly contaminated and charged with silt and suspended impurities. In reservoirs, the suspended matter tends to settle down. But in these reservoirs there is organic and vegetable growth, algae and weeds which produce bad smell. colour and taste in water. In case a river gets sweage discharge from a City or Town it becomes immediately highly contaminated. Therefore, the water supply inlets must be placed in the river upstream side of the point of entry of such discharges.

PART-B : SUB-SURFACE WATER

3.7. OCCURRENCE

A portion of the rain that falls on the ground, percolates through it and fills the interstices of the sub strata until the water re-appears at the surface to join a river, a lake or the sea.

Conditions affecting the amount of rainfall that may percolate into the ground include:

(*i*) The amount and intensity of rainfall.

(*ii*) The distribution of rainfall.

(*iii*) The topography and permeability of the ground surface.

(*iv*) And the water capacity of the ground.

Groundwater is an important source of water for industrial and public water supplies.

3.8. GROUND WATER/SUB SURFACE SOURCES

These are the sources of water which supply water from below the earth's surface. They include springs, wells and galleries etc.

3.9. QUALITY AND QUANTITY OF GROUNDWATER

Generally speaking the amount of ground water from any source is limited and cannot be compared with surface supplies. At the same time it is of a much better quality and should always be preferred for small population and moderate water supplies for colonies. They do not require costly and exhaustive purification needed for surface waters.

Ground water is utilized by drawing it by wells or Tubwells. Infiltration galleries are used in the bed or banks of a river in sandy soil. This water is however not so pure as from deep wells as it has only percolated a small distance into the ground. Springs and artesian wells give very good quality of water and occasionally in good quantities. But these sources may not be permanent.

3.10. SUB-SURFACE SOURCES OF WATER

The following are the sub-surface sources of water:

(*a*) Springs (*b*) Wells

(*c*) Artesian wells (*d*) Infiltration galleries

(*e*) Porous pipe galleries (*f*) Infiltration wells

3.10.1 Springs

Ground water reappears at the ground in the form of springs. Springs are brought about under the following conditions:

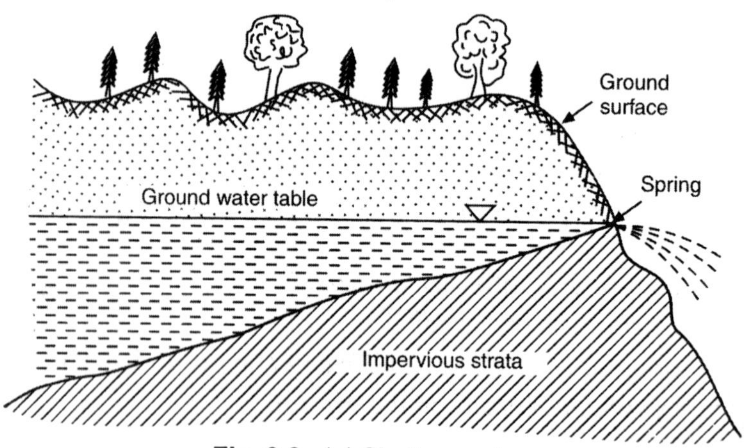

Fig. 3.3. (a) Shallow spring

(*a*) When the surface of the earth drops sharply below the normal ground-water table, the water bearing stratum (aquifer) is exposed to the atmosphere and the springs are created. The formation of such springs results from an overflow of the ground water table. This type of springs is known as 'Gravity' or 'Shallow' springs and the water table in such springs varies with the rainfall.

(*b*) When due to an obstruction ground water is stored in the form of a reservoir, and this water is forced to overflow at the surface. Springs of this type are the most common. These are formed when an impervious stratum, which is supporting the ground water reservoir, becomes out-crops. The storage capacity of these springs is very small which ceases after a drought. They can be developed by the construction of a cut off trench. A method for the development of such a spring is shown in Fig. 3.3.

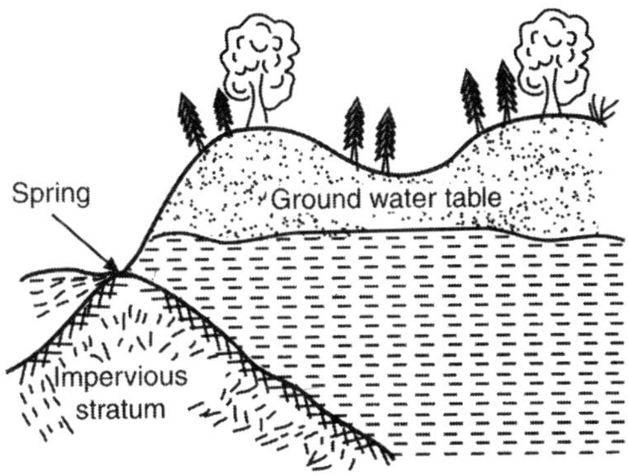

Fig. 3.3. (b) Spring

(*c*) When a fissure in an impervious stratum allows artesian water to flow in the form of springs.

Such type of springs come across when ground water rises through a fissure in the upper impervious stratum. These are also known as 'Artesian springs'. In these the water appears at the ground surface under pressure. The amount of water available is large and the rate of flow of water is constant because water comes out by a constant pressure.

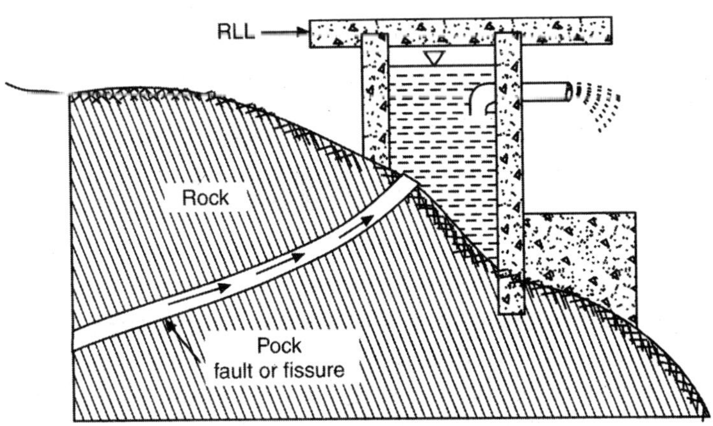

Fig. 3.4. Formation of spring due to a fault in a rock

(*d*) **Deep-seated spring:** A deep seated spring is one which is formed by exposing of a pervious stratum enclosed between two impervious strata i.e., which has impervious strata both above and below it. These springs usually have a large length of watershed and their outcrop is too much far away from the spring. The water comes out in the form of springs under pressure.

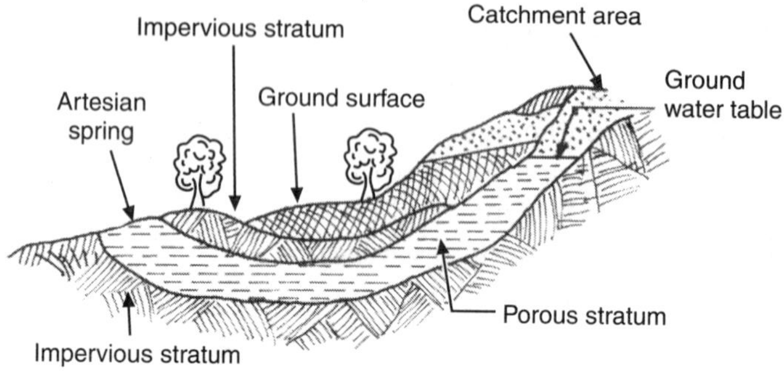

Fig. 3.5. Artesian springs

The first three springs detailed above as (*a*), (*b*) and (*c*) are called surface springs and the last (*d*) is called deep-seated springs.

3.10.2. Infiltration Galleries

(Refer Fig. 3.6). A gallery is a horizontal or approximately horizontal tunnel or open ditch constructed through water-bearing material in a direction approximately normal to the direction of flow of the underground water.

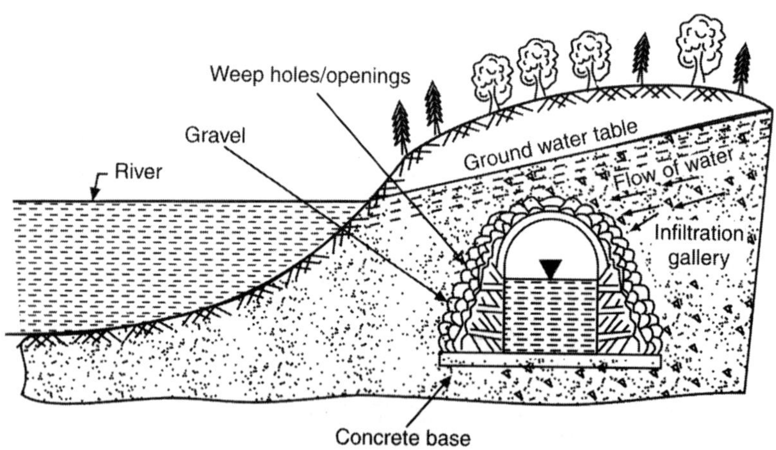

Fig. 3.6.

As we have seen earlier that sub-surface water always tries to travel towards lakes; rivers or streams. This travelling water can be intercepted by digging a trench or by constructing a tunnel with holes on sides at right angle to the direction of flow of underground water. These underground tunnels used for tapping underground water near lakes, rivers etc., are called infiltration galleries. The yield from these galleries may be as much as 1.5×10^4 litres/day/metre length of the infiltration gallery.

For maximum yield, the galleries are placed at the full depth of the aquifer. These are constructed with bricks or concrete with weep holes of 5 cm × 10 cm. Infiltration galleries are surrounded on sides and top with gravel or crushed stones to increases their intake capacity.

3.10.3 Porous Pipe Galleries

(Refer Fig. 3.7 and 3.8). If there is a much greater amount of water (ground water) available over a wide area, it can be cheaply collected by laying porous pipes or pipe with open joints in the full area at some distances. These longitudinal and cross pipes will be given a slope such that water should be brought towards a point. There intake capacity can be increased by surrounding them with rubble, gravel and sand.

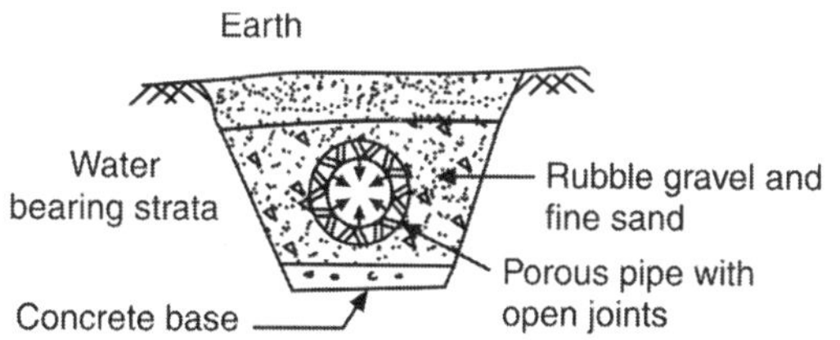

Fig. 3.7. Cross-section

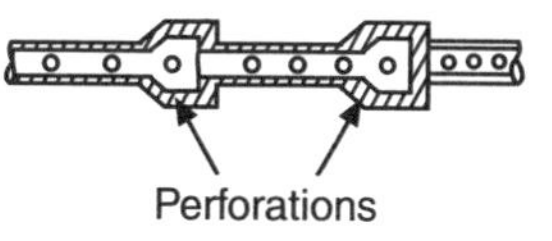

Fig. 3.8. Longitudinal section

3.10.4 Infiltration Wells

(Refer Fig. 3.9). An infiltration gallery may be a line of wells closely spaced and placed across or normal to the direction of underground flow in an aquifer. These are commonly placed close to the bank of a river or

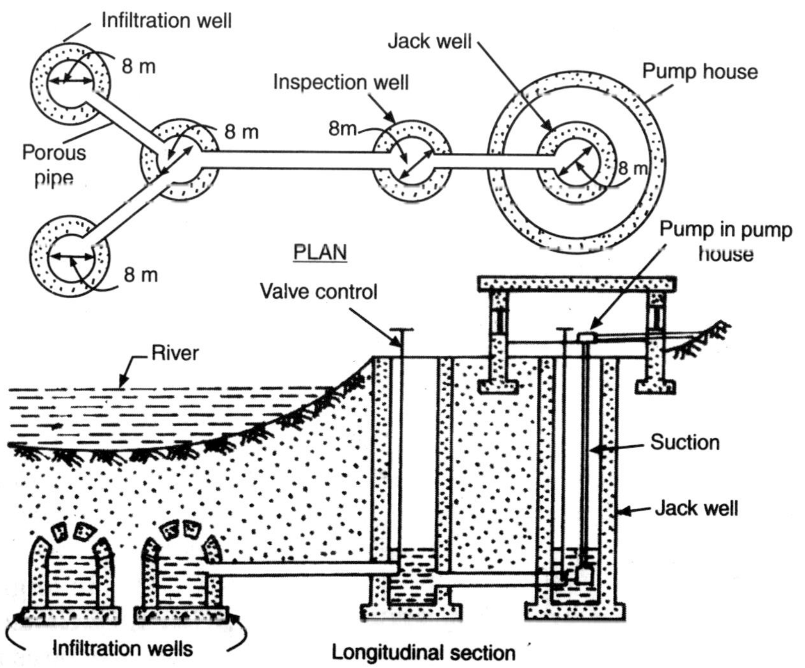

Fig. 3.9. Infiltration well

lake to intercept the underground flow towards the body of surface water. Wells so placed are called infiltration wells. It may be more economical to draw potable water from beneath a river by such means than to purify the surface water taken directly from the river.

As this water passes through layer of sand which acts as filter, so all the suspended organic and inorganic impurities are removed. Large number of wells are constructed in the river bed and then they are connected to a inspection well. The water flows under gravity from these wells to the inspection well and then towards jack well from where it is pumped.

3.10.5 Artesian Wells

(Refer Fig. 3.10). The term 'Artesian' was first applied to the flowing wells in the province of Artois in France. It is now commonly used to describe any well which derives its water from an underground source that is under pressure.

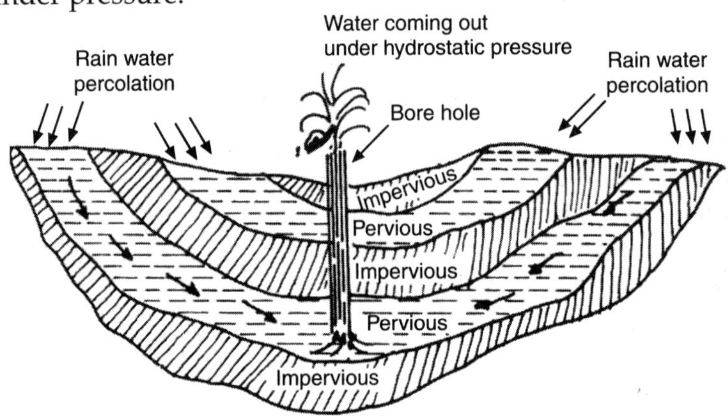

Fig. 3.10. Artesian well

Hydrostatic pressure is usually the chief moving force of artesian flow. Confined stratum aquifers are usually inclined and the pressure is greater in the lower portion. A well passing through the confining stratum and piercing the artesian aquifer releases the water, and if artesian pressure is sufficient water will be discharged as a fountain reaching far above the ground surface.

Confinement of water in water bearing strata is due to geologic structure and hydrologic properties of water conducting and water-confining formations. Water is confined by an overlying body of relatively impervious material that retards outcoming of water from the aquifer to the surface. Pressure is created by the weight of continuous water body to the collecting area.

3.11. CHOICE OF SOURCE FOR WATER SUPPLY SCHEME

The selection of the source of water supply to a town or city depends on the following factors:

(*i*) **Location of the Water-source:** The source of water should be as near to the town or city as possible. The source may be a surface or a sub-surface. If the surface sources like river, stream or reservoirs are not in the vicinity then sub-surface source is the only alternative. The water demand of the people is met from these sources. If surface sources like, river, stream or reservoirs is available near the city then the city should be developed along the source like Kolkata, Allahabad, Kanpur etc.

(*ii*) **Quantity of Water:** The source of water should have ample amount of water to meet up all the demands of the city as domestic, industrial, fire-resistance, public etc., throughout the year. It should also meet the extra quantity of water to be used in future if the city is expanded. It should also be able to meet the maximum demand in dry weather.

(*iii*) **Quality of Water:** It is an important factor. The quality of water should be such that it can be purified easily and cheaply. The water should be free from any harmful bacteria. As far as possible, the water from the source should be whole some, safe and free from any pollution.

(*iv*) **Cost of the Water Supply Scheme:** While selecting the source of water, the cost of water supply project should also be taken into consideration. This cost depends on many factors such as system of supply, ground levels of the town, distance between source and distribution system etc. The water flowing under gravity will be cheaper than lifting by pumps.

The selection of the source of water is done on the above factors but that source should be selected which is more in quantity and good in quality at the comparative least cost.

3.12. WELL

The vertical hole dug into the ground to get sub-surface water is called a well.

Most of the population of our country live in villages and wells are the main source of water supply. Yield of wells depends upon the following factors:

(*i*) Geological conditions which indicates the slope of after bearing strata.

(*ii*) Amount of water stored in different layers.

(*iii*) Porosity of various layers.

3.13. CLASSIFICATION OF WELLS

Wells may be classified into the following ways:

(*i*) According to the nature of aquifer tapped.

(*ii*) According to condition of flow.

(*iii*) According to method of construction.

3.13.1 According to the Nature of Aquifer Tapped

According to the nature of aquifer tapped wells are classed as:

(*i*) Shallow wells

(*ii*) Deep wells.

(*i*) Shallow Wells: The wells which are left in a pervious strata without meeting any impervious strata below the water table are known as shallow wells.

These well derived water from the underlying pervious strata. The quantity of water in such wells is uncertain as there is large variation of water table throughout the year. Due to the variation in water table, shallow wells, sometimes dry up in summer. The yield of these wells cannot be more than 5 litres/sec and so such types of wells are not suitable for public water supply schemes.

The water in these wells is likely to be contaminated by septic tanks, privies etc. and so the quality of water is not good. Shallow wells should be avoided for domestic purposes. If used water should be purified by bleaching powder and chloromine at an interval of two of four weeks.

(*ii*) Deep Wells: Wells which rest on impervious strata and get the supply from the underlying pervious strata are known as deep wells.

Large quantity of water can be obtained from deep wells. The yield of deep well is constant as there is no fluctuations in the water level.

The quality of water is good due to the natural straining of sub-surface water when it percolates through sands and gravels. The water obtained from deep well is purified by chlorination.

As compared to the deep well, a shallow well has the following drawbacks:

(*a*) The yield of the shallow well is uncertain due to large variations in the ground water level throughout the year. Against this, the yield of a deep well is greater and constant as there is no quick fluctuation of the water level.

(*b*) The quality of water from shallow well is mainly poor because only the uppermost sub-soil stratum is tapped, the well water may be open to contamination due to surface wash or close occurrence of septic tanks, privies etc. On the other hand, water from the deep well would be more reliable as the pollution would have been removed by longer travel of water and the straining action provided by the intervening porous strata.

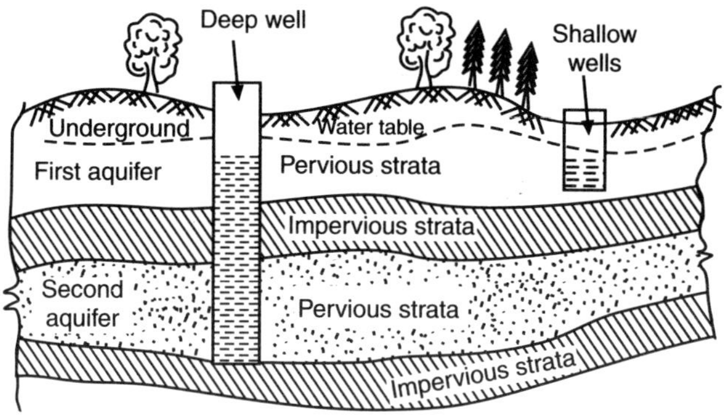

Fig. 3.11. A shallow and deep well

3.13.2. According to the Condition of Flow

Accordingly, wells may be classified as: (*a*) Gravity wells and (*b*) Pressure wells.

When the surface of the water in the water-bearing stratum (aquifer) surrounding the well is at atmospheric pressure, the well is called *gravity well*. The water flows into the well under gravity and rises to the height of saturated material surrounding it.

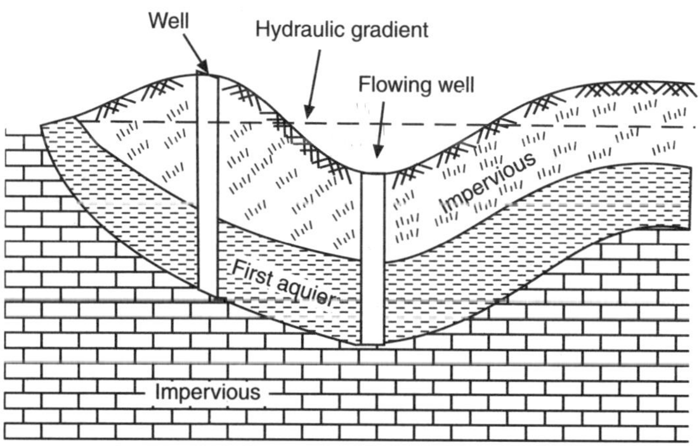

Fig. 3.12. Flowing well

When the acquifer is confined between two impervious layers, one above and the other below, and the water in this aquifer is at a pressure greater than atmospheric, the well is called *pressure well* or an *artesian well*. In some cases, the pressure is so great that the water rises from the ground surface and flows from the well. This is also termed as a flowing well.

3.13.3. According to Method of Construction

According to the method of construction, wells are of following types:

1. Dug wells
2. Bored wells or tube wells
3. Drilled wells

These may be

(*a*) Percussion drilled or driven wells

(*b*) Hydraulic rotary wells (Jetted wells)

(*c*) Core drill wells

1. **Dug Wells:** Small dug wells are generally excavated by hand. In loose soils, they are lined with brick, rubble or concrete. In rock, they are commonly left unlined. This lining is termed as 'curb'. In the case of a well used for domestic purposes, the upper portion of the curb is made impervious for a depth of 1.80 m to 2.40 m. Excavation is carried out till water flows in more rapidly than it can be bailed out. Dug wells should be completed when the water table is at its lowest level.

2. **Bored Wells or Tube Wells:** This type of well can be bored by using hand or power augers into cohesive soils. Above the water table, the excavated soil is collected in the auger, which is raised time to time to remove the collected soil. When the boring is done below the water table, sand may wash out of the auger and have to be removed from the bore hole by bailer or hand auger. The holes varying from 5 to 75 cm in diameter and 9 to 12 m in depth can be bored by hand auger, whereas power auger will drill holes of diameters ranging from 20 to 75 cm and depth 7.5 to 90 m.

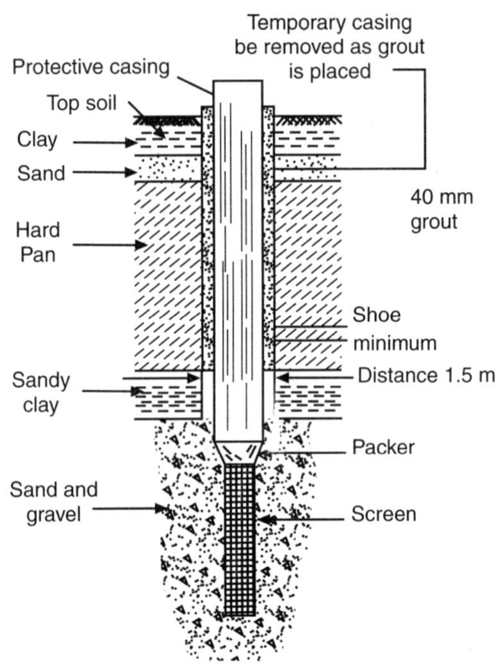

Fig. 3.13. Bored or tube well

3. **Drilled wells:**

(*a*) **Percussion Type (Driven Well):** For domestic use, driven wells develop smaller water supplies. The reason for this is that driven wells are shallow and of smaller diameter. This is because of the difficulty in driving a large pipe to greater depths. A single well gives 20 to 30 g.p.m. supply. For large supply, groups of wells are to be installed. All such wells will be located round a collecting well with a suction pump at a distance not exceeding 9 m.

Driven wells are adapted to soft, granular formations.

Percussion well consists of a drive point (pointed screen) and a pipe as shown in Fig. 3.14. A piece of pipe (casing) pointed at one end and perforated for remaining length is driven into a water-bearing formation by a wooden hammer or hydraulic ram. The diameter of this casing varies from 2.5 cm to 15 cm. In the case of sandy strata, the perforated portion of the pipe is covered with wire gauge to keep the sand away. To prevent damage to the gauge or screen and also clogging of the perforations during the driving operation, the end point is usually made larger than the upper casing. The whole pipe should be driven perfectly vertical as far as possible.

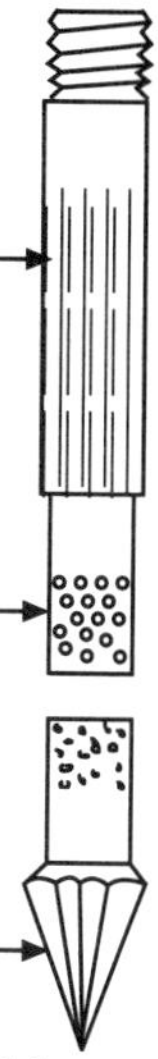

Fig. 3.14. Percussion well

(*b*) **Hydraulic Rotary Well (Jetted Well):** (Fig. 3.15.). The equipment consists of an inner pipe, which is the jetting tube; an outer pipe, which is the well casing; and a power pump and connection to supply water under pressure continuously through the jetting tube. The actual drilling consists of a fishtail, disk, or special rock bit. These are attached to the lower end of the jetting tube to facilitate breaking up of the soft materials like gravel, sand and clay. The loosened material mixed with water comes up through the annular space between the outer pipe and the jetting tube. This muddy water is collected in a settling pool and the settled water is reused by the pump. The construction of jetted wells is quicker under favourable conditions and can provide an emergency water supply.

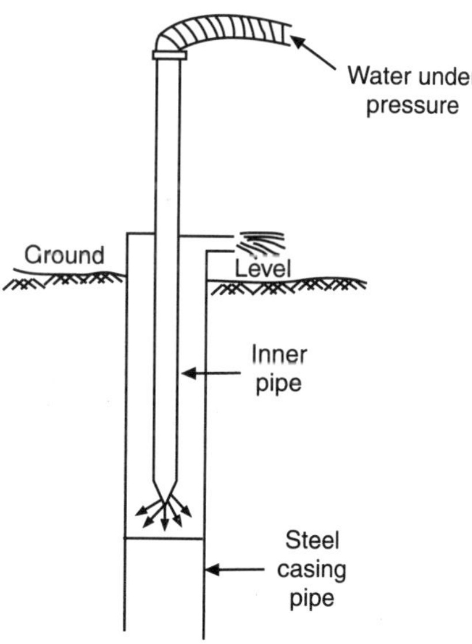

Fig. 3.15. Hydraulic rotary well

3.14. YIELD MEASUREMENT OF A WELL

Yield of a well is the amount of water flowing into the well per unit time. It is usually expressed in litre per second (lps) or sometimes in cubic metres per day. The yield of well may vary. For dug wells this may be 2-3 l/sec., for shallow tube wells this may be 10-40 l/sec and for deep wells it may vary from 50 to 200 litres/sec.

3.15 WATER TABLE WELL OR GRAVITY WELL

The various factors affecting the yield of a well are:

(*i*) D, the static head or the vertical distance from the water table to the bottom of the well.

(*ii*) d, the depth of water in the well when it is being pumped.

(*iii*) $D-d$, the difference of level in water before and at the time of pumping, called the drawdown.

(*iv*) R, the radius of the circle of influence.

(*v*) r, the radius of the well.

(*vi*) Fineness and porosity of the soil.

(*vii*) Inclination of the aquifer.

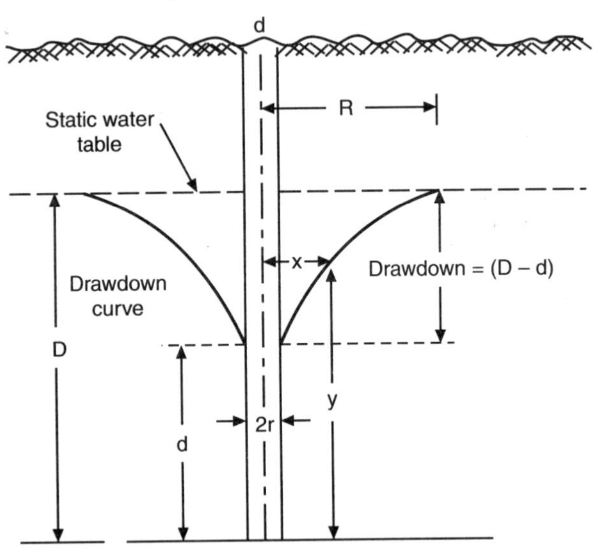

Fig. 3.16.

It is found that the velocity of flow of the ground water depends on the head consumed on the slope of the hydraulic gradient (Darcy's Law). This expressed mathematically as:

$$V = ks$$

where V = velocity of flow in metres per day.

s = slope of the hydraulic gradient.

k = transmission constant.

Yield of the well = Effective sectional area × velocity

$$Q = A \times p \times ks$$

where p = porosity ratio

A = cross-sectional area of substratum in sq. meters.

For any point on the curve with co-ordinates (x, y)

$$A = 2\pi xy$$

$$s = \frac{dy}{dx}$$

$$\therefore \quad Q = 2\pi xy \cdot p \cdot k \cdot \frac{dy}{dx}$$

or

$$Q \cdot \frac{dx}{x} = 2\pi p \cdot k \cdot y \cdot dy$$

Integrating and solving on the condition that

when $x = R, y = D$

$$Q = \frac{\pi p k(D^2 - y^2)}{\log e \frac{R}{x}}$$

Again when $x = r : y = d$

$$Q = \frac{\pi p k(D^2 - d^2)}{\log e \frac{R}{r}} \text{ cu metre/day}$$

Converting to the common log with ten as base and expressing Q in the metric system we have

$$Q = \frac{\pi p k \times 1000(D^2 - d^2)}{24 \times 60 \times 2.303 \log_{10} \frac{R}{r}} l.p.m$$

D, d, R and r being in metres

or

$$Q = k_m \frac{D^2 - d^2}{\log_{10} \frac{R}{r}} l.p.m.$$

where

$$k_m = \frac{\pi p k \times 1000}{24 \times 60 \times 2.303}$$

is the transmissibility constant.

3.16. CONFINED WATER WELL OR ARTESIAN WELL

The formula for yield of the artesian well is derived in a similar way, modified by the fact that the water in this case flows through the confined aquifer having a thickness t.

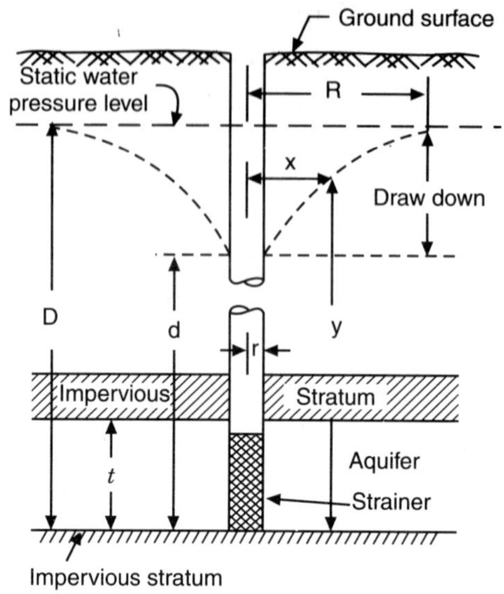

Fig. 3.17

Thus, the area of substratum becomes

$A = 2\,\pi\,x\,t$ and final expression of yield

$$Q = k'_m \frac{D-d}{\log_{10}\frac{R}{r}}\, l.p.m.$$

and $$k'_m = \frac{\pi\, p.k.2t \times 1000}{24 \times 60 \times 2.303}$$

and with various parameters expressed in their respective units.

3.17. SPECIFIC YIELD

Specific yield of a well is its yield per metre of drawdown. It is also known as specific capacity of the well. It is useful in comparing yields from different wells. A well having higher specific yield would be found to give larger yield for the same depth of drawdown and hence more acceptable.

3.18. NUMERICAL EXAMPLE

A 30 cm gravity well is being pumped at a rate of 1350 1.p.m. Measurements made in nearby test wells at the same times are as follows:

Drawdown in a test well 6 m away = 4.5 m.

Drawdown in a test well 30 m away = 0.9 m.

Distance of the ground water table above bottom of the well = 75 m.

Determine: (*a*) the drawdown in the well during pumping

(*b*) the specific capacity of the well

Solution: Given: $r = 15$ cm $= 0.15$ metre

$$Q = 1350 \text{ litres per metre}$$

$$Q = k_m \frac{D^2 - d^2}{\log_{10} \frac{R}{r}}$$

$$1350 = k_m \frac{75^2 - (75 - 4.5)^2}{\log_{10} \frac{R}{6}} = k_m \frac{75^2 - 70.5^2}{\log_{10} \frac{R}{6}} \quad ...(i)$$

$$1350 = k_m \frac{75^2 - (75 - 0.9)^2}{\log_{10} \frac{R}{30}} = k_m \frac{75^2 - 74.1^2}{\log_{10} \frac{R}{30}} \quad ...(ii)$$

From (*i*) and (*ii*), we get

$$k_m \frac{75^2 - 70.5^2}{\log_{10} \frac{R}{6}} = k_m \frac{75^2 - 74.1^2}{\log_{10} \frac{R}{30}}$$

Solving for R by trial and error method, we get

$$R = 45.6 \text{ m}$$

Now solving for k_m

$$1350 = k_m \frac{75^2 - 70.5^2}{\log_{10} \frac{45.6}{6}}$$

$$\therefore \quad K_m = 1.816$$

(*a*)

$$1350 = 1.816 \times \frac{75^2 - d^2}{\log_{10} \frac{45.6}{0.15}}$$

This gives, $d = 61.45$ m

$\therefore$ drawdown $= 75 - 61.45 = 13.55$ meters

(*b*)

$$Q = k_m \frac{D^2 - d^2}{\log_{10} \frac{R}{r}} \quad ...(iii)$$

We also know that for equilibrium conditions, the radius of the circle of influence varies directly as the yield *i.e.*, $R \propto Q$

or $R = c\,Q$

where c = a constant

$$c = \frac{R}{Q} = \frac{45.6}{1350} = 0.0337$$

$\therefore \qquad R = 0.0337\,Q$

Putting the value of $R = 0.0337\,Q$ in (*iii*)

$$Q = 1.816\frac{75^2 - 74^2}{\log.\dfrac{.0337Q}{0.15}}$$

By trial and error, $Q = 171$ l.p.m

3.19. WELL TESTS

These are: (1) Pumping Test, (2) Recuperation Test

The Pumping Test: It requires the determination of the rate of pumping water from a well while the drawdown remains constant for a number of hours. This gives the yield of the well at the particular draw-off. This test is usually carried out in the driest part of the year as otherwise it is difficult to maintain the condition of a constant drawdown.

The Recuperation Test: It needs pumping water at the maximum drawdown. Pumping is then stopped and the well water allowed to recover its original level. The gradual rise in the water level is noted down at regular brief intervals. Knowing the diameter of the well, the yield of the well at different drawdowns can be calculated.

3.20. COMPARISON BETWEEN SURFACE AND SUB-SURFACE/GROUNDWATER

Item	Groundwater	Surface water
(*i*) Available from	Springs, wells and infiltration galleries.	Rivers, Streams, Lakes etc.
(*ii*) Quantity	Small.	Less in summer but more in rainy season.
(*iii*) Quality	Less suspended impurities and free from bacteria.	More suspended impurities and contain harmful bacteria.
(*iv*) Organic Impurities	Less	More
(*v*) Pollution	Less	High.
(*vi*) Treatment	Less treatment required.	Lot of treatment needed
(*vii*) Storage problems	Water available all the times.	Reservoirs to be constructed for meeting demand during dry period.
(*viii*) Suitability	For colonies and small town.	For big cities and industrial towns.

PART–C : INTAKE WORKS

3.21. INTAKES

After the source has been fixed up, in any water supply scheme, the next problem is:

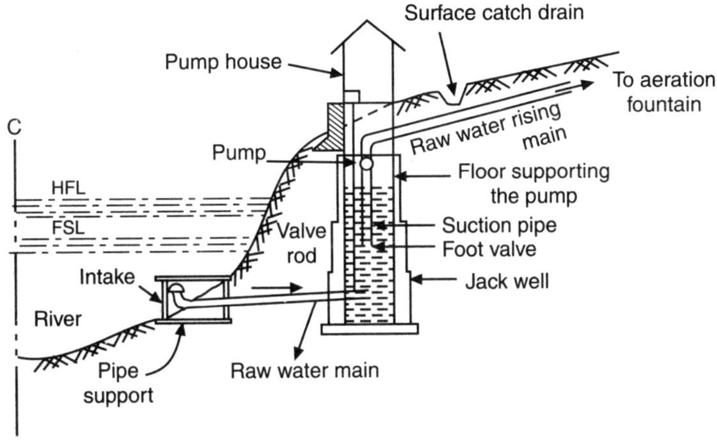

Fig. 3.18. Intakes

To draw water from this source including the provision of intake devices and head works.

A device or structure placed in a surface water source to draw water from this source and then discharge into an intake conduit through which it wili flow into the water works system, is called intake.

It consists of:

(*i*) a conduit with protective works.

(*ii*) screen at open ends

(*iii*) gates and valves to regulate the flow

3.22. LOCATION OF INTAKES

The following points should be kept in mind while selecting a site for intake works:

(*i*) At the site, the best quality of water should be available which will save time and cost of purification.

(*ii*) There should not be heavy current of water which might threaten the safety.

(*iii*) The site should provide maximum water from the source during the whole year.

(*iv*) Site should be easily accessible.

(*v*) Site should never be selected at the downstream or near the disposal of waste water.

(*vi*) Site must be free from the effects of floods as far as possible.
(*vii*) At the site, the velocity of flow in the source must be gentle.
(*viii*) Site should be near the treatment plant to minimise cost of conveying water.

3.23. TYPES OF INTAKES

Intakes are used to collect water for water works from various sources. The sources may be rivers, lakes, reservoirs or canals. Common types of intakes are:

(*a*) Reservoir intakes (*b*) River intakes
(*c*) Canal intakes (*d*) Lake intakes

3.23.1 Reservoir Intakes

These consist of intakes tower having no water inside other than in the intake pipes. The interior of the tower is thus made available for inspection and operation.

Fig. 3.19 shows a reservoir intake which is usually located either along the upstream of an earthen dam or within the body of a masonry dam. There are a number of inlets protected by screen and different levels to draw in clear water from near the surface.

3.23.2 River Intakes

In river intakes, water is drawn from the upstream side of the town to get comparatively better quality of water. They are located sufficiently inside the rivers so that required demand of water is met in all the seasons.

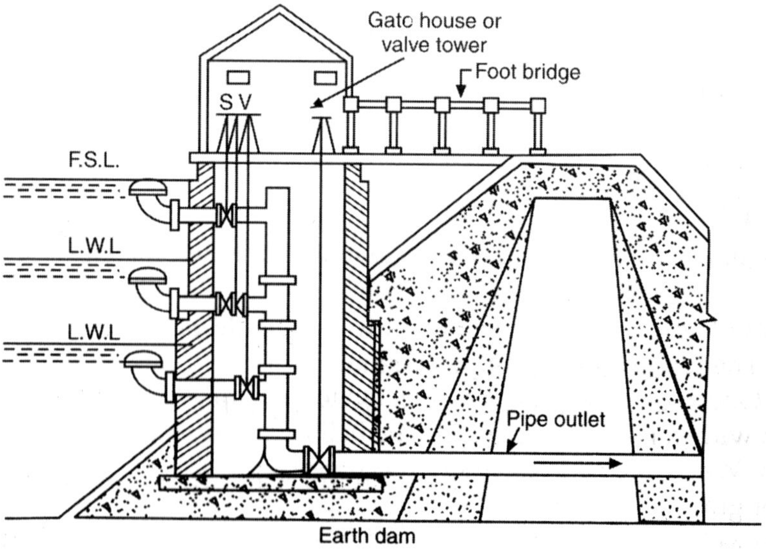

Fig. 3.19. Reservoir Intakes

Types of River Intakes

(*i*) **Intake Wells:** The water flows inside the intake wells through the penstocks located at different levels to draw water in all seasons. The water is then pumped in purification works (Fig. 3.20).

(*ii*) **Weir Intakes:** In this case small weirs are built across the river to store water for use during summer.

(*iii*) **Pipe Intakes:** Here two or three pipes are laid horizontally across the river and firmly fixed in masonry blocks. Ends are provided with strainers. The water is drawn into a jack well and then pumped to purification units. It is a cheap and simple method.

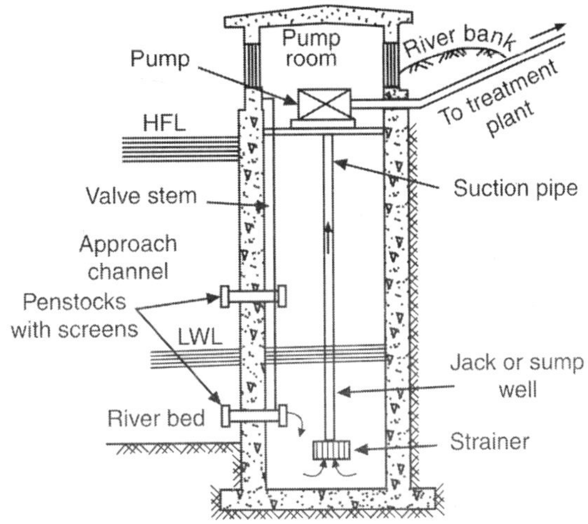

Fig. 3.20. River Intake (Intake well)

3.23.3 Canal Intake

Canal intake is a very simple structure constructed on the bank of a canal and consists of a RCC or brick masonry chamber built partially in the canal bank. Fig. 3.21, shows a canal intake. It has a side opening fitted with coarse screen which excludes heavier matter, from entering the conduit. The entry of the water into the conduit is controlled by gate valves which can be operated from the top. There will be obstruction in a portion of the water way due to intake, resulting in an increase of velocity, it is often necessary to pitch both the upstream as well as the downstream bed in order to protect it against scouring.

3.23.4 Lake Intake

Water from lakes is mostly drawn by constructing submersible intakes and so it is also known as submersible intake. Fig. 3.21 shows constructional details of submersible type intake. It consists of a pipe laid in the bed of lake. The top level of the pipe is fixed below the lowest water level of the lake. One end of the pipe is connected to a sump well and the other end is fitted with bell mouth opening. The bell mouth opening is covered with a mesh. Water of the lake enters the pipe through the opening and flows under gravity and collected in the sump well. Water is then pumped from sump well to the treatment plant for purification.

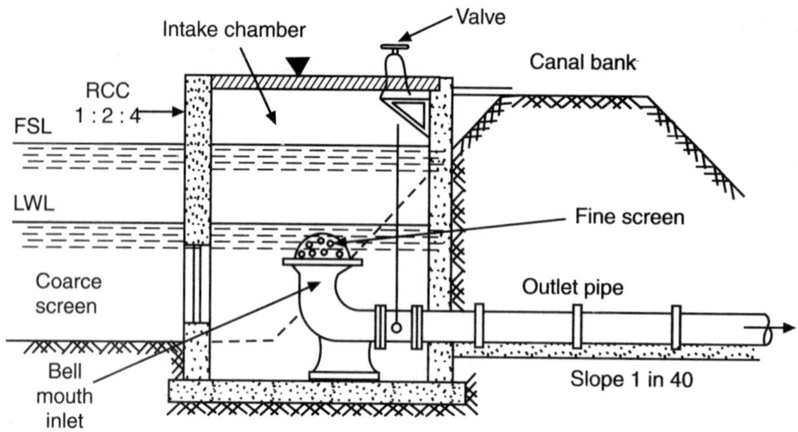

Fig. 3.21. Canal Intake

MODEL QUESTIONS

1. **Fill in the blanks:**
   (*i*) Surface and ground water sources get their supply from ______.
   (*ii*) The quality of water available in streams is good after ______.
   (*iii*) The water of the river has an important property called ______.
   (*iv*) The Source of water must be ______ to the area which is to be served with water supply.
   (*v*) Sub-surface sources are ______ affected by draught.
   Ans. (*i*) River water, (*ii*) First run off, (*iii*) Self purification, (*iv*) Near, (*v*) Not
2. Explain sources of water supply.
3. Explain the classifications of surface sources of water supply.
4. What are the main sources of water supply in India? On what basis the selection of surface source is done for water supply to a town?
5. Write a note on quality and quantity of surface water.
   (*i*) Surface and ground water sources get their supply from ______.
6. (*i*) Artesian well is that well which runs under ______.
   (*ii*) Shallow well obtains its water reserve below the ______. (Permeable/Impermeable rocks)
   (*iii*) Deep well gets its water supply below the ______. (Permeable/Impermeable strata)
   (*iv*) Spring is a ______. (Sub-surface/Surface source)
   (*v*) The condition of constant drawdown is needed in the case of ______. (Recuperative test/Pumping test)
   (*vi*) Shallow well is a surface source ______. (Correct/Incorrect)
   (*vii*) The amount of ground water directly depends on ______.
   (*viii*) Porous pipe galleries should be surrounded with ______ to increase therein take capacity. (Earth/gravel/clay)
   (*ix*) Give the formula for calculating yield of an Artesian well.

7. (*a*) What is a spring? How is it formed? What is the difference between deep seated spring and surface spring?
   (*b*) What is an artesian well? What geological conditions are necessary for its formation? Describe the properties of water obtained from it.
8. Write short note on:
   (*a*) Infiltration galleries
   (*b*) Infiltration wells
   (*c*) Drawdown cone
   (*d*) Porous pipe galleries
   (*e*) Persian Well
9. What sources of water supply would you recommend for water-supply scheme for: (*i*) Hilly area, (*ii*) Small town, (*iii*) Large city, (*iv*) Rural areas, (*v*) Backward areas with small population. Give reasons in support of your recommendations.
10. (*a*) Distinguish between shallow and deep well.
    (*b*) Sketch a shallow well showing therein constructional details to protect water from contamination.
    (*c*) Explain how would you determine the yield from a deep well.
11. What are the various sources of ground water-supply schemes? Discuss their merits and demerits from quality impurity point of view.
12. What are probable impurities in: (*i*) River water, (*ii*) Shallow well, (*iii*) Deep well.
13. In what circumstances you will conduct constant pumping test and recuperation test. Explain their procedure.
14. (*a*) Discuss the characteristics of ground water.
    (*b*) Derive an equation for the flow of ground water into an artesian well where flow is limited to sides only.
    (*c*) Describe the percussion method of drilling tube wells.
    (*d*) Write a note on development of tube wells.
15. What do you mean by:
    (*i*) Shallow well
    (*ii*) Deep well
    (*iii*) Artesian well
    (*iv*) Infiltration gallery.
    Explain with the help of sketch if necessary.
16. Sketch an infiltration gallery.
17. What is the function of a strainer in a tube well? Describe any one strainer (give sketch).
18. Describe the water jet boring method of drilling tube wells.
19. Discuss the advantages and disadvantages of underground and surface sources of public water supply.
20. What are the common sources of water? Which source is most suitable and economical and why?
21. (*a*) Define "Tube well".
    (*b*) Define "Yield of well". Name the factors upon which yield depends.
22. Describe the various sub-surface water resources and their yield.
23. Discuss the quality, characteristics of ground and surface water.

24. What are the various sources of water in water supply scheme? Discuss their merits and demerits from the quality and quantity point of view.
25. (*a*) Intake are necessary for _____ sources. (surface/sub-surface)
    (*b*) Intake is a device to _____ water. (Treat/Draw)
    (*c*) Water from river is always drawn from _____ because it is free from contamination. (down stream side/up stream side)
    (*d*) The sources for intakes may be _____.
    (*e*) For reservoir intake water should be drawn _____ the banks of shores. (near/away from)
26. What are canal intakes? Describe with a sketch.
27. (*a*) What are intakes?
    (*b*) Mention the points which should be taken into consideration in deciding the location and design of an intake for the water supply of a large town, the source being perennial river. Draw a detailed sketch of the intake and describe it fully.
28. Describe a suitable intake from an impounding reservoir.
29. What are various types of intake works? Describe side weir river intake with the help of neat sketch.
30. State the factors to be considered for the selection of site of river intake. Describe any one kind of river intake.
31. Write short note on "A Canal Intake" with the help of neat sketch.
32. (*a*) What are the principles on which intakes are designed?
    (*b*) By drawing a neat sketch, describe the details and functioning of "River Intake".
33. Describe the reservoir intakes.
34. Describe in detail the various sources of water giving quantity and quality of each.
35. Explain the following with neat sketches:
    (*i*) Canal intake. (*ii*) River intake (*iii*) Tube well
36. Discuss the merits and demerits of river water source and ground water source for water supply scheme for a town.
37. Describe surface water sources giving suitability and quality of water of each source.

❑❑❑

CHAPTER

4

QUALITY OF WATER

Learning Objectives

- ❒ Wholesome Water
- ❒ The Wholesome Water should Exhibit the Following Requirements
- ❒ Potable Water
- ❒ Polluted Water
- ❒ Contaminated Water
- ❒ Infected Water
- ❒ Impurities in Water
- ❒ Examination of Water
- ❒ Physical Standards of Water
- ❒ Chemical Standards of Water
- ❒ Biological Standards of Water
- ❒ Standards for Drinking Water
- ❒ Waterborne Diseases

4.1 WHOLESOME WATER

Water to be used for a public supply must be potable i.e., drinkable. It should bot be chemically pure because water without dissolved and suspended matter is both unpalatable and unhygienic. Chemically pure water is that water which contains only two parts of hydrogen and one part of oxygen by volume. An absolutely pure water is not recommended for human health as this water does not contain ceratin minerals which are very essential to health. Thus the wholesome water is neither chemically pure not it contains anything harmful to human health.

4.2. THE WHOLESOME WATER SHOULD EXHIBIT THE FOLLOWING REQUIREMENTS

(*i*) It should be free from pathogenic bacteria.

(*ii*) It should be free from colour, odour and turbidity.

(*iii*) It should be tasty and cool.

(*iv*) It should not corrode the pipes.

(*v*) It should be devoid of all objectionable matter.

(*vi*) It should contain dissolved oxygen and free carbonic acid so that it may remain fresh.

4.3. POTABLE WATER

It is safe to drink, suitable for domestic use and pleasant to taste.

4.4. POLLUTED WATER

It contains undesirable substances, and is not fit for drinking and domestic use.

4.5. CONTAMINATED WATER

It contains disease bacterias. It is always polluted and unfit for use.

4.6. INFECTED WATER

Infected water is water contaminated with pathogenic organisms.

4.7 IMPURITIES IN WATER

The common impurities present in water can be classified as follows:

4.7.1 Suspended Impurities

The solid particles which are dispersed in water are called suspended impurities.

The solid particles may have their specific gravity equal, less or more than the specific gravity of water. The particles of the same sp. gravity are mixed in the water. Those of less sp. gravity are lighter and float on the surface of the water. The solid particles of higher specific gravity are heavier than water and they settle down after some time. The last two types of solid particles are removed from water by simple filtration and sedimentation respectively.

The suspended imparities includes the following:

(*a*) Bacteria-pathogenic or non-pathogenic.

(*b*) Algae, fungi, protozoa.

(*c*) Clay, site and other inorganic matter.

These are microscopic and cause turbidity in the water. The load (concentration) of suspended mater in the water is measured by its turbidity.

4.7.2. Colloidal Impurities

These are very finely divided particles of solids dispersed in the water.

These particles are so small in size that these cannot be removed by ordinary filters and are not visible to the naked eye. All the colloidal impurities are electrically charged and remain in continuous movement.

Colloidal particles vary in sizes form

$$1\mu \ (= 1 \text{ micron} = 10^{-3} \text{ mm}) \text{ to}$$

$$1\mu\mu \ (= 1 \text{ millimicron} = 10^{-6} \text{ mm})$$

The existence of such particles can be detected by the Tyndall effect and the brownian movement. The colour imparted to water is due to colloidal impurities.

4.7.3. Dissolved Impurities

When water flows over soil or rock, impurities like inorganic salts, organic compounds etc are dissolved in it. These impurities are known as dissolved impurities. The concentration of dissolved solids can be measured by weighing the residue left after evaporation of water and can be expressed in ppm.

Dissolved impurities include the following:

(*a*) Salts:

(*i*) Carbonates, bicarbonates, chlorides, sulphates, flourides of sodium salts.

(*ii*) Carbonates, bicarbonates, chlorides sulphates of calcium and magnesium salts.

(*b*) Metals

Iron, manganese, lead, arsenic etc.

(*c*) Gases

Oxygen, carbon dioxide, ammonia, chlorine etc.

(*d*) Vegetables and animal matters.

Table 4.1. Effect of impurity

S.No.	Type of impurities	Causes	Effects
1.	Suspended	(*a*) Bacteria	Disease
		(*b*) Algae, Protozoa, funge	Colour, odour turbidity
		(*c*) Clay, silt, minerals	Turbidity
2.	Colloidal	(*a*) If associated with organic matter having bacterias	Chief source of epidemic
3.	Dissolved		
	(*a*) Salts	**Calcium and magnesium**	
		Bicarbonate	Hardness and alkalinity
		Carbonate	Hardness and alkalinity
		Sulphate	Hardness
		Chloride	Hardness and corrosion
		Sodium	
		Bicarbonate	Alkalinity, softening
		Carbonate	Alkalinity, softening
		Flouride	Over 1.5 ppm causes staining of teeth
		Chloride	Taste

(Contd...)

	(*b*) Metals	Iron taste, corrosion	Red colour, hardness,
		Manganese	Brown colour, teste
		Lead	Lead poisoning
		Arsenic	Poisoning
	(*c*) Gases	Chlorine	Taste, odour
		Oxygen	Corrode the metals
		Carbon dioxide	Acidity, corrode the metals
		Hydrogen sulphide	Rotten egg odour, acidity and corrode the metals.

4.8. EXAMINATION OF WATER

To ensure safety to public health it becomes essential that the water to be supplied to public through public water supply scheme should be throughly examined. The process of examining water through a series of tests in systematic manner is known as water analysis.

Water analysis help us in determining the following:

(*i*) Quality of raw water and amount of impurities present in water.

(*ii*) Various stages of purification required.

(*iii*) Daily operations of the purification plant

(*iv*) Suitability of water for feeding boilers, hot water pipes.

It consists of the following:

(*i*) Sampling of water

(*ii*) physical examination

(*iii*) Chemical examination

(*iv*) Bacteriological examination

4.8.1 Sampling of water

Because of the probable daily and seasonal changes in the quality of natural water, it is desirable that samples for analysis should be collected frequently and over a long period of time.

Samples ıre collected in sterilised bottles having stoppers. The sterilisation is done by heating the bottles or by mixing them with concentrated nitric acid or sulphuric acid. All these acids should be completely removed by washing with sterilised water.

The water or bottle should not be touched at the time of filling. The data, time, place and number should be written on the bottle. The nature of supply i.e., whether taken from well, stream, sedimentation tank, filter etc. should also be mentioned on the bottle.

Both the raw water from the source and the purified (treated) water examined to find out the actual condition of water before and after treatment. This decides the degree of treatment required and whether the treated water being supplied to the consumers from day-to-day is of the specified standard laid down by the public health department.

4.8.2 Physical Examination

The qualitative properties of water which are discussed under physical examination include temperature, odour, turbidity, colour and taste.

(*a*) **Temperature:** The most desirable range of temperature for a public water supply is seldom below 40° to 50°F. The temperature of natural waters is seldom below 40°F. The temperature above 50°F makes water less palatable. The temperature above 80°F is undersirable and above 90°F, the water is unfit for water supply schemes.

The temperature of water is measured by an ordinary thermometer.

(*b*) **Colour:** The colour imparted to water is due to decaying of vegetable or organic substances dissolved in it. The water containing natural colour is yellowish or brownish in appearance. The shade of colour present in water is normally determined by comparing it with a standard unit consisting of 1 milligram of platinum per litre of water. The colour for potable water is 10-20 ppm on platinum cobalt scale.

(*c*) **Turbidity:** In common language, it may be defined as a measure of the muddiness of water.

Turbidity of water is caused by suspended and colloidal impurities in ppm (parts per million) as ascertained by optical observation. For drinking water, a turbidity of less than 10 ppm is desirable. The optimum is 5 ppm.

Determination of Turbidity :

(*i*) **By Turbidity Rod (Fig 4.1.):** It consists of a 120 cm. long graduated non-stretchable tape attached to the upper end of 20 cm. long graduated aluminium bar. A screw with 25 mm. platinum wire at one end and a nickel ring at the other end is passed through the lower end of the bat. In order to hold the turbidity rod rigidly and vertically, a stick is inserted in a nickel ring and the whole apparatus is gradually immersed in water. While doing so, an eye is

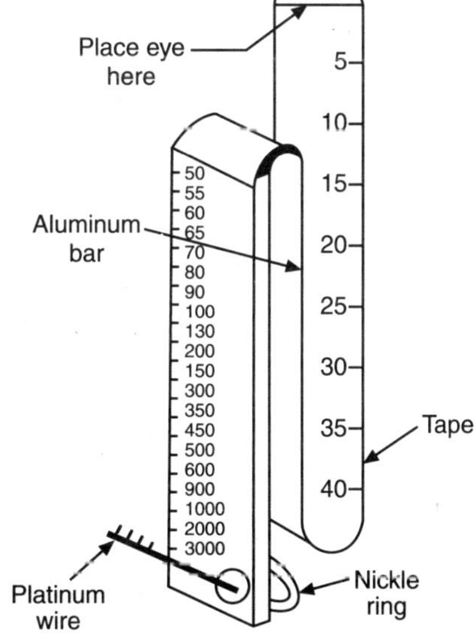

Fig. 4.1. Turbidity rod

placed at the zero graduation mark and the platinum wire, which is at right angles to the bar, is observed all the time. Lowering of the bar in water is continued until the wire becomes invisible. The reading on the bar at this instant gives the water turbidity in parts per million (ppm).

(*ii*) **By Jackson Turbidity Meter (Fig. 4.2):** It consists of a metal stand, a metal container with the glass tube and a standard candle arranged as shown in the figure. A small portion of the water sample is taken in the tube and the candle flame image is observed form top through the water in the tube. More and more water is then added until the image of the flame disappears. When this takes place, the graduation on the tube upto which water is poured, indicates turbidity. Turbidity in the range 25-1000 ppm can be recorded by this instrument.

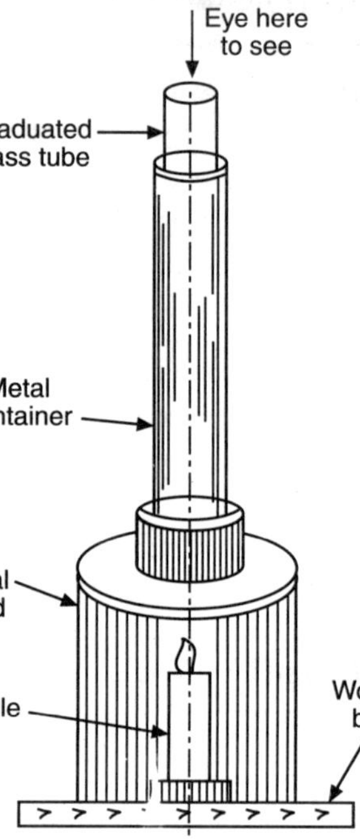

Fig. 4.2. Jackson turbidity meter

(*d*) **Odour:** The odour in water is due to organic substances, vegetable matter, microscopic organism of gases which are still undergoing decomposition.

The water having foul smell or odour should not be supplied to the public. The measurement of odour intensity is made by diluting the sample with odour-free water to the threshold value. The dilution ratio is generally 1: 4. For example, if 50 ml sample of water is diluted and is made 200 ml (till it just loose its odour), then the threshold number will be 4. For Public supplies, the threshold number should be 1 and should never exceed 3.

(*e*) **Taste:** Taste in water may result from any one or a combination of micro-organisms either dead or alive, dissolved gases such as H_2S, CH_4, CO_2 or oxygen combined with organic matter; mineral substances such as sodium chloride, iron compounds, carbonate and sulphates of other elements; and phenols and other tarry or oily matter especially after chlorination.

4.8.3 Chemical Examination

(*i*) **Total Solids:** The impurities of water in the suspended, colloidal and dissolved forms are termed as total solids in water. The quality of solids present in water can be found by evaporating a sample of water and then weighing the dry residue left. To find the suspended solids, the water sample is filtered and the residue left on the filter paper is weighed. The difference between the total solids and suspended solids will give the dissolved solids.

The amount of total solids in water for domestic use is generally restricted to 500 ppm, higher amounts upto 1500 ppm are also permitted but these may lead to some psychological effects on human body.

***(ii)* Hardness:** It is property of water which prevents the lathering of soap. It is due to the presence of carbonates, bicarbonates, sulphates, chlorides and nitrates of calcium and magnesium. It leads to scale formation in the boilers and consumptions of large amount of soap for domestic use.

Hardness is usually expressed in gms./litre or p.p.m. of calclium carbonate in water.

Table 4.2: Degree of hardness

Hardness in p.p.m.	Degree of Hardness
(i) up to 50	Soft
(ii) 50 to 100	Moderately soft
(iii) 100 to 150	Slightly hard
(iv) 150 to 200	Moderately hard
(v) More than 200	Very hard

***(iii)* pH Value:** It is the logarithm to the base 10 of reciprocal of hydrogen ion concentration expressed in gms./litre.

pH value : $$pH = \log_{10}\left[\frac{1}{H^+}\right]$$

The higher value pH means lower hydrogen ion concentration, representing alkaline solution. The lower value of pH mean higher hydrogen ion concentration, representing acidic solution.

In neutral water concentration of hydrogen ion (H^+) and hydroxyl ion (OH^-) are same. The pH of neutral or pure water is 7.

When acids or alkalies are dissolved in water, they dissociate into hydrogen and hydroxylions respectively.

$$HCL \rightleftharpoons H^+ + Cl^-$$

$$NaOH \rightleftharpoons Na^+ + OH^-$$

If the water is acidic its pH value will be less than 7 and if it is alkaline, its pH value will be more than 7.

The control of pH is essential in water treatment and this control is done by automatic devices. The device controls the dosage of chemical automatically to the required value of pH.

Determination of pH value: *Electrometric method:* This is quick and automatic method of noting pH value. In this method prepared electrodes of colomel or glass are dipped in water sample and then connected to dry cell or electric mains. A meter is also connected to the electric circuit which directly reads the pH value of the water.

***(iv)* Chlorides:** Chlorides are usually present in water in the form of sodium chloride (NaCl) and may have access to water due to : pollution from sea water, industrial and domestic wastes etc. These impart a salty waste in water if the concentration exceeds 250 gm/l. Hence chlorides content in water above 250 gm/l are not allowed.

The chloride content can be measured by titrating the water with standard silver nitrate ($AgNO_3$) solution using potassium chromate (K_2CrO_4) as indicator.

***(v)* Chlorine:** Dissolved free chlorine is not found in natural waters. It is present in the treated water resulting from disinfection with chlorine. The chlorine remains as residual in treated water for the sake of safety against pathogenic bacteria. The residual chlorine is limited from 0.05 to 0.2 p.p.m. in water to kill the disease borne bacteria. The free chlorine can be detected by two methods: the starch iodide method and orthotolidine method.

***(vi)* Iron and Manganese:** The presence of iron and manganese causes discolouration of clothes when washed in such water. Hence the presence of these metals are undersirable in portable water. When water contains iron and managanese in large concentration then they may cause incrustration in water mains due to deposition of ferric hydroxide and manganese oxide. Tests are carried out to determine their amount and as far as possible restricted to 0.3. ppm in potable water.

***(vii)* Dissolved Gases:** The solubility of gases in natural waters is a function of the temperature of the water and the pressure and concentration of the gas in the atmosphere. The following are some of the gases mostly found in the natural water:

(a) **Oxygen:** Surface waters contain large amount of dissolved oxygen, because they absorb it from the atmosphere. Dissolved oxygen keeps water fresh. More amount of oxygen corrodes the pipe materials.

Biochemical oxygen demand normally known as BOD, is a measure of the amount of oxygen demanded by living organisms in the water to oxidize organic matter present as food for the organisms. BOD is a property, not a constituent of the water.

(b) **Carbon Dioxide:** The water obtains carbon dioxide from the atmosphere. When this carbonated water comes across calcium and magnesium salts, carbon dioxide acts on them and converts them into bi-carbonates and cause hardness in the water.

***(viii)* Fluorides:** A minimum amount of fluorides is desirable in portable water to prevent dental cavities and the maximum amount is limited by the need for the prevention of fluorosis, a defect of the enamel of the teeth. Fluorides in water are added in the form of sodium silica fluoride.

***(ix)* Organic Matter:** It is produced by decaying vegetable or plant matter and waste products from animal and human beings. In general, when a chemical analysis of a sample of water shows the presence of large amounts of chlorides, ammonia and nitrates or nitrites, contamination with sewage is indicated. An examination of the water-shed must then be made to determine whether such contamination is possible.

4.8.4. Bacteriological Examination

A bacteriological analysis is a safer criterion for determining the fitness of water for drinking than a chemical analysis and consequently is of prime importance to the water works engineer. There are two standard tests performed.

***(i)* Total count of bacterias:** In this method total number of bacterias present in a millilitres (c.c.) of water is counted. One ml. of water sample is diluted in 99 ml. of sterilized water. Then one ml. of the diluted water is added to 10 ml. of agar (culture medium). This mixture is then kept in incubator at 37°C for 24 hours. It is taken out from the incubator and colonies of bacterias are counted by means of microscope. Then the product of the total number of colonies and the dilution factor will indicate the total number of bacterias per milli litre of undiluted water sample.

(ii) The determination of the presence or absence of the coliaerogenes (intestinal) bacteria – (B-coli):

In B-coli test the presence of B-coli indicates that excreta of men and animals have mixed with water and have not yet been destroyed (decomposed). The absence of B-coli represents the absence of any other bacteria of grams.

The examination for bacteria of the Coli-aerogenous group consists of:

(a) Presumptive test

(b) Confirmed test

(c) Completed test.

(a) In presumptive test a definite amount of diluted water sample is incubated in a standard fermentation tube containing lactose broth for 24 or 48 hours at 37° C. If there is evolution of gas in 24 hours, it means the test is positive (bacteria are present). Absence of gas in 24 hours indicates that test is negative and water is safe.

[Coli-aerogenes group of bacteria which includes all aerobic, anaerobic and facultative bacteria ferment lactose with gas formation.]

(*b*) In confirmed test, a small material from the positive presumptive test is transferred to the second fermentation tube containing brilliant green lactose bile as a liquid confirmatory medium. Contents are incubated for 48 hours at 37°C. Absence of gas means test is negative and water is safe. Evolution of gas means completed test is essential.

(*c*) The completed test is performed to show whether or not the gas forming organisms are Gram-negative, non-spore-forming bacilli.

4.9. PHYSICAL STANDARDS OF WATER

The following physical standards are expected in the case of water to be used for domestic purposes :

(*a*) Temperature __________ 50° to 60°F.

(*b*) Odour __________ unobjectionable.

(*c*) Colour __________ 10–20 (on platinum cobalt scale)

(*d*) Turbidity __________ 5–10 p.p.m. (Silica scale)

4.10. CHEMICAL STANDARDS OF WATER

The following chemical standards are expected in the case of water to be used for domestic purpose.

(*a*) Total Solids __________ up to 500 p.p.m.

(*b*) Hardness __________ up to 200 p.p.m.

(*c*) pH value __________ 6.5 to 8.

(*d*) Carbonate Alkalinity __________ upto 120 p.p.m.

(*e*) Chlorides __________ upto 250 p.p.m.

(*f*) Iron and Manganese __________ 0.5 to 1.0 p.p.m.

(*g*) Organic Matter-free ammonia __________ upto 0.1 p.p.m.

(*h*) Oxygen dissolved __________ 5 to 6 p.p.m.

4.11. BIOLOGICAL STANDARDS OF WATER

In India the following interpretation is usually put on the bacteriological examination of water.

(*a*) The total count of bacteria per c.c. __________ 0 to 100– good potable water.

(*b*) B–Coli test __________ No. B–Coli in 100 c.c. safe.

(*c*) Most Probable Number (M.P.N)–one number in 100 c.c.

4.12. STANDARDS FOR DRINKING WATER

Substance or property	Maximum acceptable	Max. allowable
1. Colour	5 (on Platinum cobalt scale)	50 on platinum cobalt scale
2. Taste	Unobjectionable	–
3. Odour	Unobjectionable	–
4. Turbidity	5 units on silica	20 units on silica
5. Temperature	10°C to 15.6°C	–
6. Total solids	500 mg/litre	1500 mg/lit.
7. Iron	0.3 mg/lit.	1.0 mg/lit.
8. Manganese	0.1 mg/lit.	0.5 mg/lit.
9. Copper	1.0 mg/lit.	1.5 mg/lit.
10. Zinc	5	15
11. Calcium	75	200
12. Magnesium	50	150
13. Sulphate	200	400
14. Chloride	200	600
15. pH-range	7.0–8.5	6.5–9.2
16. Mag. and Sod. sulphate	500	1000
17. Phenol	0.001	0.002
18. Carbon chloroform extract	0.2	0.5
19. Alkyl benzyl sulphonates	0.5	1.0
20. Pathogenic bacteria	Nil	3
21. B. Coli index	3	10
22. MPN	1 in 100 ml	–

4.13. WATERBORNE DISEASES

Diseases which spread mainly through contaminated waters are called waterborne diseases. Waterborne diseases are infectious diseases and may spread through direct contact or through filth or flies etc. It may be caused due to the presence of following:

(*a*) Microorganisms
(*b*) Parasitic ova
(*c*) Inorganic matter
(*d*) Organic matter

The important waterborne disease are:

1. Diseases caused by **Microorganisms**
   (*i*) Cholera
   (*ii*) Typhoid fever
   (*iii*) Paratyphoid fever
   (*vi*) Diarrhoea
   (*v*) Bacillary dysentry
2. Diseases caused by **Parasitic ova**
   (*i*) Amoebic dysentry
   (*ii*) Yellow fever
3. Diseases caused by **Inorganic matter**
   (*i*) Constipation
   (*ii*) Brittle bones
4. Diseases caused by **Organic matter**
   (*i*) Diarrhoea
   (*ii*) Gastric disturbances

Waterborne diseases are prevented by observing the following points:

1. Water should be checked throughly and disinfected before supply to the public.
2. Water should be safe and wholesome.
3. Water must be supplied to the public after sterilization and purification.
4. The water lines and sewer lines should not be laid closer to each other. Otherwise water may get polluted.
5. The water supply pipes should be checked throughly. The points of leakage of water should be sealed properly.
6. In case of any doubt in the water supply the water should be boiled before using.
7. The dose of chlorine should be increased as it may change odour or taste but will help people free from any possible serious waterborne disease.
8. The city should be kept neat and clean.
9. If any of the above diseases is reported the people should be immunised and inoculated.

MODEL QUESTIONS

1. (*i*) Turbidity is an indication of _____ of water due to suspended inorganic matter such as silt, clay and mud particles.
(*ii*) For drinking water, turbidity may be between _____ and _____ p.p.m.

(*iii*) The instrument used for determining odour is known as _____.
(Microscope, Osomoscope)

(*iv*) Hardness is due to _____ impurity. (suspended, colloidal, dissolved)

(*v*) Water is _____ when pathogenic bacteria is present. (Safe/harmful)

(*vi*) Turbidity in water is permitted upto _____ p.p.m.

(*vii*) Total solids in water should not exceed _____ p.p.m.

(*viii*) pH value of potable water is between _____ and _____.

(*ix*) Hardness is usually expressed in gms/litre or p.p.m. of _____ in water.
(calcium sulphate, magnesium chloride, calcium carbonate)

(*x*) The taste of water due to chlorides is _____.

(*xi*) Total chlorides in water should be permitted upto _____ p.p.m.

(*xii*) Alkalinity is commonly found in natural waters in the form of carbonate of _____ and as bicarbonate _____.

(*xiii*) Salinity is due to the presence of neutral salts such as the _____ and _____ of calcium, magnesium, sodium and potassium.

(*xiv*) The pH value of water is _____.

2. Give only the names of six water borne diseases and defined pathogens and carriers.

3. (*a*) What is the necessity of supplying pure water ?

(*b*) Name the various impurities present in water. What is their effect on water supply ?

4. Write the quantity of water and its suitability for a water supply scheme in the following cases:

1. Rivers, 2. Canals, 3. Reservoirs, 4. Lakes

5. What is pH value? How is it determined? What is its importance in the water analysis in water treatment?

6. What are usual standards of water supply?

7. Why biological examination of water is necessary and how is it done?

8. What do you understand by total solids in water? How are these determined?

9. Write a note on the impurities in water. Describe the standard of quality of treated water.

10. Write short notes on the following.

(*i*) pH valu and its significance

(*ii*) Carbonate hardness

11. State the permissible limits for fluorides in water to be supplied for domestic consumption. Mention the ill-effects when they are not in the permissible limits.

12. Explain the significance of the following from the point of view of water quality criteria:

(*i*) Turbidity (*ii*) Chlorides

(*iii*) Nitrates (*iv*) Sulphates

(*v*) Iron

13. What are the highest permissible limits for iron and concentrations in drinking water as per international standards? What are there ill-effects when they exceed the maximum permissible limits.

14. Give the maximum permissible limits of the following for drinking water.
    (*i*) Zinc
    (*ii*) Lead
    (*iii*) Nitrates
    (*iv*) Chlorides
    (*v*) Sulphates
15. Give a brief account of examination of water and state its necessity in connection with the water supply to a town.
16. What are the "Minimum standards" (Chemical, Physical and bacteriological for drinking water?
17. What are the various qualities for which a sample of water is tested?
18. (*a*) What is B.O.D. Demand and how is it determined.
    (*b*) Define:
    (*i*) Total count
    (*ii*) Bacterial coli index.
19. What is the necessity of examination of water for drinking purposes? How would you classify the same with important tests.
20. Describe the various physical tests carried out in water analysis, giving the significance of each.
21. Write short notes on pH value, its determination and significance.

❑❑❑

CHAPTER

5

WATER TREATMENT

Learning Objectives

- Water Treatment
- Objects of Treatment
- Location of Treatment Plants
- Layout of Water Treatment Plant
- Treatment Methods
- Comparison of Slow Sand and Rapid Gravity Filters
- Disinfection
- Chlorination
- Special Method of Chlorination
- Aeration Fountain
- Public Health Laboratory
- Removal of Tastes and Odours
- Copper Sulphate Treatment
- Water Softening

5.1. WATER TREATMENT

The treatment given to the water to bring the quality of water to the required standard before supplying it to the public is known as *water treatment*.

The amount of treatment depends upon the quality of available water and qualities required for different purposes.

5.2. OBJECTS OF TREATMENT

The objectives of water treatment are:

(*i*) To remove dissolved gases, colour, taste and odour.

(*ii*) To remove harmful pathogenic bacterias.

(*iii*) To completely remove corrosive properties of water.

(*iv*) To make water fit for drinking and industrial use.

5.3. LOCATION OF TREATMENT PLANTS

The location of treatment plants depends upon the source of supply. If the source of water is river or reservoir, the treatment plant should be located near the source to avoid pumping of muddy water which may cause wearing of pipes. If the source is underground like tube well, the treatment plant should be located in the central part of the town, so that water may

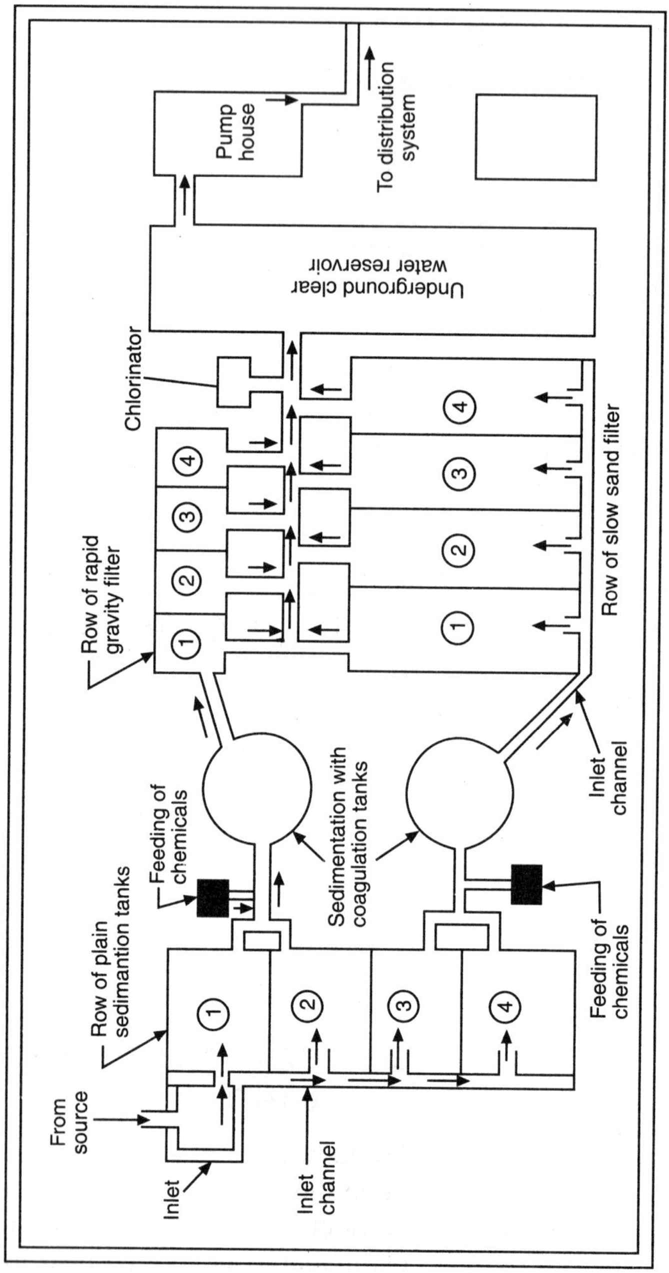

Fig. 5.1. Typical layout plan of water treatment works

reach the public as early as possible and also the water head at the farthest point will be more.

5.4. LAYOUT OF WATER TREATMENT PLANT

Fig. 5.2 shows typical sequential layout of water treatment plants.

The order of various treatment plants is given below:

1. Intake works i.e., pumping units.
2. Plain-sedimentation.
3. Sedimentation with coagulation.
4. Filtration.
5. Disinfection.
6. Clear water reservoir (C.W.R.).
7. Pumping sets to feed service reservoir through rising main.
8. Service reservoir (Underground or Elevated).

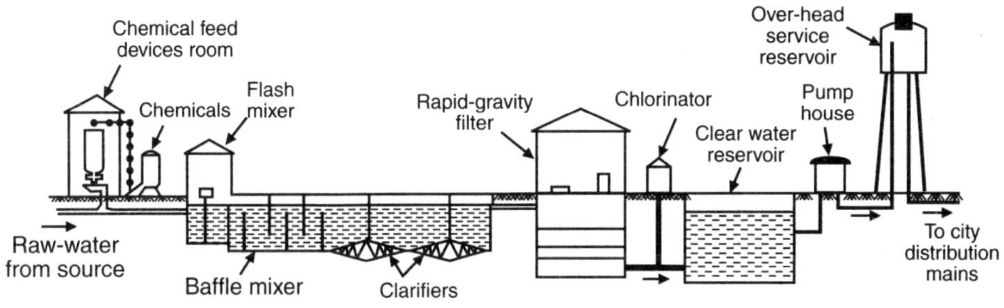

Fig. 5.2. Typical sequential layout of water treatment units of water works.

The following points should be kept in mind while deciding the layout of a water treatment plant:

(*i*) Various parts of treatment plant should be located in order of sequence.

(*ii*) The entire plant should be located at such elevation that water should flow from one unit to other under gravity.

(*iii*) The layout should be such that it requires minimum area.

(*iv*) Sufficient area should be occupied for future extension.

(*v*) Staff quarters and offices should be constructed near the treatment plant for proper functioning and inspection.

(*vi*) The site of treatment plant should be neat and clean.

(*vii*) The site should give very good aesthetic appearance

5.5. TREATMENT METHODS

The various methods of treatment and impurities removed by these methods are given in Table 5.1:

Table. 5.1. Treatment methods

S.No.	Method of treatment	Purpose
1.	Screening	To remove floating matter *e.g.*, bushes, dead animals, leaves branches of trees etc.
2.	Plain sedimentation	To remove suspended impurities e.g., sand, silt, clay etc.
3.	Sedimentation with co-agulation	To remove fine suspended matter.
4.	Filtration	To remove micro organism.
5.	Aeration and chemical treatment	To remove dissolved gases, taste and odour.
6.	Disinfection	To remove bacteria.

5.5.1 Screening

Screening is the process in which large sized particles such as branches of trees, bushes, dead animals are removed from the water to be treated.

For screening coarse and fine screens are placed in front of pump or intake works. Coarse screens consists of flat bars with openings 2.5 – 7.5 cm. These screens are kept inclined at angle of 45° to 60° to the horizontal. The main purpose of keeping the screens inclined is to reduce the velocity of flow and to increase the opening area. The fine screens remove fine suspended solids. These screens are cleaned at frequent intervals to facilitate the free movement of water.

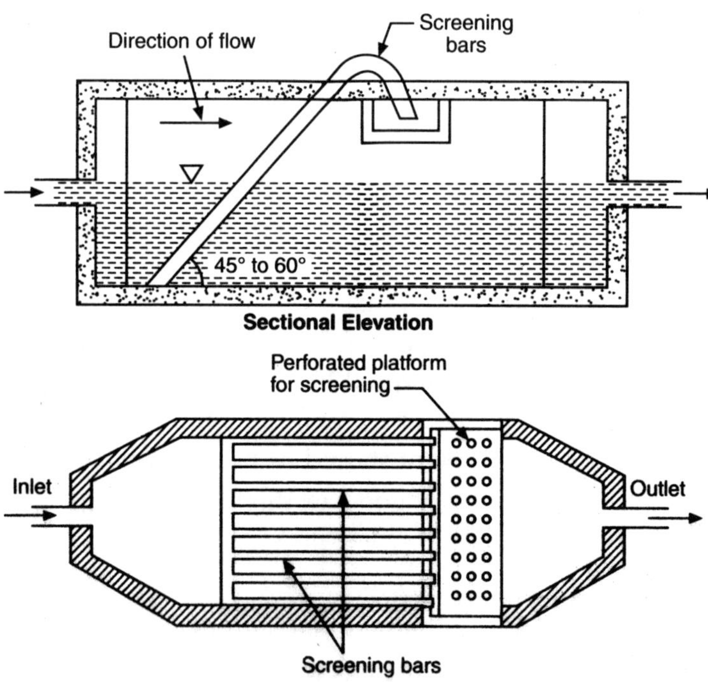

Fig. 5.3. Screens

5.5.2. Sedimentation

Sedimentation is the process of removal of suspended particles by gravitational settling. It occurs when particles are heavier than water. This process takes place in a tank, known as sedimentation tank or settling tank.

5.5.2.1 Type of Sedimentation

Sedimentation can be classified into two types:

(*i*) Plain sedimentation

(*ii*) Sedimentation added with coagulation.

When the impurities are separated from the suspending fluid by gravitation and natural aggregation, the operation is called plain sedimentation.

When chemical or other substances are added to increase segregation and settling of finely divided suspended matter and colloidal substances, the operation is called sedimentation with coagulation.

5.5.2.2 Advantages of Plain Sedimentation

(*i*) It is a preliminary process which lightens the load on subsequent processes.

(*ii*) It gives less variable quality of water and so the operation of subsequent purification process is done in a better way.

(*iii*) Cleaning cost of chemical coagulating basins is reduced.

(*iv*) No chemical is lost with the sludge discharged from the plain settling tanks.

(*v*) Less quantity of chemicals are needed in the subsequent treatment processes.

5.5.2.3 Types of Sedimentation Tanks

It is of two types :

(*i*) Intermittent tank (or quiescent type tanks)

(*ii*) Continuous flow type tanks.

(*i*) Intermittent Tank: These are simple settling basins which store water for a certain period and bring it in complete rest. After 24 hours, during which the suspended particles settle down to the bottom of the tank and clear water is drawn off. All the settled material is then removed from tank and is again filled with raw water to continue the next operations. Thus this type of tank functions intermittently, *i.e.*, after a gap of atleast 30 to 36 hrs. This necessitates the commissioning of at least two tanks. Such tanks are generally not preferred these days because lot of time and labour is wasted.

(ii) Continuous Flow Type Tank: In this type of sedimentation tank, the flow velocity is only reduced and the water is not brought to complete rest. The water enters from one end and comes out from the other end. The velocity is reduced by providing large length of travel. The velocity is so adjusted that the time required by the particle to travel from one end to another is slightly more than the time required for settlement of that particle.

Sedimentation tanks are generally made of reinforced concrete and may be rectangular or circular in plan. Long narrow horizontal tanks are generally preferred to the circular tanks. A plain sedimentation tank under normal conditions can remove as much as 70% of the suspended impurities present in water.

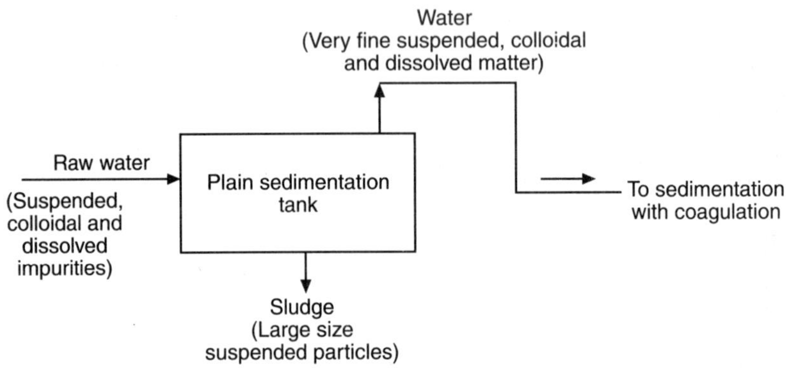

Fig. 5.4. Flow diagram of plain sedimentation plant

5.5.3. Sedimentation with Coagulation

The process of adding certain chemicals to water in order to form floc (insoluble gelatinous substance) for quick sedimentation and rapid removal of fine particles is called sedimentation with coagulation.

If the water contains large quantities of very fine and light colloidal impurities whose hydraulic settling value is very small, it is practically very difficult to eliminate them by plain sedimentation within a reasonable short period of detention. This is done by adding a desired amount of chemical compounds to water. Then this is followed by mixing. Due to the addition of these chemicals, the fine and thin particles grow in size and become heavier and then settle down.

The insoluble gelatinous substance obtained by adding chemicals to water is called floc, and the process is called flocculation.

These chemical compounds are called coagulants and their process or reaction is termed as coagulation.

5.5.3.1 Common Coagulants

The common chemicals generally used for coagulation are:

(*a*) Aluminium sulphate [$Al_2(SO_4)_3\ 18H_2O$] called Alum.

When alum is added to raw water, it reacts with the bicarbonate alkalinities, which are generally present in raw water and forms a gelatinous precipitate (floc) of aluminium hydroxide. This floc attracts other fine particles and suspended matter and thus grow in size and settles down finally to the bottom of the tank. The chemical reaction is:

$$Al_2(SO_4)_3 + 3Ca(HCO_3)_2 = 3CaSO_4 + 2Al(OH)_3 + 6CO_2$$

The insoluble and colloidal aluminium hydroxide forms the floc. The dose of alum varies from 5 mg/l for relatively clear water to 85 mg/l for highly turbid water. The average dose is about 17 mg/l. Alum for water-works:

(*a*) produces crystal clear water.
(*b*) gives better floc formation.
(*c*) promotes better settling and longer filter runs.
(*d*) aids in reduction of tastes and odours.
(*e*) has no chlorine demand.
(*f*) is a low cost coagulant.
(*g*) is superior in tests against other coagulants.
(*h*) is high in quality.

Difficulties with Alum

(*i*) slight increase in sulphate hardness.
(*ii*) an increase in CO_2.
(*iii*) formation of a colloidal precipitate in the presence of sodium or potassium.

(*b*) Iron Salts

(*i*) Ferrous sulphate –$FeSO_4 \cdot 7H_2O$
(*ii*) Ferric chloride –$FeCl_3$
(*iii*) Ferric sulphate –$Fe_2(SO_4)_3$

The reactions with ferrous sulphate and ferric chloride are:

$$FeSO_4 \cdot 7H_2O + Ca\,(HCO_3)_2 = Fe\,(HCO_3)_2 + CaSO_4 + 7H_2O$$
$$Fe(HCO_3)_2 + 2Ca(OH)_2 = Fe(OH)_2 + 2CaCO_3 + 2H_2O$$
$$4Fe(OH)_2 + 2H_2O + O_2 = 4Fe(OH)_{3\downarrow}\ \text{(floc)}$$

The ferrous hydroxide $Fe(OH)_2$ forms a desirable, heavy gelatinous precipitate which is further oxidized to ferric hydroxide $Fe(OH)_3$ – a satisfactory gelatinous precipitate. The amount of lime added varies from 1 to 5 p.p.m.

$$FeCl_3 + Ca(OH_2) = 2Fe(OH)_3\downarrow + 3CaCl_2$$

The ferric hydroxide is precipitated, forming a desirable coagulant which is heavier than $Al(OH)_3$ and requires a shorter detention period and less careful adjustment of pH.

Ferric chloride being hydroscopic cannot be applied through dry feed machines. It is applied in aqueous solution having strength between 2% or more in large well-equipped plants.

Ferric sulphate is quite effective in the pH range of 4 to 7 and above 9, whereas Ferric chloride is effective in pH range of 3.5 to 6.5 and above 8.5.

The combination has, therefore, proved to be quite effective coagulant for treating low pH waters.

***(c)* Chlorinated Copperas:** It is a mixture of ferric sulphate and ferric chloride prepared by adding chlorine to a solution of ferrous sulphate in the ratio of 1 part chlorine to 7.8 parts of copper as,

$$6\,FeSO_4 + 3\,Cl_2 = 2\,Fe(SO_4)_3 + 2FeCl_3$$

This salt is highly corrosive. It produces a desirable floc formation with tough floc particles. The coagulating effect has a wide range of optimum pH from 8 or 9 to 6 and is more effective in colour removal.

***(d)* Sodium Aluminate:** It is an alkaline compound. Its advantages are the elimination of corrosive qualities of the water, a wide range of pH, quick flocculation, and no need of the addition of alkali. It does not increase the permanent hardness due to sulphate of calcium. However, sodium aluminate being a costly chemical, it is not used widely in practice.

5.5.3.2 Dosing

The process of adding of the coagulants in correct amounts is known as dosing.

5.5.3.2.1 Dosage of Coagulants

This depends upon the following factors:

***(i)* Turbidity of Water:** It is found that greater the turbidity of water, more is coagulant required. The amount per litre varies from 0.03 to 0.13 gms or more of alum or iron.

***(ii)* Its Colour:** The dose of coagulant depends upon the colour.

***(iii)* pH value:** The control of pH value is important from the point of view of maintaining the character of the flocs. Thus for aluminium hydroxide floc if the pH value is lower then the optimum, the insoluble floc becomes soluble and the floc disappears. If on the other hand, the pH value is more than the optimum, the aluminium hydroxide ionizes into an aluminate which is both negative in electrical charge and soluble in water. It is found that alum coagulates best in the range 6.0 to about 8.0 and the optimum pH range for iron salts lies between 8.0 and 10.0. For proper control of the pH value it may sometimes be necessary to lower or raise the pH; the former can be got by adding into water, H_2SO_4 acid and CO_2 while the latter by the addition of caustic soda, lime and soda ash.

With most coloured waters, the floc is formed with pH value below 6.5. Alum is quite effective in removing colour.

Coagulation is usually found necessary for water having turbidity greater than 30–50 ppm.

***(iv)* Temperature of the water:** The amount of coagulant increases with lower temperatures because of slower reaction of clock formation.

***(v)* Time of settlement:** The dose also increase with less time of settlement.

Table 5.2. Normal dose of coagulants

Coagulant	*pH*-Range for best results	Normal dose in gm/litre
Alum	5.5 to 8.0	0.05 to 0.85
Sodium Aluminate	–	0.034 to 0.34
Ferric Chloride	5.5 to 11.0	0.085 to 0.51
Ferric Sulphate	5.5 to 11.0	0.085 to 0.51
Chlorinated Coppers	8.5 to 11	0.05 to 0.51

Calculation of Optimum Coagulant Dose: Trial determination of coagulant does, known as Jar tests, are made in a laboratory with stirring device. The set up of the apparatus is illustrated in Fig. 5.5.

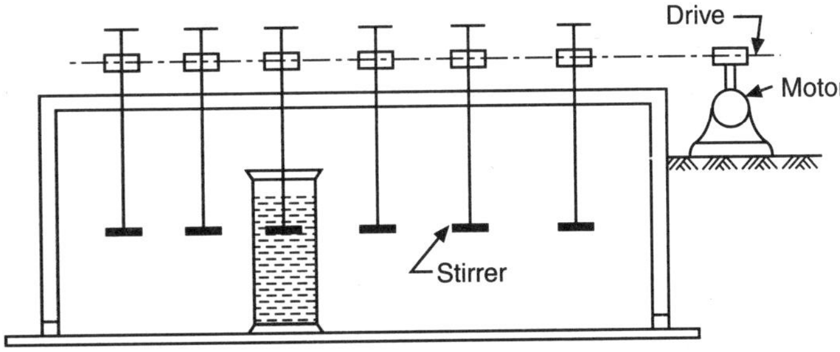

Fig. 5.5. Calculation of optimum doses of coagulant

The amount of coagulant is increased stepwise and all jars are mixed simultaneously for 10 to 30 minutes. The jar in which floc first makes its appearance is assumed to have received the most economical dose. The size of floc, its settling properties and filterability are also to be kept in mind in addition to the availability of the material.

Note: In water treatment plants, jar tests are made at regular intervals to have most efficient results and economical operations when raw water undergoes rapid changes.

5.5.3.2.2 *Feeding Devices For Adding Chemicals*

Coagulants can be fed raw water in two forms:

(*i*) Wet form

(*ii*) Dry powder form

(*i*) Solution Feed Devices: The chemicals whose solutions can be prepared easily are suitable for feeding by this method. The solution of required strength of coagulant is prepared and stored in a tank, from which it is allowed to trickle down into the mixing tank through and outlet. The level of coagulant solution in the coagulant feeding tank is maintained constant. Based on the method of control of coagulant feeding into the raw water the solution feed devices are of two types.

(*a*) **Conical Plug and Float Device:** The conical plug is attached to the pulley by means of a string. The pulley is rotated by a rack and pinion arrangement to which a float is attached. Float chamber and raw water channel are inter-connected. With increase in flow of raw water the float rises and the conical plug gets lifted because the pinion and the pulley rotate in the same direction. This adds more solution to the increased raw water flow. Thus the chemical dose gets automatically adjusted through the constant head tank.

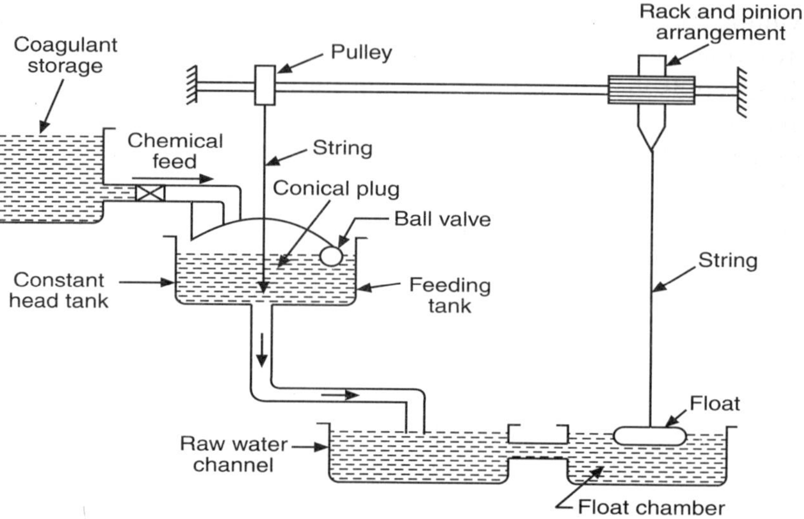

Fig. 5.6. Conical plug and float device

(*b*) **Adjustable Weir Type Device:** Rectangular holes in the sliding cylinder form the adjustable weir outlet and its position is controlled by means of a lever, the float chamber and the raw water channel are inter-connected. With increased flow of raw water the float rises and the lip of the weir gets lowered adding more chemical to raw water flow.

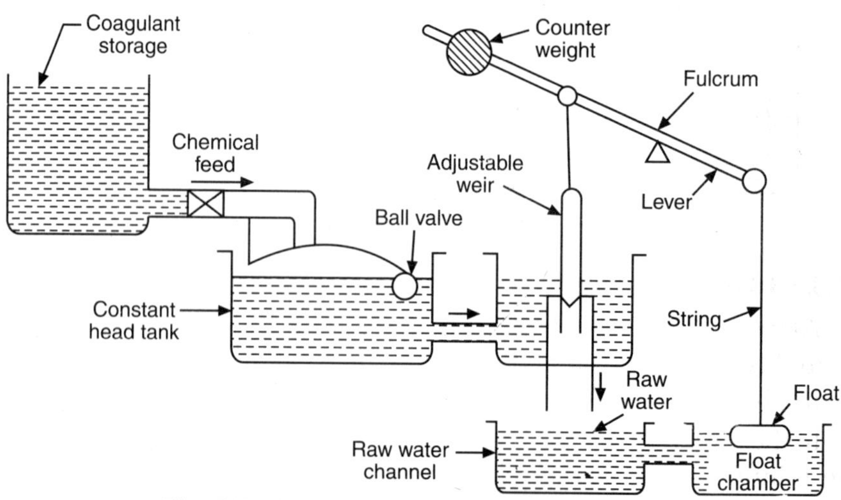

Fig. 5.7. Adjustable weir type device

The application of the correct amount of coagulating chemical is done by chemical feed machines which may be controlled manually, automatically or spaced by the flow of liquid. Proper chemical dosage is determined by coagulation control test commonly called 'jar tests'.

***(ii)* Dry Feed Device:** Dry-feed device are mainly common due to simplicity, relatively small space required, neatness and freedom from corrosion. The desirable properties of a chemical that is to be fed dry are uniformity to size of grain, homogeneity in composition, permanency under prevailing conditions of temperature and pressure, and freedom from being hygroscopic or efflorescent. Alum is fairy fine and uniform and its dry feeding avoids the corrosiveness of alum solutions of the pipes that convey it. Dry-feed machines control chemical dosage by regulated volumetric or gravimetric displacement of dry chemicals as shown in the figures. The revolving helical screw of the toothed wheel at the bottom of the conical hopper allows a small quantity of the coagulant to be drawn off continuously. The dose is regulated by adjusting the speed of the helical screw or the toothed wheel. This is done by connecting them to the venturi device in the raw water pipe. The measured chemical is then added generally in solution to the raw water.

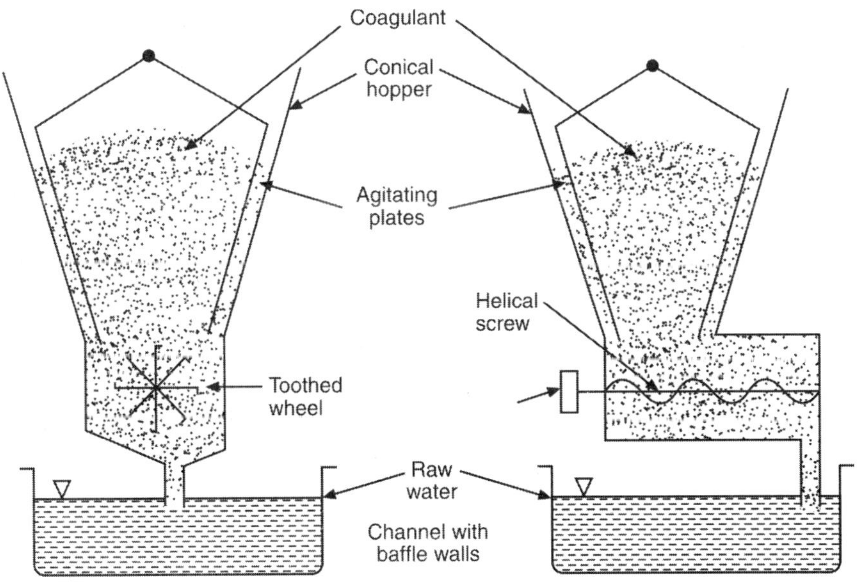

Fig. 5.8. Dry feed device

5.5.3.2.3 Mixing Device

After adding coagulants in water, the next step is to mix the coagulants throughly in water so that it gets fully dispersed in the whole water. This mixing is done by means of mixing devices.

In these mixing basins, firstly the chemicals are rapidly mixed for about one minute and then the whole mixture is gently agitated for about 30 minutes so that these coagulants may react and start coagulation. The velocity of flow of water in the mixing basins is kept between 15 to 53 cm/sec. This velocity in no case should be less than 10 cm/sec and not more than 75 cm/sec., because in the first case the floc will settle down and in the second case the floc will start disintegrating.

This mixing can be done by the following devices:

(*a*) Baffle Type Basins: These are also further divided into two classes.

(*i*) Round about the end baffles.

(*ii*) Over and under baffles.

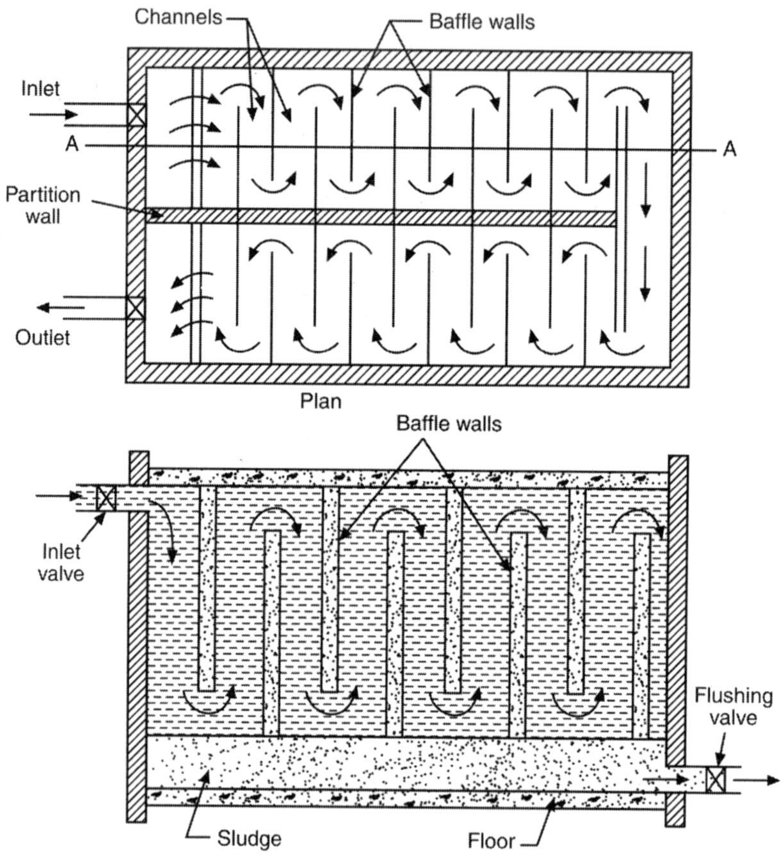

Fig. 5.9. Round about the end baffles type mixing basin

In the case of basins with round end baffles. (refer Fig. 5.9), a rectangular tank is divided into a continuous series of channels by means of baffle walls which cause the water to which the coagulant, have seen added to flow around the ends of the baffle walls through so many channels.

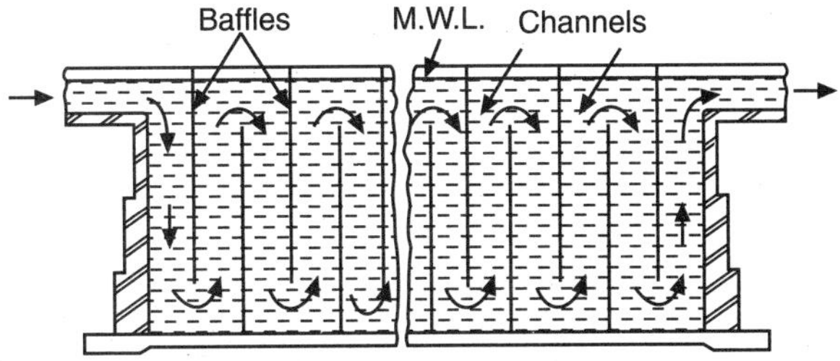

Fig. 5.10. Over and under baffle walls type mixing basin

The channels are made narrow so that the velocity of flow will be just sufficient to prevent the formation of deposits.

In the case of basins with over and under baffle walls, the chemical-mixed water poses over and under the baffles through the channels. (Fig. 5.10.)

In these basins, the baffle walls are placed 60 to 100 cms apart. The velocity of flow of water is kept in the channels between 15-30 cm./sec. The detention period in these basins is kept 25-50 minutes. These are not suitable for small plants due to costly in construction, less flexible in control and greater loss of head.

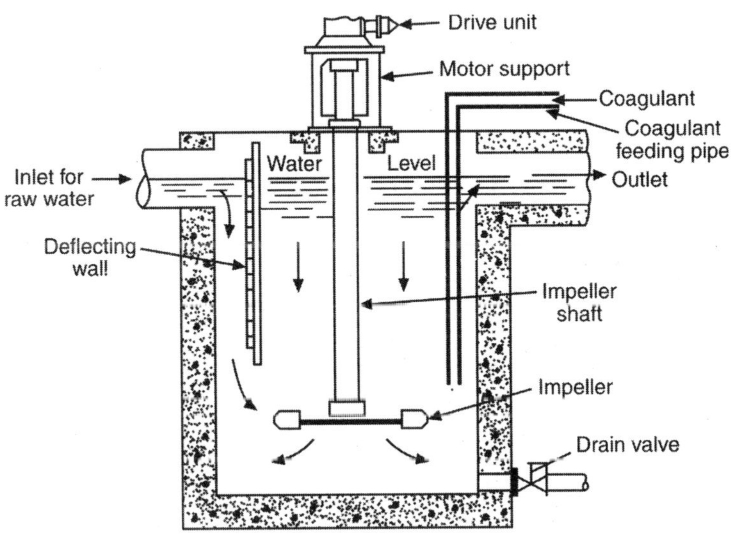

Fig. 5.11. Flash mixer

(*b*) Mechanically Agitated Mixing Basins are those in which water is subjected to a violent agitation when the chemical is first added to water and then to a gentler action when the floc is forming. The first of these operations is usually obtained in flash mixers and the second in a flocculator.

In a flash mixer the rapid mixing is brought about in a rectangular tank, by the revolutions of a propeller fixed to a properller shaft and driven by an electric motor. The coagulant is brought by coagulant feeding pipe and is discharged just under the impeller. The raw water is brought from the inlet pipe and is deflected towards the impeller by the deflecting wall. The thoroughly mixed water finally goes out from outlet.

The impeller's speed is generally kept between 100 to 120 r.p.m. and the period of detention may vary between 2 to 3 minutes.

5.5.4. Flocculator

When the thorough mixing of coagulants in the water is over, the next operation is flocculation. The device by means of which floc is formed is called flocculator and the phenomena is termed as flocculation.

Flocculators essentially consist of a circular tank equipped with paddles revolving on a vertical shaft. The paddles operate at a slow speed of 2 to 3 R.P.M. The water enters from the inlet and leaves through the outlet. The detention time allowed for best flocculation varies from 30 to 60 minutes. Fig. 5.12 shows the mechanical focculator which is mainly used in water works.

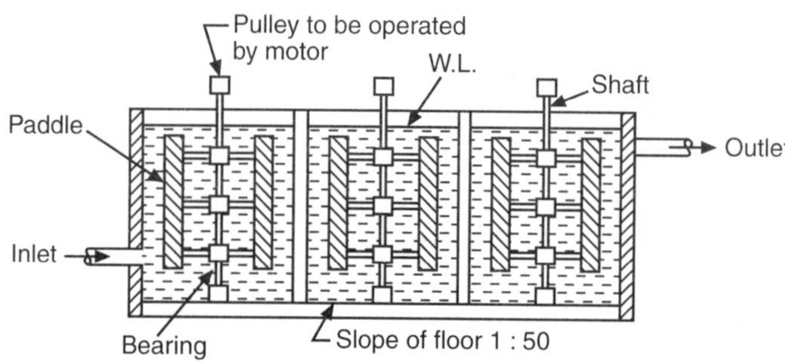

Fig. 5.12. Mechanical flocculator

5.5.4.1 *Advantages of Mechanical Mixing Devices Over the Baffle Type Basins*

(*i*) As much as 40% reduction in the amounts of coagulant due to through mixing.

(*ii*) Negligible loss of head against about 60 cms in the case of baffle-type basin.

(*iii*) Greater flexibility of operation.

(*iv*) Lower cost of installation.

5.5.4.2 *Disadvantages of Mechanical Flocculators*

(*i*) Dead spaces in corners.

(*ii*) Low velocity near the shaft of paddles.

(*iii*) Bad short circuiting.

(*iv*) Requires careful supervision and maintenance.

5.5.4.3 *Limitation of the Coagulation Process*

This process have the following limitations:

(*i*) It removes the suspended impurities of water and thereby reduces the sediment load on next process called filtration.

(*ii*) Normally it reduces the turbidity of water to 20 p.p.m. and if the process is properly controlled then the turbidity reduces to 5–10 p.p.m.

(*iii*) The floc formed in this process also removes bacteria upto 65%.

(*iv*) B-Coli index is removed by 70%.

5.5.5. Filtration

The process of passing the water through a thick layer of sand and gravel which acts as a strainer is called filtration.

5.5.5.1 Theory of Filtration

The phenomena of which filtration process removes bacterias, colour, taste, odours, iron, manganese and make water sparkling involves the following five different actions:

1. Mechanical Straining: The sand bed contains a large number of voids or interstices between the sand grains. When water is passing through these voids, the suspended matter that is too large to pass through, is retained on the surface of the sand bed. This action is called mechanical straining which removes the suspended particles.

2. Sedimentation: Sedimentation and absorption remove small particles of suspended matter, colloids and bacteria. The voids in the sand bed form a number of small sedimentation basins in which the small suspended particles settle upon the sides of the sand grains. A gelatinous coating formed round the sand grains by previously deposited bacteria and colloidal matter and the physical attraction between the suspended particles and the sand grains themselves accelerate the action of sedimentation.

3. Absorption: The huge internal and external surface area of the colloidal masses coupled with the electrical charge carried by the particles enables the material to attract large quantity of such particles. This ability to attract and hold particles at the surface is called absorption. In other words, the colloidal and dissolved matter can be made to adhere to the surface of a solid material by the phenomenon of absorption and thus impurities can be removed. The colour and odour from water is also removed by absorption to the extent of about 15 to 25 percent.

4. Biological Activity: The bacteria also contributes to sedimentation and flocculation by forming sticky, gelatinous coatings on the surface of the filter grains. Finely divided solids adhere to these coatings while even dissolved and colloidal solids are utilized for growth and energy. It gives physical as well as chemical changes in the quality of water.

5. Electrolytic Action: The sand particles of filter media and ionised matter in the water carries electrical changes of opposite nature. Therefore they attract each other and neutralise the charge of each other. In this action, the chemical constituents of the water are altered.

5. 5.5.2 *Sand for Filtration*

The sand used as filter-media should confirm the following properties:

1. It should be free from foreign material like clay, loam, lime and organic matter etc. It should be hard and resistant to atmospheric actions. If placed in hydrochloric acid for 24 hours. It should not lose more than 5% by weight.

2. Effected size of sand: The size of the sand should neither be very fine nor very coarse because the former is clogged very quickly and the latter does not retain suspended particles and bacteria. Therefore the size of sand particles is specified by the effective size which is the sieve size in mm that permits 10% of sand by weight to pass.

3. Uniformity of sand: It is specified by uniformity co-efficient which is the ratio between the sieve size that will pass 60% to the effective depth. Sand with an effective size lying between 0.30 and 0.55 and having uniformity co-efficient between 1.30 and 1.75 is used.

4. Depth of sand bed: The depth of sand should be between 60 cm. to 90 cm.

5. Grading of sand bed: The sand bed consists of graded sand in various layers (a) For slow sand filters, Sand having effective size between 0.3 and 0.25 mm and uniformity co-efficient of 1.75 is used in India practice. (b) For rapid sand filters, Sand having effective size between 0.35 and 0.5 mm, and uniformity co-efficient of 1.6 is used.

Table 5.3. Percent size distribution of filter sand grains

Percent size	Fine		Medium		Coarse	
	Min^{m}	Max^{m}	Min^{m}	Max^{m}	Min^{m}	Max^{m}
1	0.20	0.30	0.35	0.40	0.41	0.45
10	0.35	0.45	0.45	0.55	0.55	0.65
60	0.50	0.75	0.70	0.90	0.85	1.10
99	0.90	1.50	1.20	1.80	1.45	2.00

Note: Uniform grading of Sand bed is very essential because it decreases the void spaces and makes bottom and top of filter bed equally effective, and increases the rate of filtration.

5.5.5.3 *Gravel for Filter Media*

The sand bed is supported on the gravel bed. Gravel bed has many functions due to which it is provided below the Sand bed. The functions are enumerated as under:

(*i*) It supports the sand and allows the filtered water to flow freely towards the under-drains.

(*ii*) It also allows the wash water to move upward uniformly on the sand.

The gravel is placed in 5-6 layers having finest size on the top. The gravel used in filtration plant should be clean, hard, durable and rounded. It should be free from clay, loam, shells and foreign matter. Table 5.4 shows grading of gravel.

Table 5.4. Grading of gravel

Layers	Thickness of layer in mm	Size of gravel in mm
1st	5–8	2.4–4.75
2nd	5–8	4.75–12.75
3rd	7–13	12.75–19.00
4th	7–13	19.00–38.00
5th	13–20	38.00–63.00

5.5.5.4 Classification of Filters

Filters may be divided into main classes:

1. **Gravity Filters**
   (*a*) Slow Sand Filters 100–180 litre/m^2/hours
   (*b*) Rapid Sand Filters [24 to 30 Million litres/HA/day]
2. **Pressure Filters:** The difference in the above types of filters lies in the head required for filtration of water and the rate of water filtered through the filter.

5.5.5.5 Slow Sand Filters

These were first devised in England in 1829 by Mr. James Simson. Mr. James was employed in Chelsa Water Co. These filters were widely used till the last decade by 19th century when the rapid filters were invented.

5.5.5.5.1 Function

The function of slow-sand filters are listed as under:

(*i*) The suspended and settleable particles which have not been removed by sedimentation, are removed by this filter.

(*ii*) The slow sand filter removes bacteria, to an extent of 98 to 99 per cent, from the water having turbidity less than 50 ppm.

(*iii*) The slow sand filters also remove odours and tastes to a certain extent.

5.5.5.5.2 Constructional Details of the Filter

Essential parts: Shallow tank with water-tight walls and floor, under-drainage system, base materials of coarse gravel, filter media of sand and appurtenance for efficient working of the filter such as the device for measuring the loss of head and device for controlling the rate of flow through the filter.

5.5.5.5.3 Features of Slow Sand Filter

(*i*) Rate of filtration - 100 - 180 litre/m^2/hr
(*ii*) Size of Unit - 0.2 Hectare
(*iii*) Depth of filter media - 90 cm Sand; 30 cm gravel
(*iv*) Size of Sand - 0.25 to 0.35 mm.
(*v*) Uniformity co-efficient of Sand - 1.75
(*vi*) Length of run between cleanings - 20–60 days.
(*vii*) Amount of water for washing - 0.5% of filtered water

5.5.5.5.4 Construction

They consists of open water-tight tanks 2000 to 4000 square meters in plan and 3 to 4 metres deep containing 600–900 mm thick bed of sand (filtering media) supported on 300 mm thick gravel layer. The effective size of sand is 0.35 mm. with a uniformity coefficient of about 1.75.

The top of 150 mm layer of the filtering media is of a finer variety 0.25 mm. Usually sand obtained for quartz and quartzite stones is used to act as filtering media. It should be free from organic matter and other foreign material. The gravel used must be of similar good quality. It is made up of particles varying from 4.7 to 72 mm size and is placed in layers with the smallest size particles at the top and the largest on the bottom.

The sand and gravel are laid over a system of open jointed underdrains placed 3 to 6 meters apart on the bottom floor sloping towards a main covered drain constructed along the centre or aside of the filter tank.

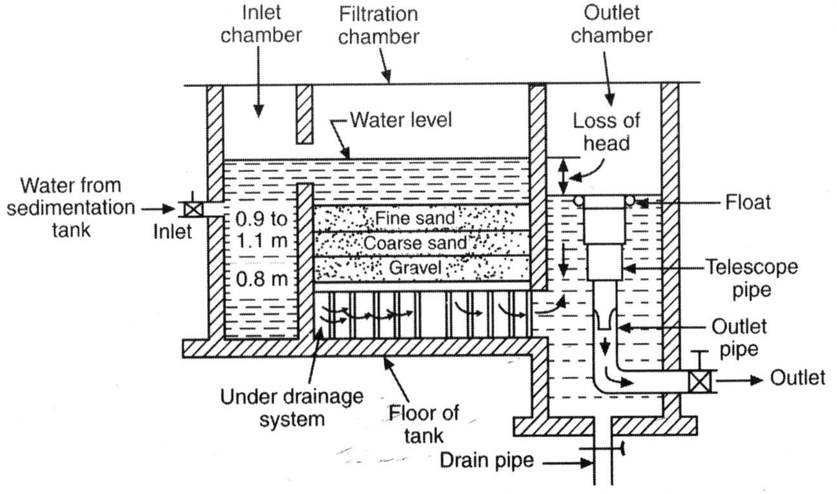

Fig. 5.13. Slow sand filter

There is a small chamber on the inlet side to admit settled water into the filter tank without disturbing the sand bed. The depth of water equal to the thickness of the filtering media has been found most suitable. On the outlet side is constructed a similar small chamber to draw out the filtered water at a constant rate by means of some controlling device known as telescopic pipe of adjusting weir.

5.5.5.5.5 ***Operation***

Settled water from plain-sedimentation tank takes entry through submerged openings or the stilling tanks and then gets distributed over the filter bed uniformly. It percolates down through the sand bed and the supporting gravel and get purified during the process of filteration. The under-drainage system collects the filtered water and passes it on the clear water reservoir.

It is essential to keep the rate of filtration constant. Different devices which are installed in the outlet chamber, can be used for the purpose. An open type of rate controller consists of a short piece of vertical telescopic pipe supported by an annular ring which floats on the water surface at a fixed distance above the submerged open end of the pipe.

Since the water falls under a constant head over a circular weir of fixed length, the flow of water falling into the outlet pipe remains constant although the water level in the outlet chamber may vary due to the gradual choking of the filter.

As the filter media gets clogged, the resistance to the passage of water and the head loss increases, the filtering media is then cleaned by throwing filter unit out of service. The water in the filter tank is first drained, the thick layer of impurities collected over the surface is then scrapped and washed clean with water jets from hoses. The filter is then put back into normal operation. The interval between successive washing will vary from 3 to 6 weeks depending upon the nature and amount of turbidity removed.

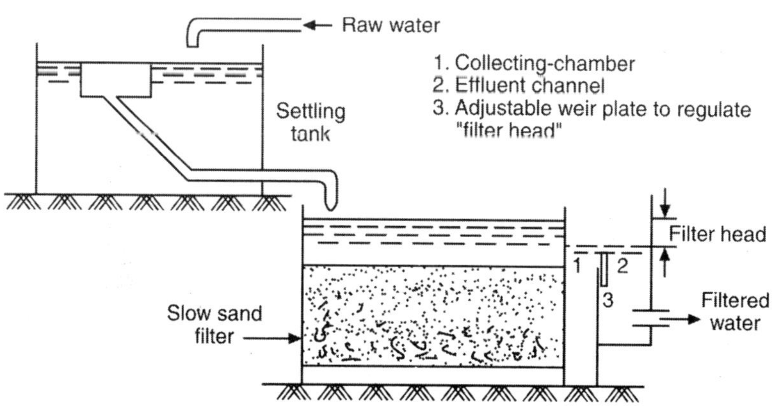

Fig. 5.14. Flow through diagram of slow sand filter

During each washing, some sand will be lost and the thickness of the sand layer will go on decreasing. New sand will be added when thickness reduces to about 0.6 m. After some years of working the impurities might penetrate deep into the filter bed and cause its complete clogging increasing thereby the resistance to the passage of water. The sand grains may also get encrusted with the absorbed impurity. The filter will then be put out of operation. The whole of the sand will be taken out, washed and replaced.

Since the filtration action of slow sand filters depends to a greater extent on the formation of "Zoogleal" (the blanket of arrested impurities) on the surface of filter media, the effluent from the filters is required to be wasted for a short while, after each washing of the sand bed till it formed again.

The filters are normally operated at a rate of 22.7 to 68 cubic metre/hectare/day by keeping the water level in the filter tank 0.9 to 1.5 metre above sand surface. The loss of head will be as low as 10 cm in the beginning of a filter run and as high as 1.3 metre just before washing.

EXAMPLE 5.1. Design a slow sand filter for NILOKHERI Town with 20,000 population assuming the water supply as 90 litres per capita per day.

Solution: Max. demand = 1.8 × 90 × 20000 l/d = 3.24 Ml/d

Assuming the rate of filtration = 45 Ml/d/ha

$$\text{We get area of filter bed} = \frac{3.24}{45}\text{h.a} = \frac{3.24 \times 10^4}{45}\text{sq m.}$$

$$= 720 \text{ sq m.}$$

Providing 2 units for the quantity to be removed and one unit more as extra:

$$\text{Area of each unit} = \frac{720}{2} = 360 \text{ sq m. Say } 15 \text{ m} \times 25 \text{ m}$$

Hence three such unit will be necessary.

5.5.5.6 *The Rapid Gravity Sand Filter*

5.5.5.6.1 *General*

The filtering media used in this type of filter is of a coarser variety and the operation heads are also higher so that the rates of loading can be about 30 times of those used in slow sand filters. The amount of filter sand and the land area required is, therefore, quite small which economises the initial cost of construction. Since filters remove large volumes of impurities in a short time and so they get clogged quickly is carried out by using some mechanical equipment which saves time and labour but at the same time it requires skilled attendance and supervision. Since these filters are washed at short interval of time, there is no opportunity left for the biological impurities to form filtering mat on sand surface as in the case of a slow sand filter. The filters have to, therefore, depend upon the unsettled floc to form the coagulation sedimentation basins to form straining layer which is essential for the filtering action.

It means that the settling (sedimentation) preceding filtration must be aided with chemical coagulation and this also helps in reducing the turbidity load on the influent to filters and ensures long filter runs.

5.5.5.6.2 *Function of Rapid Sand Filter*

(*i*) It removes colour and odour completely from the settled water from coagulated sedimentation tanks.

(*ii*) The suspended matter is removed by this filter.

(*iii*) The colloidal (*i*) impurities are removed if coagulation with chemical is done before filtration.

(*iv*) The pathogenic bacterias are not completely removed by this filter alone but if pre-chloration is done, then bacterial load is completely removed.

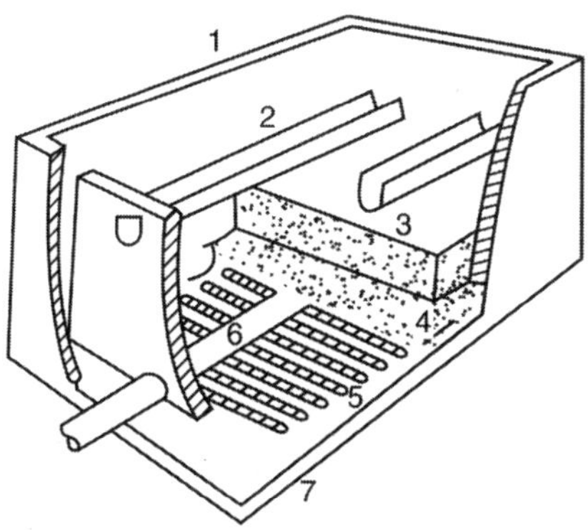

Fig. 5.15. An isometric view of a rapid filter showing details

5.5.5.6.3 *Features of a Rapid Sand Filter*

(*i*) Rate of filtration; 1200 to 2250 M. lit/ha/day or 100 litr/m^2/mt.

(*ii*) Size of one unit: $\left(\frac{1}{250}\text{ha to }\frac{1}{500}\text{ha}\right)$

(*iii*) Depth of filter media (*a*) Sand—600-750 mm (*b*) Gravel—450-600 mm.

(*iv*) Effective size of sand: 0.45 to 0.8 mm.

(*v*) Loss of head - 0.3 m initial to 2.1 m final

(*vi*) Length of run between cleanings 24 - 60 hours

(*vii*) Amount of wash-water used in cleaning: 1 to 4% of filtered water.

5.5.5.6.4 *Essential Parts of Construction*

(*a*) Filter unit: The filter units are smaller in size and are placed under a roof or in building so that the costly mechanical equipment does not get damaged due to weather action and there is also less of algae troubles. They are arranged in parallel rows leaving space for the pipe gallerias in between and for the future extension. The individual units are rectangular in plan with length breadth ratio 1.25 to 1.33. Their sizes varying from 3 m × 5 m to 6 m × 8 m. The actual sizes and shapes may however differ depending upon the site conditions.

The filter tanks are of a depth 2.9 m to 3 m and consist of an underdrainage system to the bottom and gravel and sand layers on top in across the width or lengths of the tanks for the distribution of water to be supplied and for the collection of wash-water. They remain submerged during the normal operation by about 0.6 to 0.9 m and are so placed that their lips stand 0.6 m above the normal surface of the sand so that it does not get washed and carried away through them the during back washing.

(*b*) Filter Media (Sand): The filter media used is 60 cm to 75 cm deep consisting of hard, inert and tough sand but of coarser variety and uniform size. The coarser sand allows higher ratio of filter and deeper penetration of the impurities by water but cannot effectively arrest the bacteria so that the efficiency of bacteria removal is lowered down to 95%. However nowadays due to evolution of simple, cheap and reliable methods of disinfection of water, the filters are not dependent upon for the removal of bacteria and hence the coarser sand does not prove to be disadvantage.

Normally sand is used as a filtering material because it is cheaply and easily available as a low cost at most of the places, however, alternative materials used are crushed anthracite and glass. Anthracite can give longer filter runs, requirer less amount of wash-water and gives cleaner filter beds.

(*c*) Gravel: The gravel is laid in it in 5 layers above the under drains and below the filtering media to, (*i*) support the sand and prevent it from entering into drains and thus not allow them to choke, (*ii*) Help in uniform distribution of wash water up through the filter tank.

Size of gravel may vary from 25 mm at the bottom to 5 mm at the top. The depth of gravel bed may vary from 45 to 60 cm. The faster is the rate of washing of bed, the larger the gravel size and deeper bed required. The attempts are being made to reduce the depth of gravel layers and even to eliminate them by introducing stainer of special design which do not allow it enter into the under-drains. Such elimination of gravel helps in reducing the total depth of the filter tank and gives economy in the initial cost of construction. It also reduces the loss of head in filters.

(*d*) The Under Drainage System: The function of under-drains is to:

(*i*) collect the filtered water percolating down through the sand and gravel layers.

(*ii*) distribute upwards the wash water uniformily underneath the gravel.

5.5.5.6.5 *Various Types of Under Drainage Systems*

(*a*) Perforated-pipe under-drains or the manifold and lateral pipe system.

(*b*) Vitrified clay tile block design.
(*c*) Wheeler under-drainage system.
(*d*) Porous plate bottoms.
(*e*) Concrete ridge and valley bottoms.

Out of the above five types the first type of under-drainage system is more commonly used. It consists of a cast iron main of 50 cm. dia placed longitudinally along the centre of the bottom floor with a number of 7.5 – 10 cm dia laterals placed across at 15 to 20 cm centres and connected to central manifold. The lateral have 6–12 mm dia perforated holes drilled to their bottom either vertically or inclined at 30° to 60° with the vertical or strainer of special design mounted on top at 10 to 15 cm centres. In the former case, the laterals are placed on concrete blocks 4 cm above the filter bottom and the pipe material around the perforations is protected against corrosion and erosion by lining it with brass or bronze bushings.

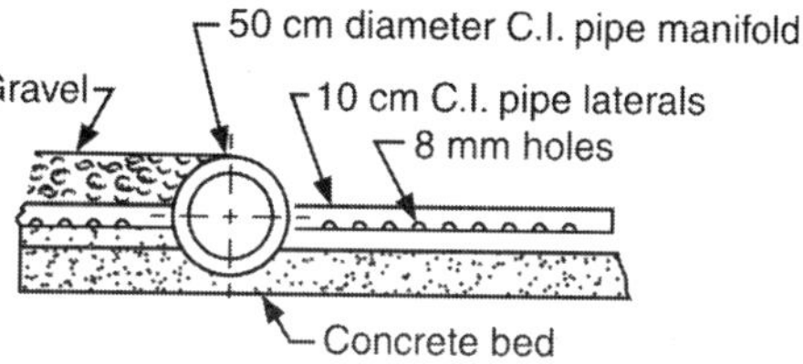

Fig. 5.16. Perforated pipe type

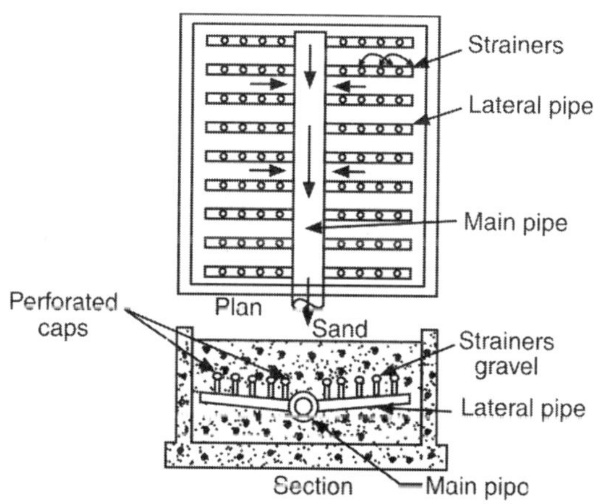

Fig. 5.17. Manifold and laterals

The strainers used in lateral case are made of brass, bronze or plastics and have the shape of an umbrella or a cylinder having small preforations. In both the cases the jet actions are prevented during back washing and the gravel or the filtering media does not get displaced. The combined area of the orifices or openings is kept between 0.2 to 0.3% of the plain area of the filter unit. The whole system is illustrated in Fig. 5.16 which gives the plan and elevation of the same.

Vitrified Tile Blocks: It essentially consists of vitrified tile blocks having orifices and holes in them for passage of water. All these tiles are laid in the bottom, over which gravel layer is laid as shown in Fig. 5.18.

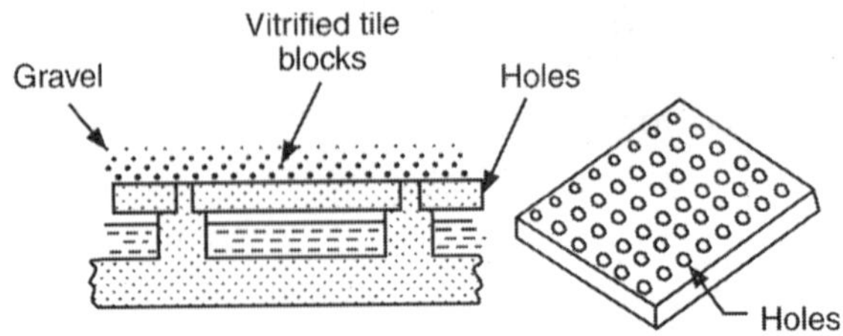

Fig. 5.18. Vitrified tile blocks

The wheeler filter bottom: This is very cheap method. In this system a false bottom is prepared above the real bottom by RCC. This false bottom consists of solid, inverted, truncated pyramids with water connections at the apex of each pyramid and the pockets filled with cement or glazed earthen-ware spheres as shown in Fig. 5.19.

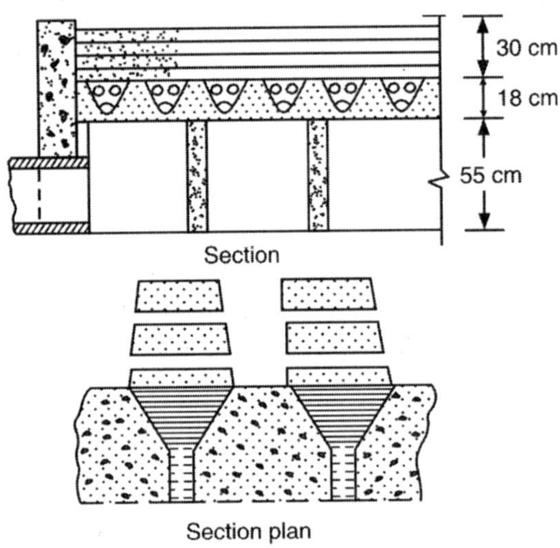

Fig. 5.19. Wheeler under drainage system

The porous Plate Bottom: In this case also a false bottom is prepared with porous plates. The sand layer is directly laid over it and there is no necessity of gravel layer. In this type of underdrainage system, the rate of filtration and wash water distribution is very uniform.

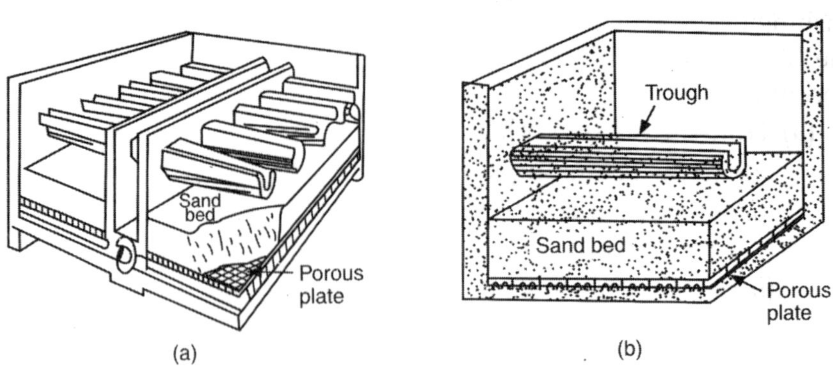

Fig. 5.20. Porous plate bottoms

5.5.5.6.6 *Operation of Rapid Sand Filter*

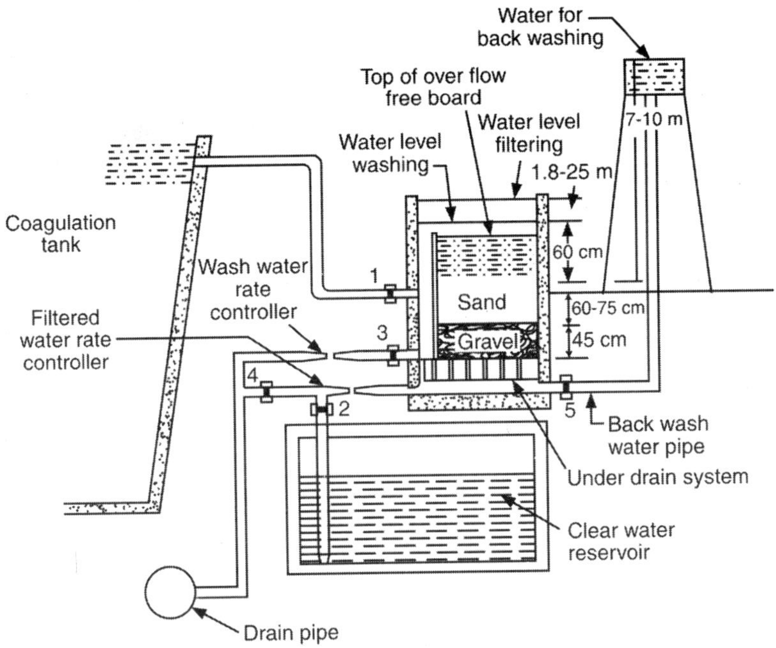

Fig. 5.21. Operation of a rapid sand filter.

The pretreated water from the coagulation sedimentation tanks through the inlet valve is admitted into the filter units. It is distributed by the troughs which remain submerged while working, over the entire bed area. The water then percolates through the sand and gravel layers and thereby the fine suspended and colloidal impurities present in it are arrested by the sand layer. After this, it enters the laterals through the strainer or holes and then leaves the filter unit via the manifold.

In the beginning the loss of head of a filter is about 15 to 22 cm. As the filtering media goes on getting clogged with the arrested impurities, the resistance to passage of water and hence the loss of head increases. Usually a maximum loss of head of 2 to 2.5 metres is permissible. If it increases further till it may pack the sand so tightly as to cause difficulties in back washing or may cause the water to break through sand without filtration. The filter unit is then washed, the clogging is removed and the normal working resumed.

(*a*) **How the Filter Works:**

(*i*) In the beginning keep all the valves closed.

(*ii*) Open the valve (1) to allow coagulated and settled water to flow to the filter.

(*iii*) Open the valve (2). This allows the water filtered through the sand, gravel and under-drains to flow into the clear water storage.

(b) How the Filter is Washed:

(i) Close valve (1) to stop settled water flowing to the filter.

(ii) Keeping all the valves closed, open the air valve (3) to pass the compressed air through the separate pipe system for 2-3 minutes which agitates the sand bed and causes loosening of the dirt, clay etc.

(iii) Open valves (5) and (4) to allow water from wash water tank to be forced up through the under-drains, gravel and sand. Due to back flow of water, the sand expands and all the impurities are carried away with the wash water to the drains.

The washing process is continued till the sand appears clean and neat. The washing of filter is done generally after 24 hours and it takes about 10 minutes and during back washing the sand expands by about 50%.

(c) How to Run Filter to Waste:

(i) Close valves, (2), (3) and (5) and open valves (1) and (4). This allows the coagulated and settled water to flow through the filter, and under-drains to run to waste through the valve (4). This is done for a few minutes only for conditioning the filter before it is put into service again.

(d) How to Resume Filtering:

Open valve (2) and close valve (4).

5.5.6 Negative Head

The loss of head in the filter operation is caused by the frictional resistance offered by the filtering medium and the under-drains to the water flow. The value of head loss is equal to the vertical distance between the water level on the filter bed and the elevation of the hydraulic grade line at the filter outlet. The loss of head immediately after washing should not be more than 10 to 15 cm. When the sand bed gets choked up with impurities the rate of discharge falls down and it is necessary to wash it to regain its original clean condition and the stipulated rate of discharge.

In the filters provided with loss of head indicators this choked up condition is shown by increase in the loss of head so that when it reaches a certain figure generally about 1–8 meters, the filter requires washing.

When the loss in the top layer of sand becomes greater than the head of water above the sand bed (due to choking) and the water column below such layers acts as a draft tube and a partial vacuum is created. This condition is known as negative head. Thus the negative head at any point in a filter is equal to the intensity of vacuum at the point and is usually a maximum at the point where the water enters the underdrainage system.

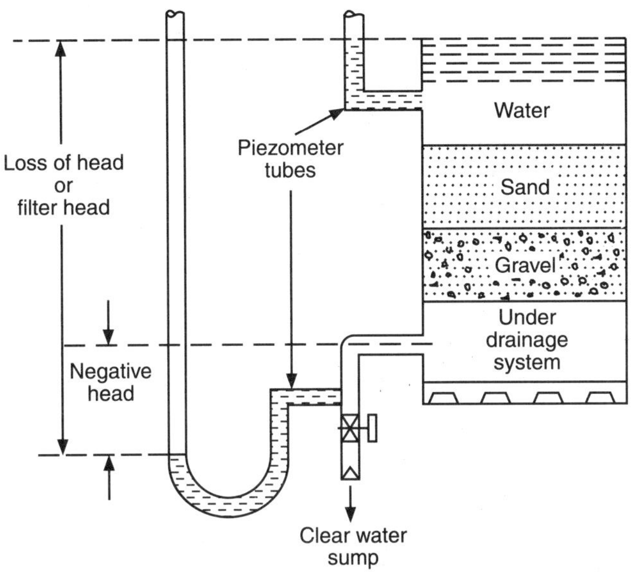

Fig. 5.22. Negative head

EXAMPLE 5.2. Design the rapid sand filters to treat 2.27 ml/d supply of a town. The filters works for 12 hrs a day. Assume rate of filtration of 1123 ml/d/ha.

Solution:

$$\text{Quantity of water to be treated per hour} = \frac{2.27}{12}\text{ml}$$

$$\text{Filtering area} = \frac{2.27 \times 24}{12 \times 1123} \text{ hectare}$$

$$= \frac{2.27 \times 24}{12 \times 1123} \times 10^4 \text{ sqm.} = 4040 \text{ sqm}$$

Providing 2 units

$$\text{Area of each unit} = \frac{40.4}{2} = 20.2 \text{ sqm.}$$

The back washing is necessary in 2 or 3 days and hence every filter may have its turn every day. Assuming washing time as 20 minutes (nearly 3% of 12 hrs.) and adding 3% of 20.2 m^2 or 0.606 m^2.

$$\text{Area of each unit} = 20.2 + 0.606$$

$$= 20.806 \text{ sqm}$$

$$= 6.0 \text{ m} \times 3.5 \text{ m, say}$$

Both the units should be placed side by side and hence the space occupied will be 7 m × 6 m plus the thickness of walls. **Ans.**

5.5.7 Pressure Filters

A pressure filter is a small rapid gravity sand filter placed in a closed cylindrical vessel and through which the water to be treated is passed under pressure. The vessel may be installed either in a vertical or horizontal position and its size may vary from 40 cm to 240 cms dia for vertical types and 2 meters long; 2–2.5 metres dia and 2.5 metres long for horizontal type.

The filter is operated like an ordinary rapid gravity filter except that the applied water may not be flocculated or conditioned after addition of the coagulants. The coagulants commonly used are alums. The alum is kept in a pressure container which is connected to the influent filters are operated at high rate of 100–250 litre/sq. metre/day of filter bed area. When the loss of head increases due to clogging effect, the filter is cleaned by back washing. The pressure filters are classified as:

(*i*) Vertical pressure filters. (*ii*) Horizontal pressure filters.

These filters are not used for treating municipal water supplies, because:

(*i*) Although the rate of filtration is high, the filter units being small the overall plant capacity is small.

(*ii*) They cannot reliably remove the bacteria and are inefficient to remove the turbidity.

(*iii*) They are costly.

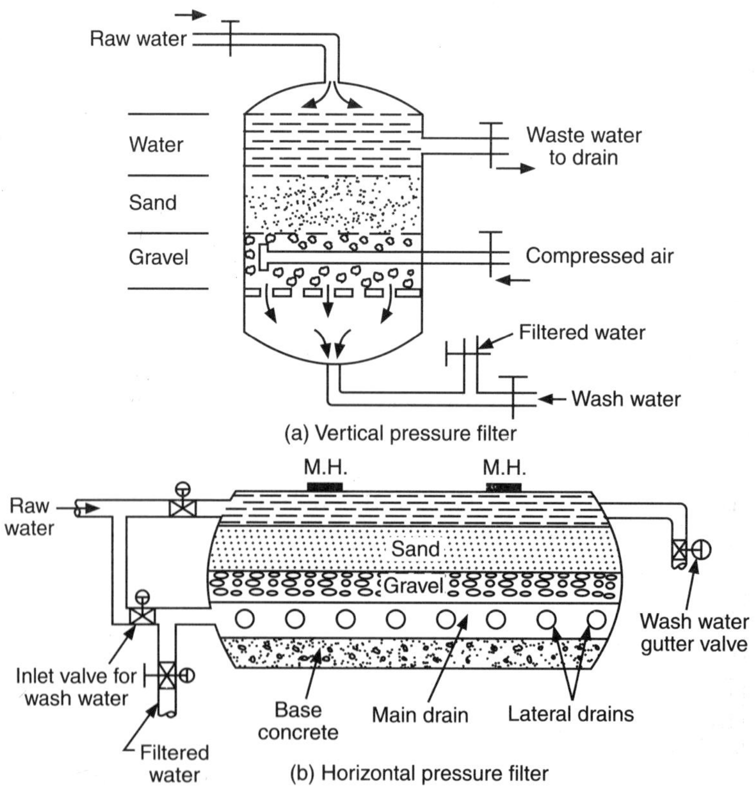

Fig. 5.23. Pressure filter

The other disadvantages are:

(*i*) The filter operation cannot be seen and inspected as they continue to be closed vessel.

(*ii*) It is difficult to inspect, clean and replace the sand, gravel and under-drains.

(*iii*) Because of shape no wash water gutter can be provided.

(*iv*) The operation of these filters being under pressure, it requires pumping of water at excessive rate.

The pressure filters are useful in small plants where automatic operations for receiving and pumping water are available. The filters are used in clarifying softened water in industries and in treating the swimming pool water that is to be recirculated.

5.6. COMPARISON OF SLOW SAND AND RAPID GRAVITY FILTERS

S.No.	Item	Slow sand filter	Rapid gravity filter
1.	Rate of filtration	100 to 200 lit/hour/sq. m of filter area	3000 to 6000 litres/hour/sqm. of filter area
2.	Size of one unit	30 m × 60 m are varies from 500 to 2000 sq. m.	6 m × 8 m to 8 m × 10 m area varies from 40 to 80 sq.m
3.	Depth of Filter Media	30 cm. gravel, 90–105 cm. sand	45 cm. gravel, 75 cm. or less sand.
4.	Effective size of sand	0.25 to 0.35 mm $C_u = 1.75$	0.45 to 0.8 mm $C_u = 1.6$
5.	Distribution of Particles	Uniform	stratified smallest at top coarsest at bottom.
6.	Under-drainage system	Open jointed pipe or covered drains	Strainer pipe, laterals discharging into C.I. main
7.	Loss of head	9 cm. initial to 1.2 m final	0.3 m initial, 2.1 m final
8.	Length of run between cleanings	20–60 days	24–60 days
9.	Method of cleaning	Scrapping of surface and washing down by hoses	Scour by back washing and removal of dislodged particles by upward flow.
10.	Penetration of suspended matter	Very small	Deep
11.	Amount of wash water use in cleaning	0.2 to 0.6% of water filter	1 to 4% of water filter.
12.	Preparatory treatment	Aeration along or with flocculation and setting	Aeration flocculation and setting.

(Contd)...

13.	Cost of construction	Higher	Lower
14.	Cost of operation	Lower	Higher
15.	Depreciation of plant	Lower	Higher
16.	Quality of raw water	It may not be treated with chemicals, but should not have turbidity more than 50 ppm.	Treatment with coagulants is essential.
17.	Area	Very large	Small area
18.	Construction	Simple	Complicated
19.	Flexibility in operation	Not possible	Possible
20.	Skilled supervision	Not required	Most essential.
21.	Efficiency	Efficient in removal of bacterias and suspended impurities	Cannot remove all bacterias, disinfaction necessary. More efficient in removal of colour, odour and taste.
22.	Size of one unit	30 m × 60 m	6 m × 8 m to 8 m × 10 m

EXAMPLE 5.3. Design slow sand filters for a town of 30,000 population, the average daily demand being 135 litres/head/day. The maximum demand may be taken as 1.5 times the average.

Solution: Max. demand per day

$$= 30000 \times 135 \times 1.5$$
$$= 6075000 \text{ litres}$$

Assuming the rate of filtration as 35 million litres per hectare per day,

the required area in sq. metres $= \dfrac{6075000}{35000000} \times 100^2 = 1735.7 \text{ m}^2$

This is a large area which cannot be provided by one filter only. If 3 beds are assumed each will have an area of about 600 m^2. Suitable size shall be 30 m × 20 m, giving this area. Adding one more filter as spare for cleaning purposes and breakdowns, total number will be 4.

5.7. DISINFECTION

The treatment of water with chemicals to kill bacteria is called disinfection.

5.7.1. Necessity of Disinfection

Necessity in the water treatment processes like sedimentation, coagulation, filteration, etc., considered so far, all the bacteria from the water cannot be removed. Moreover there is every chance of getting the water contaminated using its flow through the water distribution system especially in case of intermittent supply, where the pipes remain empty for a considerable period. Therefore, in order to ensure safety to such water it is sterilized or disinfected as soon as it leaves by chlorine or bleaching powder.

5.7.2. Requirements of Disinfectants

The requirements of good disinfectants may be:

(*i*) They should be able to destroy all the harmful pathogenic bacteria and make the water perfectly safe for use.

(*ii*) They should be economical and easily available.

(*iii*) They should be able to kill all pathogenic germs within required time at normal temperature.

(*iv*) After their treatment the water should not become objectionable and toxic to the customer.

(*v*) The disinfectant dose should be such that, it may leave some residual concentration for protection against contamination in the water.

5.7.3. Methods of Disinfection

The methods employed to disinfect water depend upon the disinfecting materials used for the purpose. The disinfection of water can be done by the various methods such as :

(*a*) **By Heating or Boiling of water:** The water can be disinfected by boiling for 10 to 15 minutes. This method is known as sterilization. By boiling water all the disease-causing germs are killed and the water becomes perfectly safe for use. This method being costly and can be used only in individual case in emergency.

(*b*) **By Light:** If the water is exposed for sufficient time in sunlight, it is disinfected. Sun light is considered to be a good natural disinfectant. The ultraviolet ray is quire effective to kill microbial bacteria in water. It can penetrate through a depth of 30 cm of clear water. The water is allowed to pass in thin layers and the ray is applied to it. As it is costlier, its use is limited to dispensaries and surgical purposes.

(*c*) **By Chemical Disinfectants:** For disinfecting water on large scale, chemicals have been found to be most efficient. These are economical too. Hence in municipal and industrial water supplies only chemicals e.g., chlorine, potassium permanganate, iodine, silver and copper ions, caustic lime etc., have been used as disinfectant. In practice, chlorine and its compounds are mostly used because they meet all the general requirements of disinfectants.

5.8. CHLORINATION

The process of applying small quantities of chlorine or chlorine compounds to water is called chlorination.

The chlorine dose applied is generally 1 ppm so as to produce residual chlorine or quantity varying from a trace and about 0.05 to 0.20 ppm. The quantity of chlorine so required is known as chlorine demand of water. Chlorine demand is defined as the difference between the amount of chlorine added and the amount of chlorine remaining at the end of a contact period of 15-30 minutes. As chlorination possesses great disinfecting powers, therefore this method is universally employed for disinfection public water supplies.

5.8.1. Action of Chlorine

The commercial product of chlorine is about 99.99% pure. Chlorine gas is soluble in water and forms a solution containing a decreasing percentage of chlorine as temperature rises. Chlorine reacts with water to produce hydrochlorous acid (HOCl) and hypochlorite ion (OCl^-), which are together known as free available chlorine. The chemical action may be given as:

$$Cl_2 + H_2O = HOCl + HCl$$

$$HOCl = H^+ + OCl^-$$

In case ammonia is present in water, other formed compounds may be monochloramine (NH_2Cl) and dichloramine ($NHCl_2$), which together are known as *Combined chlorine.*

Second theory about the destruction of bacteria to the liberation of nascent (newly born) oxygen ($HOCl = HCl + O^-$) which oxidizes the germs. This theory is nowadays considered to be inadequate due to the fact that the quantity of nascent oxygen produced is too less for the purpose.

5.8.2. Application of Chlorine

This may be applied by the following methods :

1. **Dry Chlorine Gas:** It may not be drawn from the liquid chlorine, it is applied directly to the water supply through submerged diffusers. This method is not satisfactory because it results corrosion in pipes.
2. **Chlorine Solution:** The chlorine solution formed by mixing chlorine gas with small quantity of water is applied to the water supply through a discharge line. This method is very commonly used.
3. **Powder Form:** In the form of hypochlorite or bleaching powder. Bleaching powder is very commonly used for smaller water-plants.

5.8.3. Application of Gaseous Chlorine

Where water is to be treated in large quantity (as in chandigarh water works) chlorine gas is used. Some bleaching powders are unstable. They originally contain about one-third by weight of available chlorine, this small proportion is very rapidly reduced during storage, particularly in damp and warm climates. Hence accurate and precise treatment is essential.

Pure chlorine available in gaseous form is entirely free from objections raised above. Chlorine is available in 50 kg and 100 kg cylinders stored under a pressure of 8 to 10 kg/cm^3 in seamless steel cylinders. The chlorine gas may be dissolved in small quantity of water and the solution so prepared is fed to the point of application. An injector type chlorinator may be used for this purpose.

5.8.4 Application of Liquid Chlorine

Liquid chlorine is found more effective in disinfecting filtered water, if liquid chlorine is applied at such point where adequate mixing is done. It can be applied to any pressure. The rate of application of chlorine may be manually controlled or mechanically controlled at various points.

The chlorine applying device is known as chlorinator. Fig. 5.24 shows liquid chlorinator of M/s Wallace and Tiernan Co.

Chlorinators are of two types:

(*i*) Solution feed

(*ii*) Direct feed

In the first type, a measured quantity of chlorine is mixed with a fixed quantity of water before it is conducted to the point of application in the water main or filtered water reservoir. In case of direct feed type chlorinator, a measured quantity of chlorine is fed directly into the water main or the water reservoir. The use of direct feed chlorinator is limited to 2 to 2.5 kg/sq-cm pressure in the delivery tube, whereas the solution feed equipment may carry to any pressure.

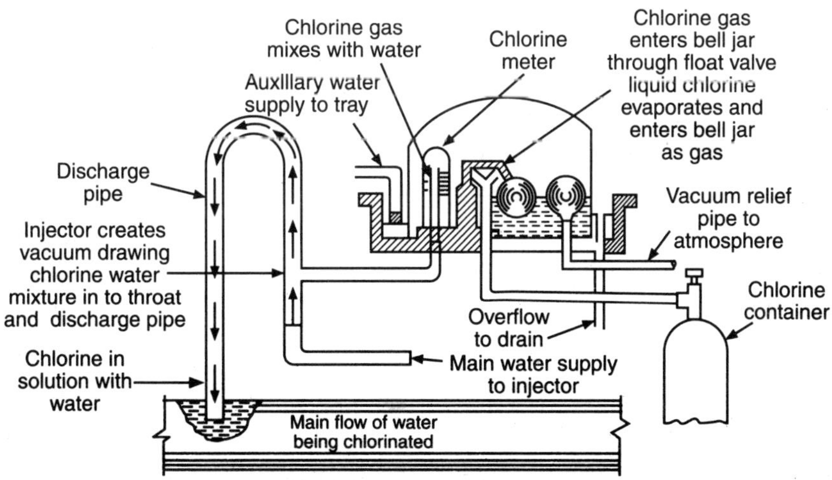

Fig. 5.24. Liquid Chlorinator

5.9. SPECIAL METHOD OF CHLORINATION

Chlorine is generally applied after all other treatment have been given to the water supply. The special methods of chlorination may be as follows :

(*i*) Post-chlorination: As already explained above, when the chlorine is added in the water after all treatments, it is known as post chlorination, it is generally done after filtration. The chlorine is commonly added in the clear water reservoir. The minium contact period should be half an hour, before use of water.

(*ii*) Plain chlorination: When only chlorine treatment is given to raw water, the process is called plain chlorination. The amount of chlorine required is 0.5 mg/l.

(*iii*) Prechlorination: It is the application of chlorine before filtration. It may be added in the suction pipes or in the mixing basins. It reduces bacterial load on filters, this results increased filter runs and oxidizes excessive organic matter. This helps in removing taste and odour and makes the water fit for use.

(*iv*) Super-chlorination: It is application of excessive amount of chlorine to water. The amount of chlorine may vary from 5 to 15 mg/l of water. This is not ordinarily employed but is practised only during the epidemic of waterborne diseases.

(*v*) Double-chlorination: It is application of chlorine at two points in the treatment process. It is also prechlorination with an added treatment to the final effluent from the filters.

Advantages of double chlorination

(*a*) greater removal of bacteria-higher bacterial efficiency.

(*b*) decrease in the load in filters.

(*c*) more factor of safety due to maintenance of two chlorinating plants.

(*d*) control of algae and slimy growths in coagulating basins and filters.

(*vi*) Break point chlorination: (Fig. 5.25)

This term gives an idea of the extent of chlorine added to water. It represents a dose of chlorination beyond which any further addition of chlorine will appear as free residual chlorine.

When chlorine is added to water it reacts with ammonia present in water and forms chloramines. Some of them form chlorine combination. When chlorine is slowly added in water the residual chlorine go on increasing. However, some chlorine is consumed in killing bacterias and so the amount of residual chlorine is slightly less than the added amount as shown by line *oa* in Fig. 5.25. If the addition of chlorine is continued beyond point *a*, the organic matter present in water gets oxidised and so the residual chlorine content suddenly falls from *a* to *b*. Beyond point *b* any further addition of chlorine will remain in water as free chlorine. The point *b* is called the *break point*. The addition of chlorine beyond break point is called *break point chlorination.*

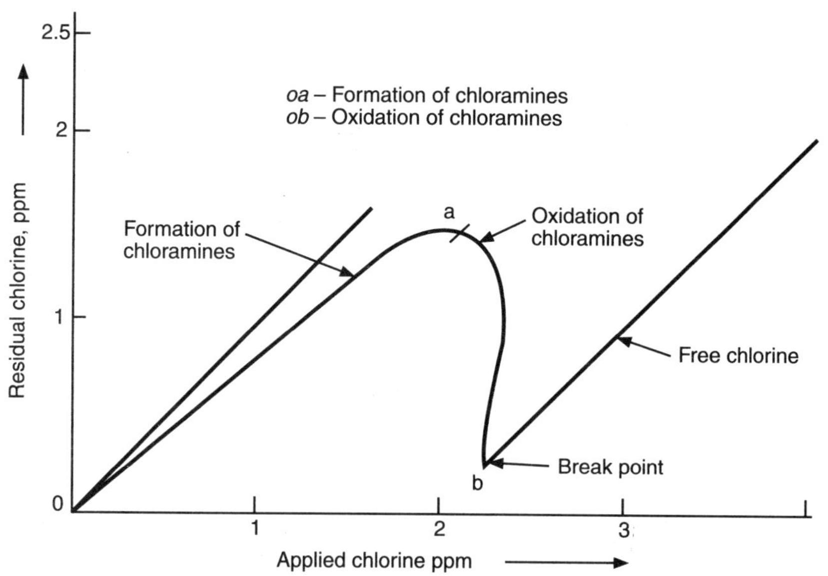

Fig. 5.25. Break point chlorination

Advantages of Break Point Chlorination:

(*a*) It destroys all the disease bacterias.

(*b*) It removes completely taste and odours from the water.

(*c*) It completely oxidizes impurities of water e.g., ammonia etc.

(*d*) It helps in removing colour or water which is due to organic matters.

(*e*) It also prevents weeds growth in water.

(*vii*) Dechlorination: The process of removing excess chlorine from water is called dechlorination.

It must be done in such a way that some residual chlorine remains in water dechlorinating agents or chemicals used are:

(*a*) Potassium permanganate

(*b*) Sodium bisulphate

(*c*) Sodium thiosulphate

(*d*) Sodium sulphite

(*e*) Sulphur dioxide etc.

5.10. AERATION FOUNTAIN

Function: The function of an aeration fountain is to remove the tastes and odour which are unpleasant and the dissolved gases other than oxygen and nitrogen from raw water.

5.10.1 Methods of Aeration

Water can be aerated by the following methods:

1. Exposing it to the atmosphere as an open aqueducts and reservoir.
2. Flowing over cascades, weirs, steps, roughs etc.
3. Flowing through trickling devices such as coke beds and perforated trays.
4. Spraying it through nozzles;
5. Diffusing air through it;
6. Aspiring air through the water.
7. Mixing air and water under pressure to increase the solubility of the air. The first four methods involve the exposure of water to air and are most commonly used.

5.10.2. Construction of Aeration Fountain

It essentially consists of water pipe, nozzle and stop-cocks. When water under pressure is forced through the nozzle, it sprays and the droplets of water come in contact with the atmospheric oxygen. The water absorbs oxygen from atmosphere and gets aerated. The dissolved gases like CO_2, H_2S etc. escape into the atmosphere and the oxidation of organic matters takes place which reduces the odour. The aeration fountain works efficiently at a pressure of 10 to 14 m head of water.

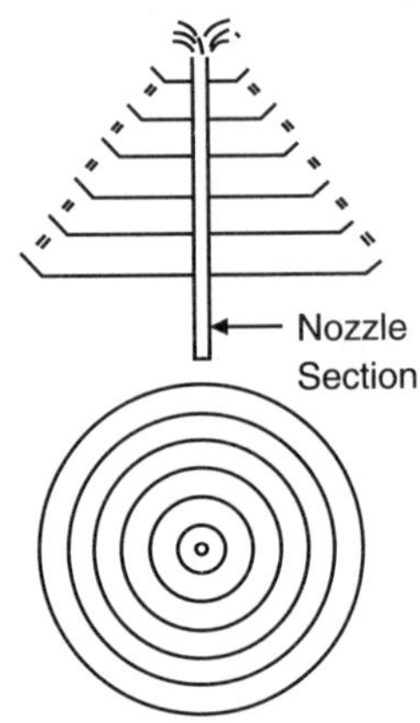

Fig. 5.26. Aeration fountain

5.11. PUBLIC HEALTH LABORATORY

A well equipped public health laboratory should be provided at the treatment plant for checking the quality of raw water and treated water.

The following Physical, Chemical and Biological tests are carried out in the laboratories of water works:

1. **Physical Tests:**
   (*i*) Temperature
   (*ii*) Colour
   (*iii*) Turbidity
   (*iv*) Taste and odour
2. **Chemical Tests:**
   (*i*) Total, volatile and fixed solids
   (*ii*) Hardness
   (*iii*) pH value, acidity and alkalinity
   (*iv*) Free carbon-dioxide

(*v*) Nitrogen—including Nitrite, Nitrate, Albuminoid and free ammonia

(*vi*) Chloride

(*vii*) Active chlorine (Free and combined)

3. **Biological Tests:**

(*i*) Presence of members of B-coli group

(*ii*) Plate counts for bathing and swimming pool water

(*iii*) Microscopic organism and amorphous matter—their examination and enumeration.

5.12. REMOVAL OF TASTES AND ODOURS

Tastes and odours in raw water are due to the presence of industrial wastes, dissolved gases, algae, dissolved mineral matter etc. These can be removed by any one of the following methods:

(*a*) Aeration

(*b*) Pre-chlorination

(*c*) Using-chlorine dioxide

(*d*) The application of activated carbon

(*e*) The application of potassium permanganate

(*f*) Using copper sulphate

(*g*) Applying ozone

(*h*) By coagulation, sedimentation and filtration.

5.13. COPPER SULPHATE TREATMENT

Copper sulphate treatment is done to destroy algae from the water. Powdered copper sulphate is sprinkled in water or sometimes it is applied by dragging racks of crystals through water with the help of boats. The amount of copper sulphate should not be more than 3 ppm. Copper sulphate is also effective in killing many types of aquatic weeds.

5.14. WATER SOFTENING

The reduction or removal of hardness from water is known as water softening. The advantage of water softening is in the reduction of soap consumption, a lowered cost in maintaining plumbing fixtures and improved taste of cooked food.

5.14.1 Necessity of Water Softening

The water containing sulphates, chlorides and bicarbonates of calcium and magnesium is known as hard water and causes the following troubles:

(*i*) More quantity of soap is consumed in homes and laundries.

(*ii*) The fabric of clothes are spoiled while washing to remove precipitate formed by soap in hard water.

(*iii*) In industries, hard water forms scales in the boilers, due to which more fuel is wasted and boiler metals are over-heated causing danger to the boiler plant.

(*iv*) Hard water creates serious troubles in the manufacturing processes of textile finishing, paper making, dyeing, rayon industry, tannery, ice-making etc.

(*v*) Hard water choke and clog plumbings due to deposits of salts in them.

Due to the above difficulties its removal from water becomes necessary.

5.14.2 Types of Hardness

Hardness is of two kinds : (*i*) Temporary and (*ii*) Permanent.

Temporary hardness is due to bi-carbonates of both calcium and magnesium and can be removed by boiling.

$$\underset{\text{Cal. Bicarbonate}}{Ca\,(HCO_3)_2} + \text{Heat} = \underset{\text{Calcium Carbonate}}{Ca\,CO_3} + CO_2 + H_2O$$

Permanent hardness is due to the presence of sulphates of chlorides or both of calcium and magnesium. The addition of caustic soda or soda ash will remove permanent hardness.

$$MgSO_4 + Ca(OH)_2 = Mg\,(OH)_2\downarrow + CaSO_4$$

According to Clark's scale the degree of hardness is expressed in terms of grains of calcium carbonate ($CaCO_3$) dissolved per gallon of water. Thus 6 grains of $CaCO_3$ dissolved in gallon of water will cause 6° hardness. The desirable figure of hardness for domestic water supply is limited to 10° or 100 ppm 1° = 14.3 ppm. Imperial units or = 17.1 ppm. – U.S. units. In metric scale the standard of hardness is expressed in parts and is obtained by multiplying the Clark's scale by 0.7.

In a muncipal water-softening operation it is now a practice to reduce the carbonate hardness to 35 to 40 ppm, and the total harness to 50–100 ppm. In general, it is good practice to carry a calcium carbonate alkalinity of about 25 ppm. This calls for a pH of 8.2 to 8.5.

5.14.3 Water Softening Process

The following processes are used in water works:

(*a*) Lime process

(*b*) Lime and soda ash process

(*c*) Base-exchange process

5.14.3.1 *Lime Process*

When water contains only carbonate hardness, it can be removed by adding milk of lime in the water. This process is known as Clark Process. The chemical reaction are;

(*i*) $CO_2 + Ca(OH)_2 \rightarrow CaCO_3\downarrow + H_2O$

The carbon dioxide is neutralized with milk of lime. If water contains bicarbonate hardness the chemical reactions are:

(*ii*) $Ca(HCO_3)_2 + Ca(OH)_2 \rightarrow 2CaCO_3\downarrow + 2H_2O$

(*iii*) $Mg(HCO_3)_2 + Ca(OH)_2 \rightarrow CaCO_3\downarrow + MgCO_3 + 2H_2O$

(*iv*) $MgCO_3 + Ca(OH_2) \rightarrow Mg(OH)_2\downarrow + CaCO_3\downarrow$

Calcium carbonate and magnesium hydroxide are insoluble and can be removed by sedimentation.

5.14.3.2 *Lime and Soda-Ash Process*

If water having hardness to sulphates of calcium and magnesium, only lime cannot remove it. But if soda-ash is added in lime, the non-carbonate hardness can be easily removed. The chemical reactions are:

(*v*) $MgSO_4 + Ca(OH)_2 \rightarrow Mg(OH)_2\downarrow + CaSO_4$

(*vi*) $CaSO_4 + Na_2CO_3 \rightarrow CaCO_3\downarrow + Na_2SO_4$

(*vii*) $MgCl_2 + Ca(OH)_2 \rightarrow Mg(OH)_2\downarrow + CaCl_2$

(*viii*) $CaCl_2 + Na_2CO_3 \rightarrow CaCO_3\downarrow + 2NaCl$.

When lime [$Ca(OH)_2$] acts with sulphates and chlorides of magnesium, insoluble magnesium hydroxide are formed which settle in the bottom. But in these reactions sulphates and chlorides of calcium are formed which further increase the hardness. When soda ash (Na_2CO_3) is added it reacts with this hardness and removes it.

5.14.3.3 *Basic Exchange Process*

In this process hard water is passed through a bed of Zeolite Sand (Complex silicates of aluminium and sodium). Which the passage of water, Ca and Mg cation get replaced by Sodium from the exchanger and water becomes soft. The sodium from the zeolite sand goes on getting exhausted and after some time it cannot remove the hardness of water. But the reactions are reversible and the zeolite can be recharged by passing through it a solution of common salt. The chemical reaction are:

(*a*) For softening

$$Na_2R + \begin{Bmatrix} Ca(HCO_3)_2 \\ CaSO_4 \\ CaCl_2 \\ Mg(HCO_3)_2 \\ MgSO_4 \\ MgCl_2 \end{Bmatrix} \rightarrow \begin{Bmatrix} Ca \\ Ca \\ Ca \\ Mg \\ Mg \\ Mg \end{Bmatrix} R + \begin{Bmatrix} 2NaHCO_3 \\ Na_2SO_4 \\ 2NaCl \\ 2NaHCO_3 \\ Na_2SO_4 \\ 2NaCl \end{Bmatrix}$$

(*b*) For Regeneration

$$\begin{matrix}Ca \\ Mg\end{matrix}\left[R + 2NaCl \rightarrow Na_2R + \begin{matrix}Ca \\ Mg\end{matrix}\right]Cl_2$$

where R = Zeolite

MODEL QUESTIONS

1. (*i*) Permanent hardness is due to presence _____ and of calcium and magnesium.
   (*ii*) _____ oxygen is only effective in killing bacterias. (molecules/nascent)
   (*iii*) The degree of hardness is expressed in terms of _____.
   (*iv*) The iron and manganese are removed from the water due to _____.
2. Write short notes on the following:
   (*i*) Aeration fountain
   (*ii*) Removal of silica from water
   (*iii*) Activated carbon
3. (*i*) Alum is used for _____ (Coagulation/Disinfection).
   (*ii*) Write the processes in succession.
   (Filtration, Sedimentation, Chlorination and Coagulation).
   (*iii*) Plain sedimentation removes _____ impurities from raw water.
   (*iv*) Coagulation is needed when the size of the particles is very ___.
   (Coarse/fine)
   (*v*) Factors affecting good coagulations are _____.
   (*vi*) Define floc and flocculation.
   (*vii*) Why alum is mainly used as coagulant?
4. Name the different coagulants used in a water-treatment plant. How these remove suspended impurities?
5. Write short note on:
   (*i*) Flash Mixer, (*ii*) Dorr-Clasifier,
   (*iii*) Flocculator, (*iv*) Detention period.
6. What are the various device for feeding the coagulants? On what factors the choice of feeding device is made?
7. State the ob ect of adding alum to water before filtration. When is ferrious sulphate pr eferred to alum? Mention the function of a sedimentation tank and explain the various points to be remembered in its design.
8. Draw neat sketches of the following:
   (*i*) Rectangular sedimentation tank
   (*ii*) Mechanical mixing basin used for coagulation
9. Give the chemical reaction when ferrous sulphate is used as coagulant along with lime-to-remove turbidity.

10. Make necessary assumptions. Sketch the inlet, outlet and the sludge removal arrangements.
11. What is meant by coagulants? What coagulants are commonly used? Please explain in detail.
12. Explain the terms:
    (*i*) Coagulation (*ii*) Coagulants
    (*iii*) Clarification (*vi*) Flocculation
    (*v*) Clariflocculator
13. What do you understand by sedimentation and coagulation? How do they help impurifying water.
14. (*i*) Discharge rate of slow sand filter is _____ litres/m^2/hour.
    (*ii*) Discharge rate of rapid sand filter is _____ litres/m^2/hour.
    (*iii*) Back washing is necessary when loss of head exceed _____ metres.
    (*iv*) Slow sand filter is more efficient in removal of _____. (Bacteria/colour)
    (*v*) Size and uniformity co-efficient of sand used in slow sand filter are _____ and _____ respectively.
    (*vi*) Coagulants are essential in case of filtration with _____.
    (Slow and filter/ rapid sand filter)
    (*vii*) Skilled labour is needed for _____.
    (*viii*) What impurities are removed by filtration?
    (*ix*) What is meant by Double Filtration?
15. (*a*) What is the function of sand and gravel in the filters?
    (*b*) Why is the under-drain system necessary in sand filters?
    (*c*) What precautions should be kept in mind while washing rapid gravity filter?
16. Sketch a rapid gravity sand filter. Compare its working with that of a slow sand filter.
17. Answer with respect to slow and filters.
    (*i*) Rate of filtration
    (*ii*) Effective size and uniformity coefficient of sand and depth of sand bed
    (*iii*) Size filter unit
    (*iv*) Method and frequency of cleaning the filter
18. What are pressure filters? Where are they used? Please explain their working with a neat sketch.
19. Explain the constructions and working of a Rapid Gravity Sand filter with the help of a neat sketch. What is the importance of back washing of these filters. How is it done?
20. Describe in detail the function of a slow sand gravity filter with a neat diagram.
21. Explain the detailed working of the following:
    (*i*) Slow Sand Filter
    (*ii*) Rapid Sand Filter
    (*iii*) Pressure Filter
22. Describe the construction (give sketch), working principle and operation of slow sand filters with rapid gravity sand filter.

23. Differentiate between "Loss of head" and "Negative head" in rapid sand filters.

24. Draw neat sketches of the following:
    (*i*) Pressure filter
    (*ii*) Rectangular sedimentation tank

25. Answer with respect to rapid sand filters:
    (*i*) Maximum permissible head loss
    (*ii*) Filtration of rate
    (*iii*) Function of rate of flow controller
    (*iv*) Effective size and uniformity coefficient
    (*v*) Length to width ratio of filter bed
    (*vi*) Rate and duration of back washing with air and water.

26. Draw a flow diagram showing Treatment Units in the case of rapid sand filtration water supply scheme. Detail out briefly the functions of each of the components of the scheme.

27. Explain briefly with the help of sketches the difference between a slow filter and a rapid sand filter in the matter of their constructional details function operation and utility. How do we make up for the draw backs in each?

28. (*a*) What is break-point chlorination and under what conditions is it advocated?
    (*b*) Write short notes on chloromines

29. (*a*) Write detailed note on chlorination of water supplies? What do you mean by Pre-chlorination, Post-chlorination, super-chlorinations? Under what circumstances are these resorted to any why?
    (*b*) Explain the following terms:
    (*i*) Pre-chlorination
    (*ii*) Post-chlorination
    (*iii*) Super chlorination
    State under what condition they are recommended

30. (*a*) What is meant by chlorination of village wells? Describe the method adopted for chlorinating wells in villages.
    (*b*) What is usual quantity and function residual chlorine?

31. Explain briefly the following processes:
    (*i*) Break Point chlorination
    (*ii*) Super chlorination

32. (*i*) The process of killing infective bacteria from the water is known as (Sterilization/Disinfection)
    (*ii*) The chlorine added in the water after all treatment is known as _____. (Pre-chlorination/Post-chlorination)
    (*iii*) The substances added for disinfection of water are called _____.

33. Write short notes on:
    (*i*) Post-chlorination.
    (*ii*) Pre-chlorination.
    (*iii*) Double-chlorination
    (*iv*) Super-chlorination
    (*v*) Plain-chlorination

34. With the help of neat sketches, describe how liquid chlorine is applied in water.
35. Describe briefly with the help of a neat sketch a chloroscope.
36. (*a*) Name at least two waterborne diseases.
    (*b*) What is bleaching powder? How is it used for chlorinating village wells? What do you mean by available chlorine in a sample of bleaching powder?
37. (*a*) Why is water disinfected.
    (*b*) Describe any method adopted in practice to disinfect water.
38. What do you mean by 'Hardness of water'? What troubles it will cause if not removed? Describe lime-soda process for softening of water.
39. Describe briefly the Zeolite process of softening hard water. What are the limitations of the zeolite process?
40. What are iron exchanges? What are their functions in softening of water.
41. Differentiate between lime-soda process and zeolite process in water softening.
42. What is the proposes of aerating water? How is it carried out?
43. Differentiate between carbonate and non-carbonate hardness. Mention the various method of removing harness and describe any one method in detail. Discuss its advantages and disadvantages.
44. (*a*) Why disinfection is necessary after filteration?
    (*b*) What is the function of gravel and sand in filter.
    (*c*) Discuss the construction and working of clarifier with the help of neat sketch.
45. (*a*) What is filteration? Give its significance?
    (*b*) Draw the sketch for a slow sand filter and name the various components.
    (*c*) What is disinfection of water?
    (*d*) Explain break-point chlorination.
46. Compare slow sand and rapid sand filter in details.
47. Differentiate between the following:
    (*a*) Pre-chlorination and Post-chlorination.
    (*b*) Floculation and coagulation.
    (*c*) Effective size and uniformity co-efficient.

❑❑❑

CHAPTER 6

CONVEYANCE OF WATER

Learning Objectives

- Introduction
- Method of Conveyance of Water
- Choice of Pipe Material
- Types of Pipe
- Pipe Joints
- Types of Joints
- Distribution System
- Pressures in Distribution System
- General Responsibilities of the Water Works
- System of Distribution
- Time Restricted Supply (Continuous and Intermittent Supply)
- Layout of Distribution Dystems
- Causes of Wastage of Water
- Prevention of Wastage
- Appurtenances in Distribution System
- Service/Storage/Distribution Reservoirs
- Type of Service Reservoirs
- Advantages and Disadvantages of Reservoirs
- Boosting

PART-A

CONVEYANCE OF WATER

6.1. INTRODUCTION

There are three stages in the conveyance of water:

(*a*) From the source to the treatment plant.

(*b*) From treatment plant to the service reservoir.

(*c*) From service reservoir to the consumer.

The first two stages of transportation of water from part of conveyance of water, whereas the third stage is covered under distribution of water.

In the first stage water is transported by gravity or by pumping or by combination of both.

6.2. METHOD OF CONVEYANCE OF WATER

Conveyance of water is carried out by one of the following systems:

(*a*) Free flow system

(*b*) Pressure system.

(*a*) Free Flow System: In this system, water flows through the sections under gravity. The demerits of this system are:

(*i*) transporting section required is very long.

(*ii*) percolation and evaporation loss is more.

(*iii*) chances of water gets polluted from surface drainage are more.

(*iv*) head loss due to friction is more.

(*v*) construction and maintenance costs are high.

The following sections may be used for free flow of water from source to the treatment plant.

1. Canals
2. Flumes
3. Grade aqueducts
4. Grade tunnels

1. Canals: These are generally not used for supplying water, because of losses due to percolation and evaporation are also the possibility of pollution and misuse of water. Most economical sections for canals are generally trapezoidal but rectangular section prove economical where rock cutting is involved. The canals are either lined or unlined depending upon the nature of ground and available slope.

2. Flumes: These are open channels supported on or over the ground surface particularly when water has to be taken across valleys, rivers and depressions. They are built of concrete or of metal. The sections may be rectangular or semicircular. Velocities of 1.8 to 2.4 m/s may be permitted in flumes.

3. Grade Aqueducts: These are covered waterways used to carry water from a remote source to the treatment plant or to the origin of distribution. These are designed in such a way that they flow three quarter full at required capacity of supply. In the Initial stages they were constructed of brick masonary but now concrete is extensively used for its construction.

4. Grade Tunnel: Grade tunnels are built to serve following purposes:

(*a*) to shorten the length of water channel.

(*b*) to conserve the head of water.

(*c*) to reduce the cost of aqueducts traversing uneven terrain.

They are generally of horse-shoe type and are usually lined.

(*b*) Pressure System: The flow in conduits is under pressure in this system. This system may be classified into:

(*i*) Pressure aqueducts

(*ii*) Pressure tunnels

(*iii*) Pipe lines.

(*i*) Pressure Aqueducts: They are generally of circular in section and built of reinforced concrete. These conduits are constructed *in situ* in balanced cutting and embankment. The advantage of pressure aqueduct is that they are not exposed to pollution or seepage.

***(ii)* Pressure Tunnels:** These are constructed when water has to cross river or valleys. They are generally of circular section and are usually lined.

***(iii)* Pipe Lines:** Pipe lines are circular in section and normally follow the profile of the ground surface quite closely. Gravity pipelines have to be laid preferably below the hydraulic gradient.

6.3. CHOICE OF PIPE MATERIAL

The factors which affect the final choice of pipe material are:

(*a*) Internal pressure and external loads to which the pipe is subjected.

(*b*) Type of water to be conveyed and its resistance to corrosion.

(*c*) Capital and maintenance cost.

(*d*) Availability of fund.

(*e*) Carrying capacity of pipe.

(*f*) Expected life and ease of repair and replacement.

6.4. TYPES OF PIPE

The following type of pipes are available:

(*i*) Cast iron pipe

(*ii*) Steel pipes

(*iii*) Reinforced cement concrete (RCC) pipes

(*iv*) Prestressed concrete (PSC) pipes

(*v*) Asbestos cement (AC) pipes

(*vi*) Copper, brass and lead pipes

(*vii*) Wrought iron pipes

(*viii*) Hume pipes

(*ix*) Polyethylene (low and high density) pipes

(*x*) Polyvinyl chloride (rigid PVC) pipes

6.4.1 Cast Iron Pipe

Cast iron pipes are mostly used in water supply schemes due to their durability, strength, resistance to corrosion, easy of laying etc. But the disadvantages of this type of pipes are:

(*a*) Due to its heavy weight larger diameter pipes (larger than 1200 mm) are difficult to transport in hilly and difficult terrain.

(*b*) Coating inside and outside of the pipe is required for carrying corrosive water.

Cast iron pipes are of following types:

***(i)* Vertically Cast or Pit Cast Pipes:** These types of cast iron pipes are manufactured by casting the metal into a vertical mould. Generally best grey pig irons are used for casting. Most of the old CI pipes were cast vertically. These pipes are more durable because of its heavy wall thickness.

***(ii)* Centrifugally Cast-iron (Spun) Pipes:** These pipes are manufactured up to a diameter of 900 mm. These are formed by spinning or centrifugal action. This method of CI pipe production is universally adopted as the production is faster and gives longer lengths of pipes with good metal qualities. The weight of these pipes is about 3/4th to 2/3rd the weight of the vertically cast pipes. Spun pipes are available in diameters ranging from 50 mm to 900 mm and lengths from 3.7 to 5.5 m.

***(iii)* Centrifugally Ductile Cast-iron Pipes:** These are recent development over spun pipes. They are manufactured from a material, produced by combining modern iron of suitable composition with magnesium. They are lighter in weight (about 25% less than of spun pipes) and hence easy to handle and transport.

6.4.2 Steel and Wrought Iron Pipes

These pipes are stronger than cast iron pipes. They are, however, less durable having life upto 50 years, more liable to corrosion. To increase the life of wrought iron pipes sometimes these are galvanized with zinc.

These pipes can withstand much higher pressure but are of lighter section and hence easy to transport to site. Moreover there is difficulty of easy availability of pipe specials. Joints in steel pipes may be done by welding or riveting, lap joints suitable upto 1.20 m diameter. Welded steel pipes may be used for large diameter upto 2.50 m or even more.

Advantages and Disadvantages of Steel Pipes

Advantages

(*i*) Steel Pipes are cheap.

(*ii*) These pipes are more durable

(*iii*) These pipes are light in weight, hence easy-to-transport.

(*iv*) These pipes are available in large lengths which decreases the number of joints.

(*v*) Steel pipes can resist high internal pressure.

Disadvantages

(*i*) Steel pipes are likely to be rusted which reduces their life.

(*ii*) These pipes require more time for repair and therefore, are not suitable for distribution pipes.

(*iii*) The maintenance cost is more.

6.4.3 Reinforced Concrete Pipes

These are very durable, heavier and can be used upto 1.8 m dia. Transportation costs are much reduced if the pipes are "Cast-in-situ". These pipes are resistant to corrosion and specially suitable to soft and acidic waters. These pipes have low maintenance costs but cannot withstand high pressures unless reinforced. The concrete mix normally used is 1 : 2 : 2.

Advantages and Disadvantages of Concrete Pipes

Advantages

(*i*) The pipes have low maintenance cost.
(*ii*) The pipes are not corroded from inside by normal drinkable water.
(*iii*) The pipes do not require expansion joints as co-efficient of expansion for concrete is low.
(*iv*) The pipes do not require any skilled labour.

Disadvantages

(*i*) The pipes are difficult to repair and join.
(*ii*) The pipes have tendency to leak due to shrinkage cracks and porosity.
(*iii*) The pipes are difficult to transport due to their large size and weight.

6.4.4 Prestressed Concrete (PSC) Pipes

Prestressed concrete pipes are used for intermediate pressure range, i.e., between 0.3 MPa, because at this pressure range metallic pipes are expansive and R.C.C. pipes would not be suitable. The sizes of these pipes range from 80 to 1800 mm. The lengths are 2 m for the sizes up to 400 mm dia and 2.5 m for the higher sizes. These are of two types:

(*a*) PSC cylinder pipes
(*b*) PSC non-cylinder pipes

6.4.5 Asbestos Cement Pipes

These are manufactured from a mixture of portland cement and asbestos fibre combined under pressure into a dense homogeneous structure. These pipes are very light in weight (about 1/4th that of C.I. pipes), can be easily cut, joined and handled in transit. They resist corrosion and are very smooth. They are available in large variations of sizes form 5 cm to 120 cm and are manufactured in four different grades to withstand internal pressure from 3.5 to 140 N/mm^2. Use of these pipes is restricted to minor works of the distribution system (upto 22.5 cm dia.) because of poor structural resistance to bending stresses caused during transport.

Advantages and disadvantages of Asbestos cement Pipes

Advantages

(*i*) The pipes are very light in weight (about 1/4 that of C.I. pipes), can be easily cut, jointed and transported.
(*ii*) The pipes are smooth and their carrying capacities do not reduce with time.
(*iii*) The pipes are very suitable as small size distribution pipes.
(*iv*) The pipes are flexible as such the joints are easily formed.

Disadvantages

(*i*) The pipes are costly and less durable.

(*ii*) The pipes are soft and brittle and do not have much strength.

(*iii*) The pipes cannot be laid in exposed places.

(*iv*) The pipes are likely to be damaged during transportation.

6.4.6 Hume Steel Pipes

These pipes are manufactured from a thin mild steel shell lined and out-coated with cement concrete (1 : 2 : 4) or cement mortar (1 : 3) through a centrifugal process. These are used for trunk and distribution mains and are available upto 135 cms diameter. They are difficult to handle being heavy and the concrete coating easily gets damaged while fixing branch connection etc.

6.4.7 Plastic Pipes

Plastic pipes are of three kinds for potable water supplies:

(*i*) Low density polyethylene pipes

(*ii*) High density polyethylene pipes and

(*iii*) Unplasticised PVC pipes

Polyethylene pipes are extruded from a compound consisting **virgin** polyethylene in which carbon black and a suitable non-antioxidant are **evenly** dispersed.

(*a*) Low and High Density Polyethylene Pipes: Low density polyethylene shall have a density not more than 0.93 g/ml and high density shall have a density greater than 0.94 g/ml at 27°C. Refer Table 6.3 for properties of polyethylene and PVC pipes. The anti oxidant in the material should be physiologically harmless and be not more than 0.3 percent by weight. The polyethylene pipes are classified as pressure ratings (working pressure) viz. 0.25, 0.40., 0.60 and 1.00 N/mm^2.

Table 6.1.

Working pressure in N/mm^2	Outside dia. range in mm	
	Low density polythylene	High density polythylene
0.25	40–140	63–500
0.40	32–140	40–500
0.60	20–110	32–500
1.00	12–63	20–500

Refer Table 6.3. for recommended temperature ranges, tensil strength etc.

The low density pipes are generally supplied in coils of nominal length of 25 to 200 m. The high density pipes are to be supplied either as coils measuring 25 times the mini-dia. of the pipes in straight length of 5 to 20 m.

(b) Unplasticised PVC Pipes: The material shall be substantially polyvinyl chloride plus necessary additives for getting good surface finish and mechanical strength. These shall not cause toxicity and no detrimental effect on the composition of water passing through the pipes. Refer Table 6.3 and 6.4 for properties of PVC pipe.

The pipes shall withstand a hydraulic pressure equal to 1½ times the maximum working pressure for one hr. at room temp. The pipes are classified as test ratings as in the case of polyethylene pipes 0.25, 0.40, 0.60, 1.0 N/mm^2 pipes. The range of dia. for various working pressure are given in Table 6.2:

Table 6.2. Range of diameter for various working pressure

Working pressure	Range of dia in mm
0.25 N/mm^2	90–630
0.40 N/mm^2	63–630
0.60 N/mm^2	40–630
1.00 N/mm^2	16–630

2. Jointing of PVC Pipes: PVC joints are quite simple and cheap. Remember the following four points:

(*i*) Use only that solvent compound specified by the manufactures of the unpasteurized PVC.

(*ii*) Throughly clean the joint surfaces before applying the solvent.

(*iii*) Avoid twisting the joint when you are finally assembling quickly, push it firmly and directly home.

(*iv*) Wipe off surplus solvent at once. The joint will set within fifteen minutes, but it needs twelve to twenty four hours to dry out completely. In case of PVC pipes at the final stage an adhesive jointing compound is used instead of a solvent as recommended by the manufacturers.

3. Advantages and Disadvantages of Plastic Pipes:

Advantages:

(*i*) Plastic pipes are light in weight i.e., one-sixth that of steel.

(*ii*) These pipes are cheap.

(*iii*) They are easier to instal and join.

(*iv*) These pipes are durable.

(*v*) These pipes are not damaged due to thawing and freezing of water.

(*vi*) Jointing of these pipes can be done by solvent or cementing techniques.

(*vii*) These pipes are good electric insulators.

(*viii*) These pipes are free from corrosion.

(*ix*) These pipes posses low hydraulic resistance.

(*x*) These pipes are unaffected by age, weather or sunlight.

(*xi*) These pipes are easy to handle and transport.

(*xii*) Plastic pipes are highly resistance of acidic waters.

Disadvantages

(*i*) Co-efficient of expansion of pipe material is high.

(*ii*) The plastic may impart taste to water flowing through these pipes.

(*iii*) Plastic pipes have low resistance to heat.

4. Application of Rigid PVC Pipes:

(*i*) **In Potable Water Supply:** For main lines and service lines.

(*ii*) **In industries:** For pumping chemicals and corrosive water and ventilating corrosive fumes.

(*iii*) **In Drain Waste and Vent Systems:** For drainage, ventilation and waste disposal in factories and residential buildings.

(*iv*) **In Agriculture:** For laying irrigation pipes to sprinker layouts for tube well systems and also for cattle breeding.

(*v*) **In Electrical System:** For electrical conduits, smooth surface inside facilitates pull-through of wires:

Note:

(*i*) Refer Table 6.6 for country-wise break down of PVC consumption.

(*ii*) CPHERI: Central Public Health Engineering Research Institute, Nagpur.

(*iii*) DGHS: Directorate-General of Health Services, New Delhi.

Table 6.3. Properties of PE and PVC pipes

Property	LD polythene	HD polythene	Rigid PVC
(*i*) Tensile strength (N/mm^2)	11.52	23.04	46.8–61.2
(*ii*) Compressive strength (N/mm^2)	—	17.28	69.12–82.1
(*iii*) Flexural strength (N/mm^2)	12.25–26.65	26.65–43.2	72–115.2
(*iv*) Sp. Gravity	0.92	0.94	1.35–1.46
(*v*) Softening Point 0°C.	0.91	120–128	75.82

Note: PVC = Polyvinyl chloride

LD = Low density

HD = High density

Table 6.4. Properties of plastic pipes

Property	Unplasticized PVC	LD polythene	HD polyethylene
(*i*) Ultimate tensile strength (N/mm^2)			
(*a*) at 23° C	45.5 – 60.0	11.5 – 17.0	26.5 – 28.0
(*b*) at 40° C	5.3	9.1	19.5
(*ii*) Recommended temp. range of pipe, °C pressure	– 1 to + 40	– 40 to + 60	– 18 to + 45
(*iii*) -Do-For non-pressure pipe °C	– 1 to + 60	– 40 to + 60	– 18 to + 80
(*iv*) Coefficient of thermal expansion 10^{-5} per °C	5 – 6	16 – 18	11 – 13
(*v*) Impact strength Izond m kg/mm notch			
(*a*) at 23° C	5 – 25	No Break	15 – 25
(*b*) at 0° C	0.4	–	3
(*vi*) Weathering characteristics	Excellent	Fair	Fair
(*vii*) Flammability	Self extinguishing	Slow	Slow
(*viii*) Abrasion resistance	Good	Moderate	Fair
(*ix*) Jointing methods	Solvent welding	Heat welding or Compression fitting	Heat welding or Compression fitting

Table 6.5. Principal manufactures of plastic pipes

S. No.	Name of Mfg. units (mm)	Sizes of pipes Mfd. used	Plastics raw material	Standard specification
1.	Union Carbide India Ltd. Kolkata-1	12–50 mm	PE	IS 3076–1965
2.	I.C.I. (India) Pvt. Ltd. Kolkata-16	12–50 mm	PE	IS 3076–1965, ISS & BSS for LDPE
3.	Garware Plastic Pvt. Ltd. Mumbai-18	12–150 mm	PE & UPVC	America C.S. Rigid
4.	Pioneer Plastic Works Pvt. Ltd. Kolkata-1	12–75 mm	PE	PV BSS
5.	Calico Chemical and Plastic Dvn. Mumbai-74	12–160 mm	UPVC	Din
6.	Wavin India Ltd. Chennai-58	15–200	UPVC	BSS & ISS
7.	Emco General Industries, Kolkata-15	12–50 mm	PE	ISS & BSS
8.	Chemical and Plastics India Ltd. Chennai-2	15–100 mm	UPVC	ISS & BSS
9.	Polyolefins Indusatries Ltd. Mumbai-1	32–160 mm	HDPE	Din.
10.	Atlas Industries Company, Kolkata-6	Pipe Fitting only	HDPE	–

Note: PE : Polyethylene
UPVS : Unplasticized (Rigid) Polyvinyl-chloride
HDPE : High Density Polyethylene

Table 6.6. Country-wise break down of PVC consumption (1964) 1000 tonne

S.No. country	PVC consumption			Breakdown of rigid PVC		
	Total	Flexible	Rigid	Pipe	Sheet/ film	Corrugated panel
1. UK	147.4	129.3	18.1	6.8–9.1	4.5–6.8	2.3
2. W. Germany	238.1	147.4	90.7	27.2–31.7	11.3–13.6	15.9–18.1
3. France	158.7	131.5	27.2	56.7	–	4.5–6.8
4. Italy	124.7	74.8	49.9	24.9–27.2	–	4.5–6.8
5. Benelux	45.50	29.5	15.9	11.3–13.6	–	2.3–4.5
6. US	635.0	607.8	27.2	11.3	11.3	2.3
7. Japan	340.2	170.1	170.1	90.7	34.0	22.7
8. India	13.25	12.65	0.60	N.A.	N.A.	N.A.

Report of APO " Plastic Industry" study Mission, Japan, 1966, p. 37.

Stresses in Pipes

Plastic pipes may be subject to pressures of following types:

(*i*) Temperature stresses
(*ii*) Internal water pressure
(*iii*) Water hammer stresses
(*iv*) Pressure due to foundation reaction
(*v*) Longitudinal stress due to flow around various bends.

Fig. 6.1. Plastic Pipes

4. Production of Plastic Pipes in India: The production of plastic pipes in India is not yet well organised. Refer Table 6.5 which gives comparative data for principal plastic pipe manufacturers in India. Annual production in India is 1750 tones of P E and 1550 tones of rigid P.V.C. pipes at present.

5. Suitability in India: Of plastic pipe for potable water has already been certified by the I.S.I., CPHERI, DGHS etc., and the plastic-pipes both PE and right PVC has been in use in India for more than ten years, mostly in hilly regions and for village water supply. During 1962, about 1350 meters of rigid PVC pipes for potable water in various sizes were denoted by "Wavin India Ltd". Chennai to Delhi Muncipal Corporation of installation in Chattarpur village near Mehrauli, Delhi.

6.5. PIPE JOINTS

Pipes are manufactured in small lengths of 3.5 m to 5.5 m. These small pieces of pipes are then joined together after placing in position, to make one continuous length of pipe line. The design of joint depends upon type of pipe, support condition, internal water pressure etc.

6.5.1. Requirements of Jointing Material

The requirements of jointing material are:

(*i*) Strength (*ii*) Durability
(*ii*) Imperviousness (*iv*) Elasticity
(*v*) Availability (*vi*) Workability
(*vii*) Adhesiveness (*viii*) Economy.

The various types of joints which are commonly used in practice are described in the following articles.

6.6. TYPES OF JOINTS

(*i*) Spigot and socket joints (*ii*) Flanged joint
(*iii*) Expansion joint (*iv*) Flexible joint
(*v*) Collar joint (*vi*) Joint for AC pipe
(*vii*) Screwed and socket joint

6.6.1. Spigot and Socket Joint

This type of joint is commonly used in case of cast-iron pipes. For the construction of this joint the spigot or normal end of one pipe is centered into the socket of the preceding or other pipe; hemp yarn or tarred gasket is then wrapped around the spigot, leaving unfilled the required depth of socket for lead. The hemp yarn is caulked tightly home with a yarning tool

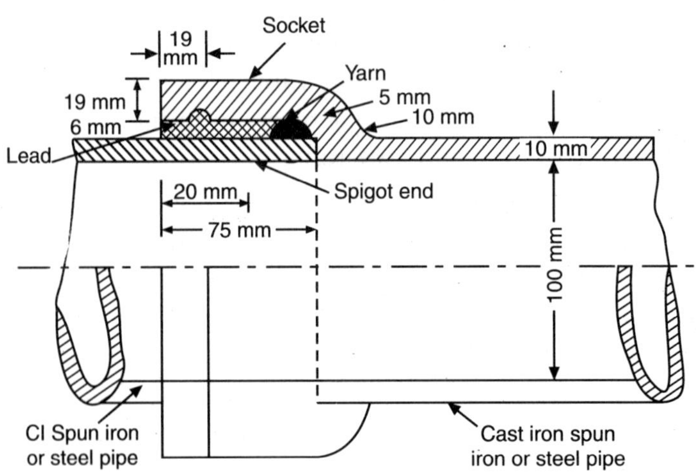

Fig. 6.2. Spigot and socket joint

[refer Fig. 6.3 (a) & (b)]. A knealed clay ring is then placed around the barrel and against the face of the socket. Refer Fig. 6.2 below of construction details.

After this Molten pig lead is poured in to fill the remainder of the socket. The lead which shrinks while cooling, is again tightened by means of 2 kg weight hammers and by caulking tool. Joint thus becomes water tight and should be finished 3 mm behind the socket face.

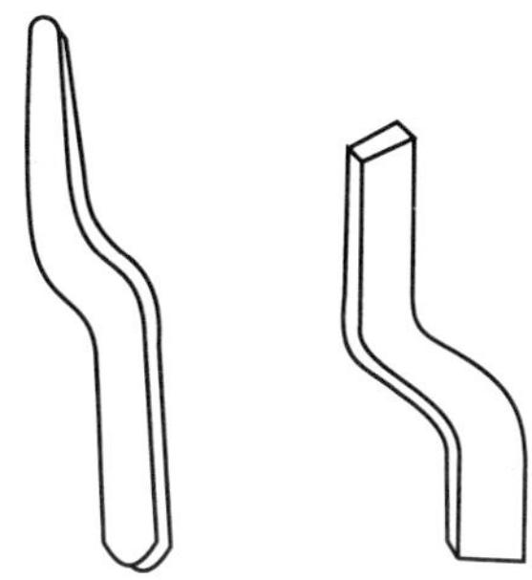
(a) Yarning tool (b) Caulking tool

Fig. 6.3.

6.6.2. Flanged Joint

These joints are rigid and are easy to disjoint; as such used where pipe-joints have to occasionally opened out for carrying out repair work, as in pumping chambers. Refer Fig. 6.4 for construction details.

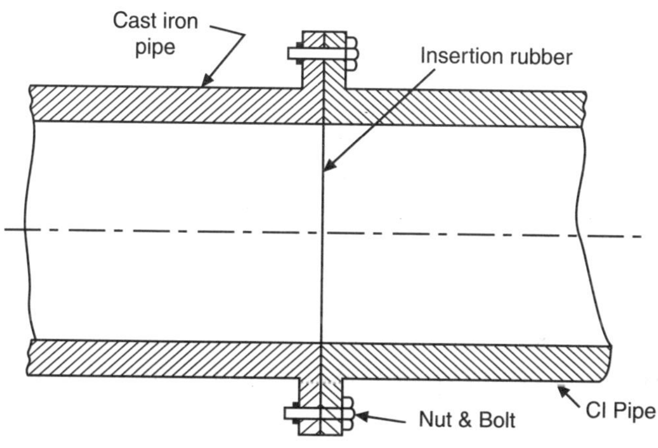

Fig. 6.4. Flanged Joint

The pipe in this case has flanges on its both ends, casted, welded or screwed with the pipe. A gasket of rubber, canvass or lead is introduced between the two flanger of cast-iron pipes, which are then tightened with bolts and nuts.

6.6.3. Special Joints

These are used for special purposes in case of cast iron-piper. These may be:

(*a*) Expansion Joint: These joints are used in pipes exposed to considerable differences of temperatures allowing for free expansion or contraction without setting up thermal

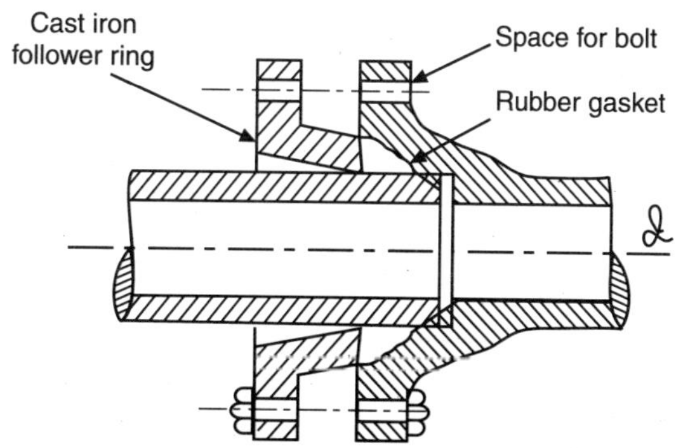

Fig. 6.5. Expansion joint

stresses in the pipes. Here when the pipe expands the socket end moves forward and when pipe contracts, it moves backward in the space provided for it as shown in the Fig. 6.5 and the elastic rubber gasket in every position keeps the joint watertight.

(*b*) Flexible Joints: Flexible joints are used for pipes to be laid submerged under water, where the bottom of the river or trench bed is uneven with the possibility of settlement and consequent damage. Details of construction have been shown Fig. 6.6. It is quite clear from the Figure that if one pipe is given any deflection the ball shaped-portion will move inside the sockets and the joint will remain water-proof in all the position. Sometimes this joint is also known as universal or Bell and Socket Joint.

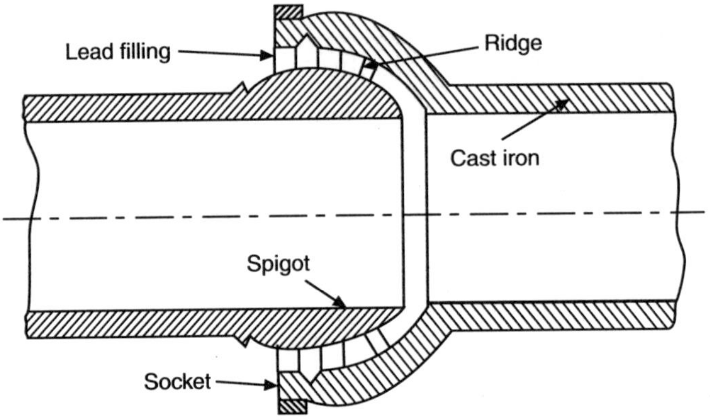

Fig. 6.6. Flexible joints

6.6.4 Joints for Concrete and Asbestos Cement Pipes

(*a*) Collar-Joint: This joint is mostly used for joining concrete and asbestos cement pipes having bigger diameter. A rubber gasket is placed between steel rings in the groove after bringing the ends of the two pipes in one level. Then the collar is placed at the joint so that it should have the same lap on both the pipes. After this cement mortar (1 : 1) is filled in the gap between the pipes and the collar as clearly shown in Fig. 6.7.

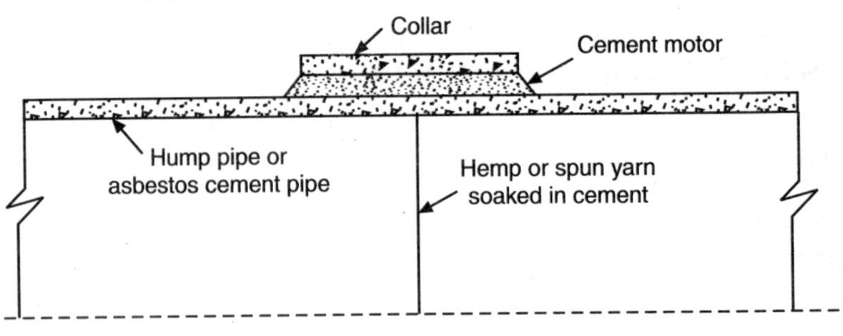

Fig. 6.7. Concrete collar joint

(*b*) **Joint for AC Pipes:** In case of the asbestos cement-pipes, the two plain ends of the pipe butt against each other, two rubber rings are then slipped over and a coupling pushed over rubber rings, thus making it a tight and flexible joint which can be easily completed in a dry trench or under water.

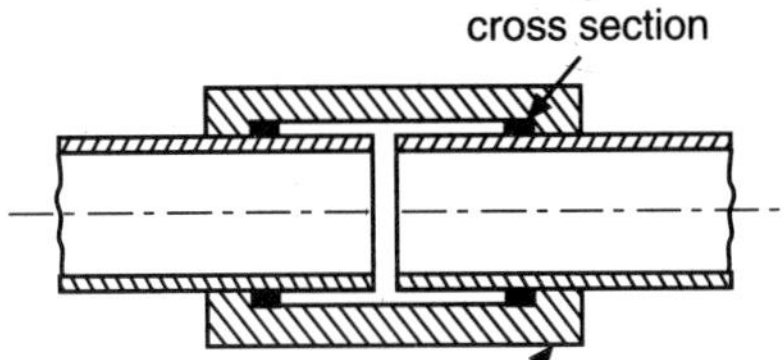

Fig. 6.8. Simplex joint for AC pipes

6.6.5. Screwed and Socketed Joint

The is a simple type of joint commonly used for jointing screwed wrought iron or galvanized iron pipes. Refer Fig. 6.9 for construction details.

Here two ends of the pipes are threaded on the outside and on them a suitable jointing compound (e.g., Zinc paint or hemp yarn) should be used before screwing socket over it having corresponding threads from insides.

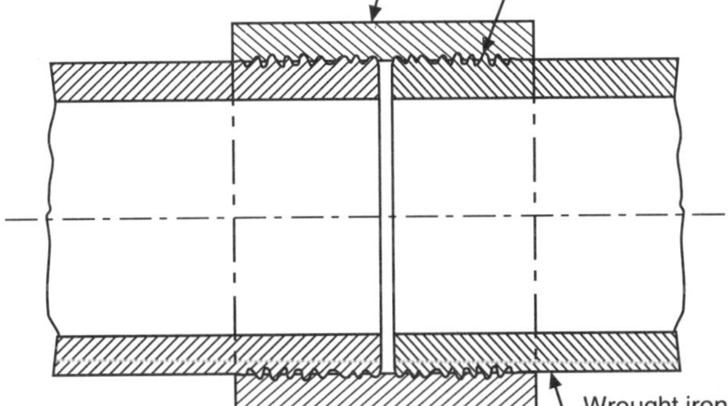

Fig. 6.9. Screwed joint/threaded joint

PART-B

DISTRIBUTION METHODS AND APPURTENANCES

6.7. DISTRIBUTION SYSTEM

The system of distributing water efficiently to consumer is called distribution system.

Because distribution system requires a large part of capital expenditure (about 60%), hence requires economical planning. The water must reach the consumer with required rate of flow.

6.8. PRESSURES IN DISTRIBUTION SYSTEM

The pressure which is required to be maintained in the distribution system depends upon the following factors:

(*i*) The distance of the locality from the service reservoir.

(*ii*) The height of highest building upto which water should reach without boosting.

(*iii*) The supply is to be metered or not. Higher pressure will be required to compensate for the high loss of head in meters.

(*iv*) How much pressure is required for fire hydrants.

(*v*) The funds available for the project work.

The following pressure are recommended for multi-storeyed structures:

Upto 3 storeys – 0.21 N/mm^2

From 3 to 6 storeys – 0.21 to 0.42 N/mm^2

From 6 to 10 storeys – 0.42 to 0.53 N/mm^2

Above 10 storeys – 0.53 to 0.7 N/mm^2

6.9. GENERAL RESPONSIBILITIES OF THE WATER WORKS

(*i*) **Rate of Supply:** To supply the required quantity of water to meet the domestic and industrial demand at the rate which it is needed.

(*ii*) **Fire Demand:** To meet the rate of Fire demand in case of configeration.

(*iii*) **Reliability of Supply:** To make available the quantity of water for the above purposes, at all time, in the distribution system. Reliability of water works is therefore a very essential requirement which is fulfilled with the help of a good distribution system.

(*iv*) **Quality of water:** The standard of the quality of purified waters should not be allowed to deteriorate, either by long storage in the Distribution Reservoir or by its long travel in the distribution system or by sub-soil water entering the mains through braking joints.

(*v*) **Equipment used:** The various materials used and the machinery employed should be of a good make and of an approved manufacture.

(*vi*) **Repairs:** In the case of repairs to any part of the distribution system, as small a section should be cut off from the supply, as possible.

(*vii*) **Maintenance:** The effective pressure and the rate of supply should be checked occasionally at various points of distribution system. Samples of water are taken from various tests points at the consumers end for examining its quality.

(*viii*) **Circulation of water:** In no part of the system the stagnation of water should be allowed to occur. A minimum number of dead ends should be provided.

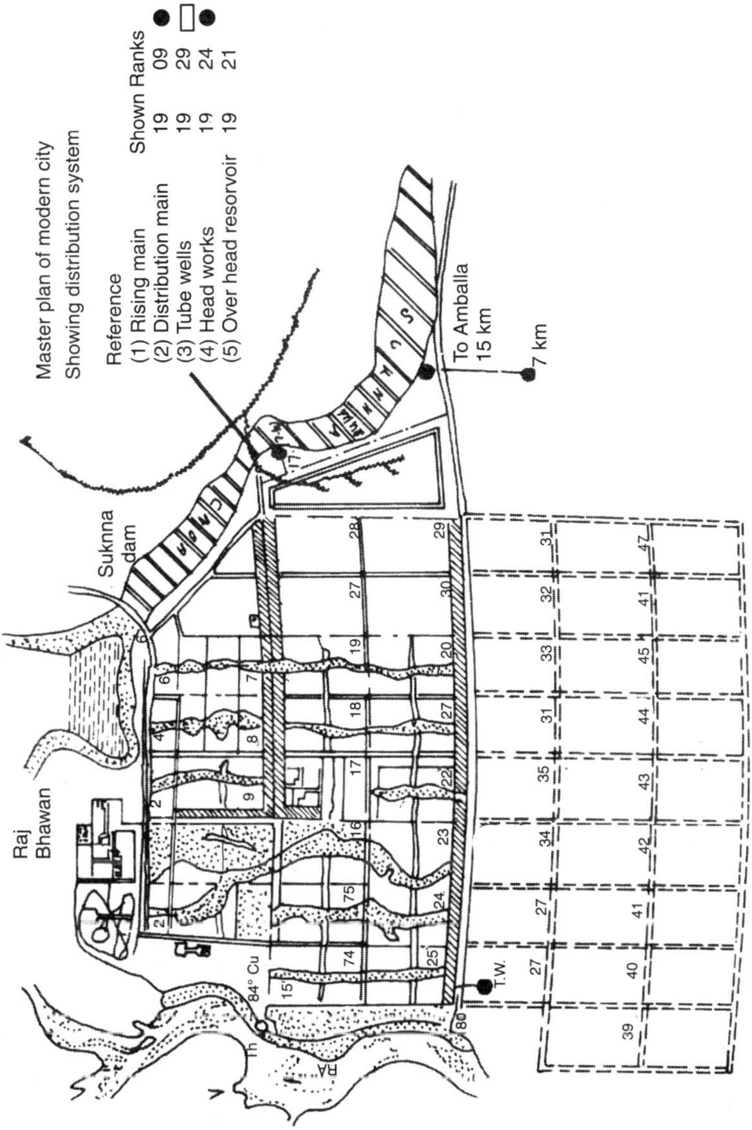

Fig. 6.10. Distribution of modern city

(*ix*) **Simplicity in Layout:** The entire system of distribution should be designed in a simple and straight-forward manner. This reduces the cost of outlay, construction and maintenance.

(*x*) **Waste detection:** Provisions should be made to detect the wastage of water. An adequate number of meters, valves and other hydraulic appliance should be fitted at suitable portions in the system.

(*xi*) **Size of pipes:** The pipes should be designed to carry 2.25 times the average discharge. But to meet exigencies of demand and also to accommodate partly the increase in demand for the future, the sizes are adopted to carry 3 times the average demand.

(*xii*) **Effective Pressure:** After overcoming the frictional loss, etc., there must be enough residential pressure in pipe to raise the water to required height in a building and to deliver the required quantity even in the period of maximum demand.

6.10. SYSTEM OF DISTRIBUTION

There are various systems of distribution. The adoption of one system or the other depends upon the relative elevation of the principal elements of the scheme and on the topography of the city.

6.10.1. Gravity System

This is the most reliable method of distributing the water. In this case, the purification works are at a higher level and the distribution system is at a sufficient lower level. The adequate pressure of distribution is maintained by the gravity force only. No pumpings is used. For a complete gravity system the source of supply is also at a higher level than the purification works, and the whole project becomes an ideal one.

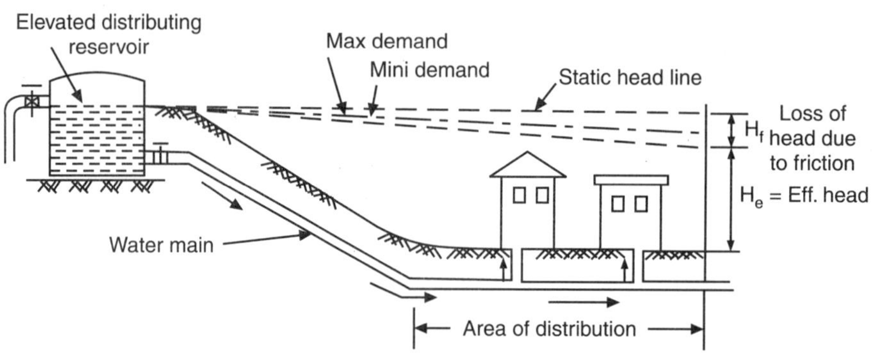

Fig. 6.11. Gravity system

6.10.2 Combined Gravity and Pumping System

(*a*) In this case the distribution of water and the pressure required for the same are effected by the gravity force. But the water works are almost at the same level and the treated water is pumped in to a distribution reservoir. The distribution reservior is located on a natural high level ground, or it may be placed on a fabricated staging of steel or reinforced cement concrete. Then it is termed as an elevated reservoir.

(*b*) In the other case the water may be also required to be pumped from the source to the purification works. Combined system is also usually reliable but involves the cost of construction and maintenance of pumps, rising mains, and the distribution reservoir.

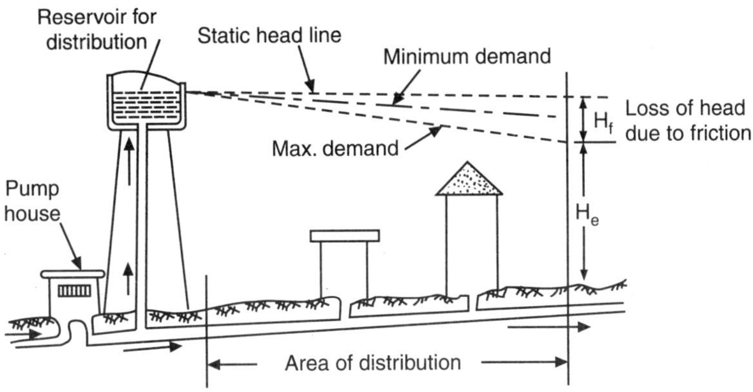

Fig. 6.12. Combined gravity and pumping system

6.10.3 Pumping System

When the distribution pressure is maintained with the aid of direct pumping in the mains, it is termed as pumping system or direct system. In an extreme case, pumping has to be done in the case of raw water supply to the purificating works also. The method increases the cost of maintenance and since the pumps have to pump at varying rates, their life is also reduced. In the case of towns where the supply is intermittent, the direct system could be recommended, since the pumping has to be done in shifts only. It has been found that steam operated reciprocating pumps with low speed, are found to give best service for pumping directly into the mains.

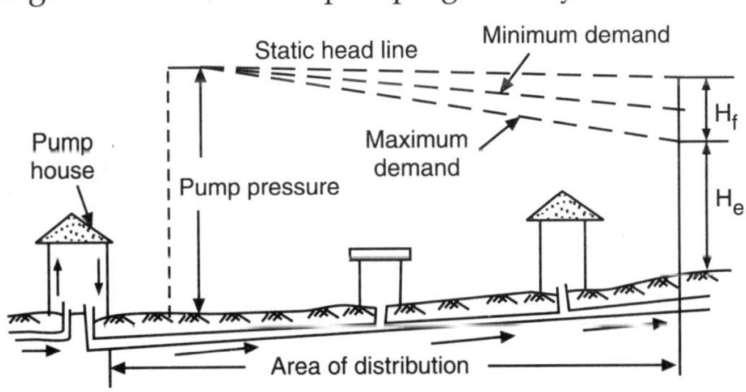

Fig. 6.13. Pumping system

6.11. TIME RESTRICTED SUPPLY (CONTINUOUS AND INTERMITTENT SUPPLY)

To minimise the quantity of consumption of water, in some towns, water is supplied to the consumers only during the periods of maximum demand i.e., from about 5 A.M. to 1 P.M. and from 4 P.M. to 10 P.M. This is not always recommended because of the following reasons:

(*i*) A fire may occur when supply is turned off.

(*ii*) It induces storage of water by a consumer and very often these storage tanks are badly leaking and water is wasted. In addition its quality also deteriorates by storage.

(*iii*) The taps are often kept open carelessly when the supply is cut off. When the supply is turned on, water starts flowing until the tap is closed. This leads to wastage.

(*iv*) When the air valves come into operation, much noise is created in the distribution system when the supply is cut off and also when the supply is turned on.

6.12. LAYOUT OF DISTRIBUTION SYSTEMS

There are principally four systems adopted for the layout of pipe lines to distribute water through mains, sub-mains, branches and service pipes.

1. Dead end system or Tree system.
2. Grid iron or Reticulation system.
3. Ring system or Circular system.
4. Radial system

6.12.1 Dead End System

Suitable for town developed in irregular manner.

In this system the trunk main starting from the service reservior and as smaller mains and branches taken off from it along roads joining the main roads along which the trunk main is laid. It is gradually reduced in size. The same principles apply to all branches and submains.

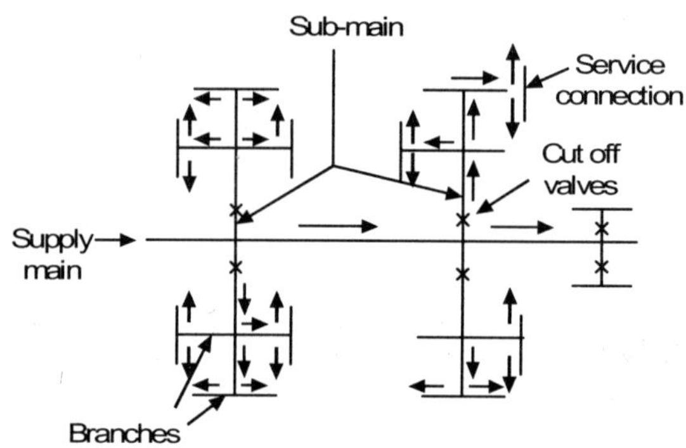

Fig. 6.14. Dead end system

Advantages and disadvantages of dead end system

Advantages

(*i*) The system is cheap.

(*ii*) Hence pressure to be allowed can be calculated easily.

(*iii*) The design calculations are simple in this system.

(*iv*) It is easy to plan the layout.

(*v*) It is easy to lay pipes.

(*vi*) It can be used in irregularly developed cities.

(*vii*) It can be extended when desired.

Disadvantages

(*i*) There are dead ends where water may decompose and contaminate the supply.

(*ii*) To provide a scour valve at every end is costly.

(*iii*) The supply will be cut off if any repair work is carried in main/sub Main.

(*iv*) Difficult to extinguish fire during repair work.

6.12.2 Grid Iron System

In this system the layout of pipelines assume the shape of a network and all the dead ends are eliminated by interconnecting. The water is kept in perfect circulation. This system is most suitable for towns that have a rectangular lay out of roads. In case of fire, water can be drawn from more than one point. This system is in common fashion these days and for all newly developed cities and model towns, greatly in use.

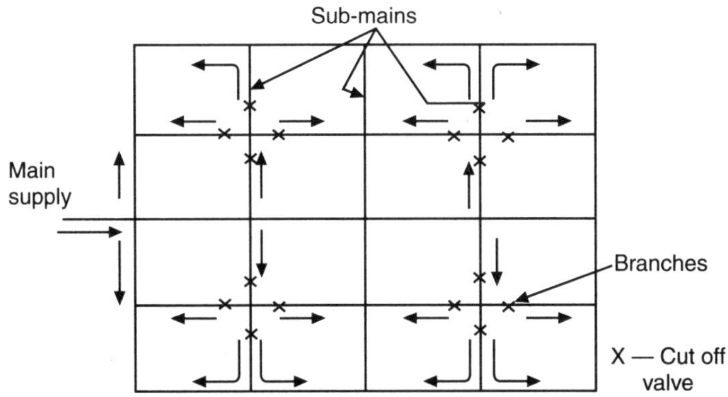

Fig. 6.15. Grid iron system

Advantages and disadvantages of grid iron system

Advantages

(*i*) In this system all dead ends are eliminated.

(*ii*) Supply is available from various points.

(*iii*) The supply is even available during repair work.

(*iv*) The water will not decompose due to continuous circulation.

(*v*) The supply available for many points for free demand.

Disadvantages

(*i*) It is costly system.

(*ii*) Design of system is difficult.

(*iii*) More number of sluice valves required.

(*iv*) Computation of pipe line sizes is difficult.

(*v*) Fittings are required to cut off any desired section from supply.

6.12.3 Ring System

This system of laying pipes is shown is the Fig. 6.16. In this system, the supply main is laid around the district of distribution. The flow bifurcates and reaches a point from both ends. Water can reach the consumer very quickly and then the service can be effected quite satisfactorily.

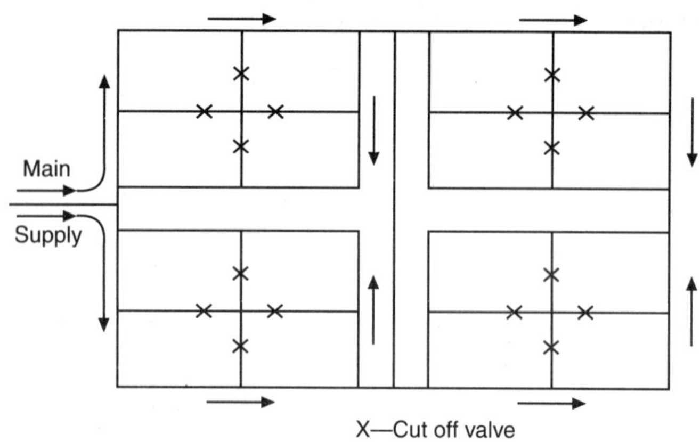

Fig. 6.16. Ring system

6.12.4 Radial System

In India this system is not so popular, because this system is suitable when the roads are laid radially. In a sense it is reverse to that of ring system. The entire district is divided into various zones and one reservoir is provided for each zone. Water is taken from water mains and pumped into the distribution reservoir. The water is then supplied through radially laid pipes as shown in Fig. 6.17.

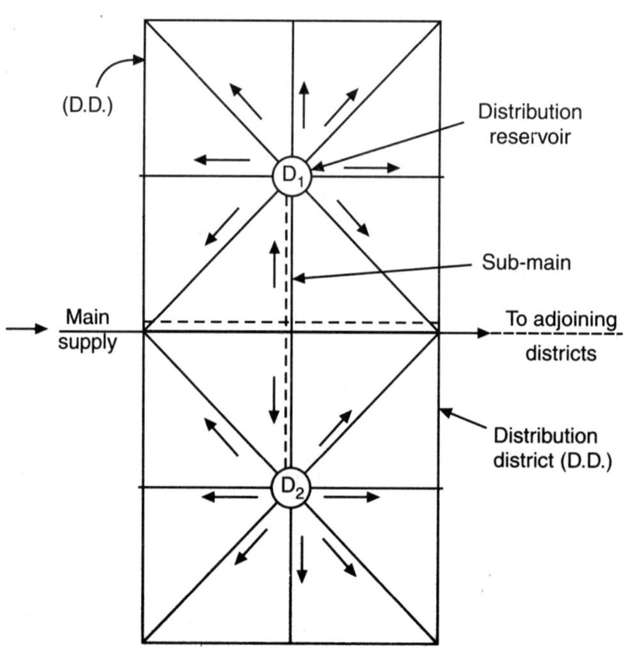

Fig. 6.17. Radial system

6.13. CAUSES OF WASTAGE OF WATER

The following are the reasons of water wastage.

1. Leakage in the mains, joints, valves and other fittings.
2. Leakage in the service connections under the control of the consumer.
3. Leakage in the cistern where a consumer stores his water for use.
4. Keeping the taps carelessly open, even during not using them.
5. Lack of general sense of responsibility for economical use and value of water.
6. General mistaken idea that a flow of water on the whole clean gutters etc.

6.14. PREVENTION OF WASTAGE

1. Insisting the use of best fittings.
2. By occasional inspection of the same and maintaining the same in good condition.
3. By installing meters in areas of large consumption and charging a consumer by the quantity of water he uses.
4. District meters should be installed to study the consumption of water from day to day. Any excess than the normal indicates wastage.
5. It is also argued that wastage could be minimum by an intermittent supply as against continuous supply. But this is not desirable. In modern practice the latter method is invariably recommended.

6.15 APPURTENANCES IN DISTRIBUTION SYSTEM

6.15.1 Appurtenances

The different devices required for controlling the flow of water, for preventing leakage and other purposes in water distribution are called appurtenances.

6.15.2 Valves

The importance of valves in pipe line should not be forgotten. Depending upon different situations and requirements different types of valves are used in the distribution system. The important ones are as follows.

1. Stop valve or sluice valve.
2. Check valve or reflux valve.
3. Air relief valve or float valve.
4. Pressure relief valve or safety valve.
5. Scour valve or blow off valve.

6. Plug valves.
7. Fire hydrant.

1. Stop valve or Sluice Valve: These are also known as gate valves and used to control the flow of water through pipes.

They are provided at street corner where pipe line intersect. Two valves are placed at each intersection. As large a number of valves should be provided as possible, so that when there is any breakdown on a pipe line, the last length of the same may be shut off and isolated for repairs by closing the valves which should be able to close down slowly. If the supply is suddenly cut off, it causes, water hammer on all joints fittings.

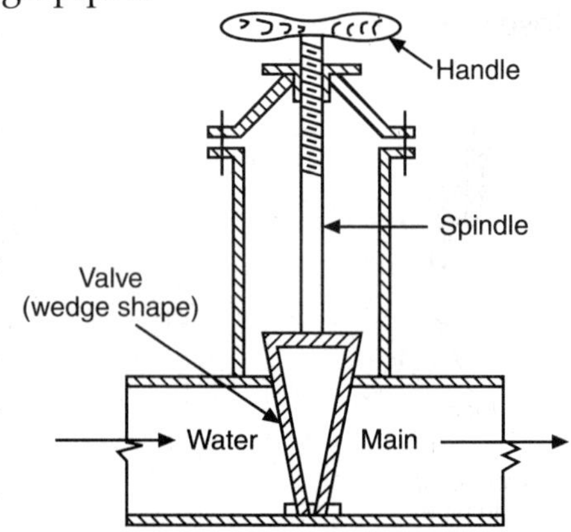

Fig. 6.18. Sluice valve

2. Reflux Valve or Check Valve: It allows the flow in one direction only and is sometimes used at the foot of a long rising water main to prevent running back of water if pipe breaks down in the lower reaches. The valve is also provided on pipe lines leading off from the pumping station, to prevent empty back of pumped water in case a pump breaks down or is removed for repairs.

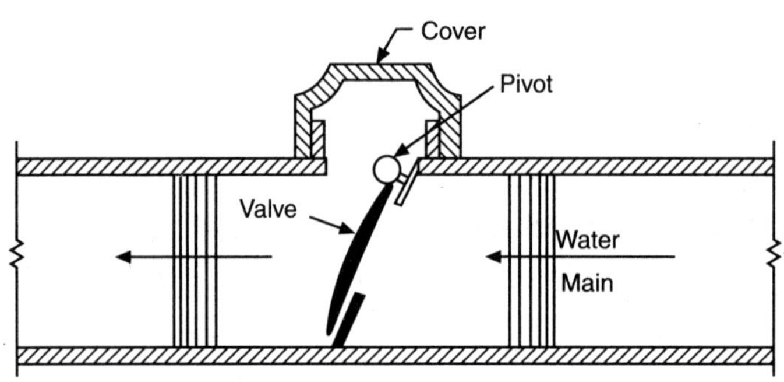

Fig. 6.19. Reflux valve

It has a metallic disc resting against a projection in the casing and to which a hinge is attached. When water flows in the direction shown in Fig. 6.19 the disc goes up and allows the flow. In case the direction of flow is reversed, the disc falls down under its own weight and shuts tightly against its bearings. A cap is provided at the top for inspection.

3. Air Relief Valve or Float valve: In long pipe lines, air accumulates at high points of the line which interferes with stream line flow of water. At such points air valves are provided which remove the accumulated air automatically. This valve has one or two hollow floats inside float chambers. There are air opening at the top and the valves are connected to the main as indicated in the figure. Float chamber normally remains full of water. When air fills it, the water level goes down and the float falls, thereby openings the air opening at the top through which air escapes out, after which the chamber again fills up with water and takes the float up, which closes the air openings.

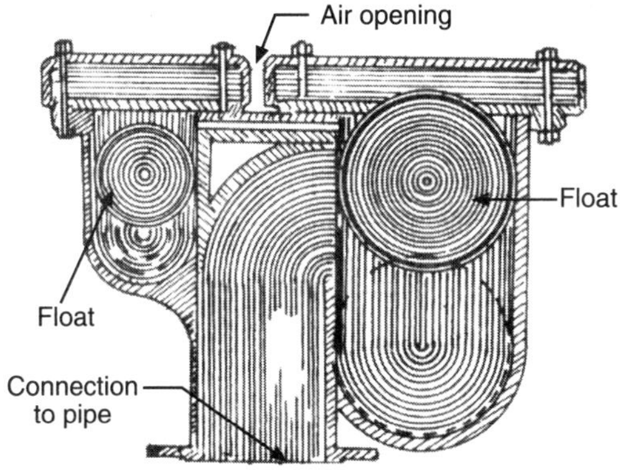

Fig. 6.20. Double acting air valve

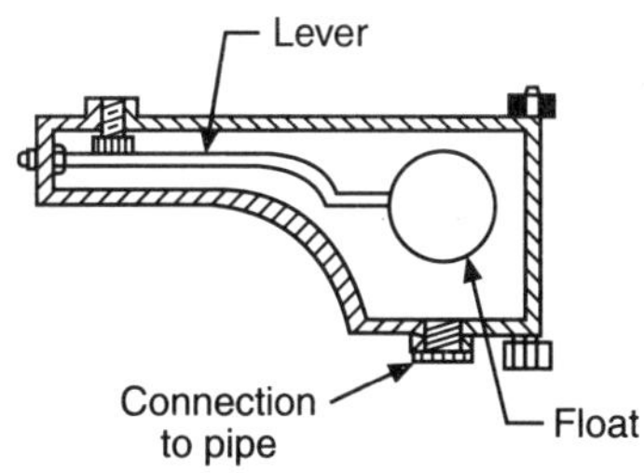

Fig. 6.21. Single acting air valve

4. Pressure Relief or Safety Valve: It relieves high pressures in pipe lines and plumbing fixtures in building. It consists of disc controlled by a spring which can be adjusted for any pressure.

As the pressure in the pipe line exceeds, the disc is forced off its seat, and the pressure is relieved through cross pipe. After this, the disc again comes down automatically due to the springs.

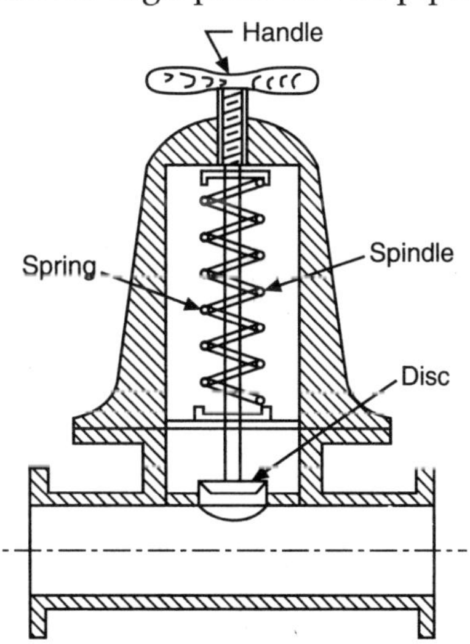

Fig. 6.22. Safety valve

5. Scour out Valve or Wash out Valve: It is placed at every depression in a pipe line. Mud and sludge get collected in the pipes at these depressions which generally occur when pipes cross valleys etc. The valve is essentially mounted on a branch of the main pipe line at the lowest point. When the valve is worked, it removes the dirt due to high velocity of water.

6. Plug Valve: These are also known as cone valves, which have a tappered plug which turns in a tappered seat.

Plug valves are used not only for water under high pressure but also for sewage, oils, gases etc.

7. Fire Hydrant

Fire fighting requires water at much higher pressure than that used for domestic purposes. Because water is used to extinguish fire at a larger rate and has to reach even upto the top of the buildings. This higher pressure is generated by fire engines at the time of fire. The fire engines are attached to fire hydrants which are connected to street mains and water is taken through them in case of fires in the vicinity. Hydrants are placed on all streets at distances of 154 m to 300 m apart, as well as at all street crossings and turnings, so that in case of a fire in a building, the nearest fire hydrant shall not be more than 75 m away.

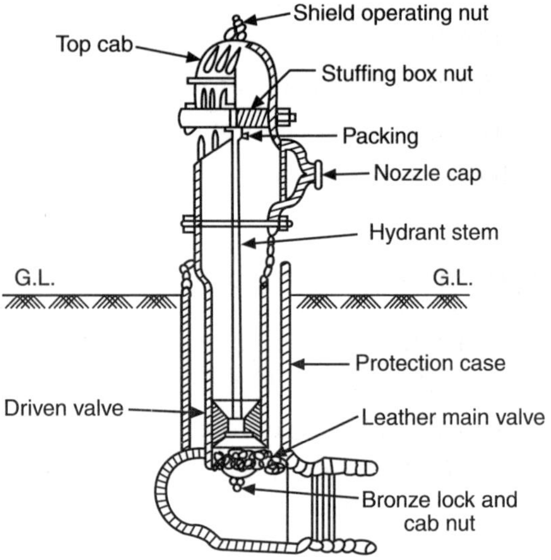

Fig. 6.23. Fire hydrant

A fire hydrant has a 10 cm connection for attaching it to fire engine and 6.25 cm connection for hose outlet. Hose pipes carry the water to the building. They have a 2.75 cm smooth nozzle through which a stream of water comes out under a pressure of atleast 3 kg/cm^2. Available pressure at fire hydrants should preferably be

With pumping engines ... 0.75 to 1.25 kg/cm^2

For direct flow from hydrants 3 to 5 kg/cm^2

Sometimes pressures upto 10 kg/cm^2 are developed.

Hydrant consists of a C.I. barrel with connection to the street main. It has a leather valve connected to valve stem which is turned by a handle at the top. By working the handle; it can be opened and closed. In the closed position the valve rests against the valve seat. The hydrant usually has two 6.25 cm hose outlets and one 10 cm upper outlet for attaching it to the fire engine. Hydrants may have up to 4 outlets, when they are placed four ways. The outlets are spaced all round the hydrant barrel. The hydrant is either partly or wholly underground. Hydrants are used to fill water carts.

6.16. SERVICE/STORAGE/DISTRIBUTION RESERVOIRS

A reservoir which stores the treated water for supplying water during emergencies is called service reservoir.

Functions:

(*i*) It stores treated water.

(*ii*) It meets variation of demand.

(*iii*) It creates necessary effective pressure in the distribution system.

(*iv*) It meets emergency of demand in case of fire.

(*v*) It avoids continuous operation of pumps.

(*vi*) It brings economy in distribution system.

(*vii*) Size of filter units, pumps and mains reduced due to storage reservoirs.

6.17. TYPES OF SERVICE RESERVOIRS

(*i*) Earth reservoirs

(*ii*) Masonry and RCC reservoirs (underground)

(*iii*) Elevated reservoirs.

(*i*) Earth Reservoirs: These reservoirs are used when large quantity of water is to be stored before the treatment. These reservoirs are constructed by excavating to the required depth below the ground surface and the excavated earth is used for the embankment construction to the required height above the ground. Core walls may be constructed to make the earthen reservoirs impermeable. The embankment should be keyed down to the bottom as shown in the diagram. Outlet pipes may be provided in the embankment walls at various places to minimise seepage of water. To prevent the loss of water and leakage, the sides and bottom of the reservoirs should be properly lined with bricks, stones, asphalts or cement concrete etc.

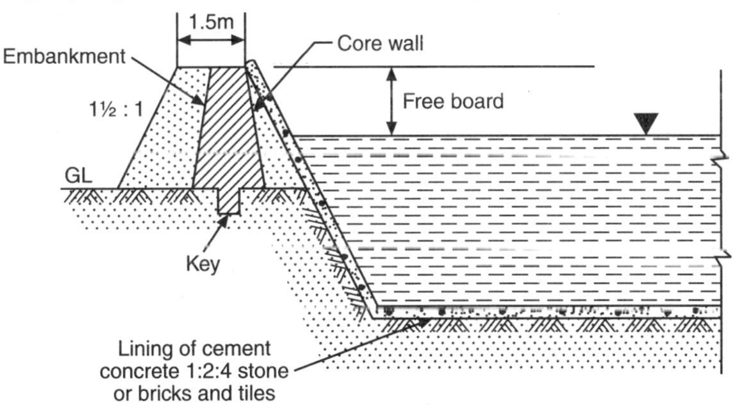

Fig. 6.24. Earth reservoirs

(*ii*) Underground Masonry and RCC Reservoir: These reservoirs are constructed on high natural grounds. It may be made of bricks, stones, plain or reinforced cement concrete. While designing these reservoirs the groundwater table is kept in mind and the side walls should be able to take up pressure of water, when the reservoir is full, and the earth pressure when it is empty. The floor of these reservoirs are usually constructed with RCC slab and bituminous compounds may be used at all construction joints for water tightness.

These reservoirs should be provided with RCC roofs on number of columns. The top of roof is covered with about 50–75 cm thick earth layer to protect the reservoir from heat and cold. Generally green grass lawns are maintained at the top. This serves as a safety measure from defence point of view. Ventilation pipes are provided for aeration and stairs are provided for internal inspection. Refer Fig. 6.25.

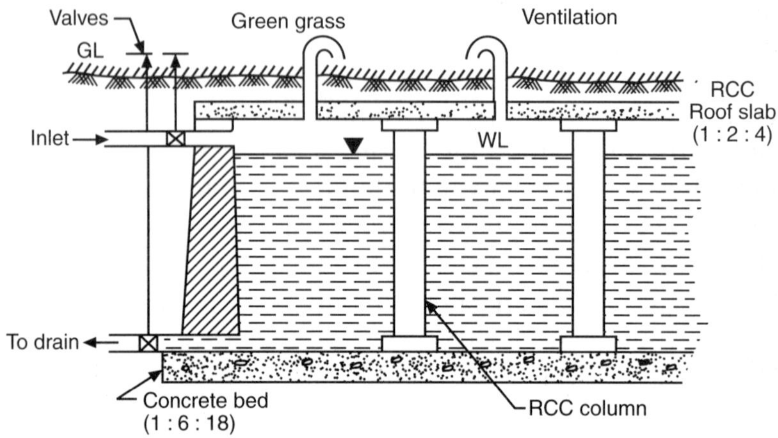

Fig. 6.25. Underground reservoirs

These reservoirs are used for storing and distributing clear water. Two such underground reservoirs having (4.3 m litres) capacity have been constructed at head works.

***(iii)* Elevated Reservoirs:** When sufficient high ground above the distribution is not available for the construction of ground reservoirs, from where water can flow under gravitational force in the distribution system, elevated reservoirs are constructed. These reservoirs reduce the initial cost of pumps, reduce the peak demands on the pumps and reduce the maximum pressure required at the pumps discharge.

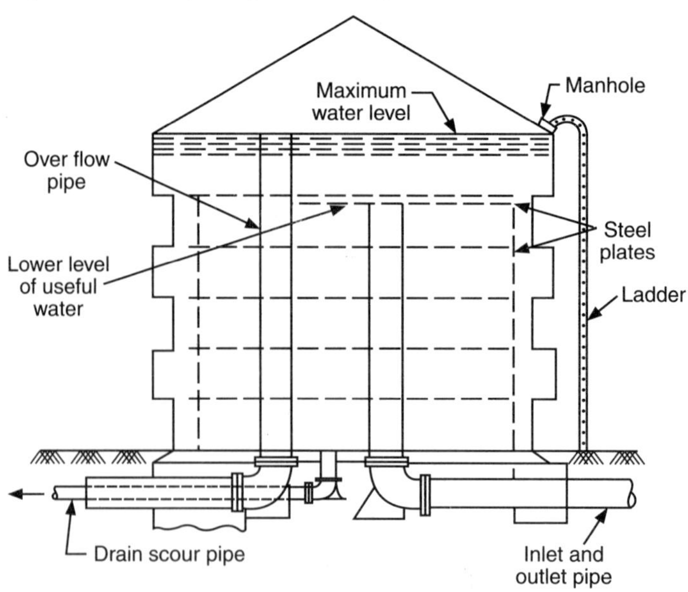

Fig. 6.26. Steel stand pipe

Generally two types of elevated reservoirs are used which are:

(*a*) Stand pipes, (*b*) Elevated tanks

(*a*) Stand Pipe: Stand pipes are generally made of RCC or steel and will be circular in plan. These are upto 12 m in height. If it is possible to locate the stand-pipe on a hill or high ground then its entire capacity can be made use of successfully. Its main function is to increase pressure on the distribution system by creating extra storage in the tank above the elevation required to give the necessary pressure for distribution. These pipes are provided with single inlet and outlet pipes terminating at the lower elevation of the useful capacity of the tank, scour or drain-pipe to drain out and flush the tank and the overflow pipe to discharge surplus water.

(*b*) Elevated Tanks: When water is to be distributed at very high pressure and the use of the stand pipes become impracticable and uneconomical. Then elevated tanks may be constructed with steel or RCC. Nowadays RCC elevated tanks are very popular, because they have long life, require very little maintenance and give good appearance in the locality. In modern practice prestressed RCC tanks are coming up, because they are even economical than plain RCC tanks.

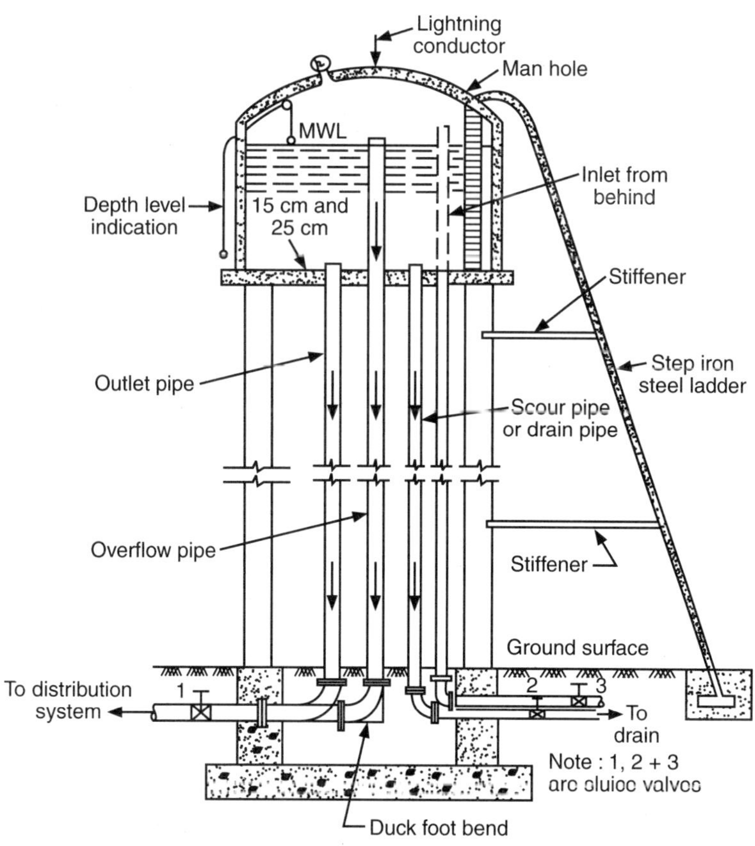

Fig. 6.27. Elevated tank

The elevated tank is provided with the usual-piping consisting of inlet, outlet overflow and scour or drain pipes as shown in the above diagram. A depth level indicator with a float arrangement is provided to measure the depth of the water in the tank. About 60 to 100 cm balcony is provided around the tank for inspection and maintenance of the tank and step-iron steel ladders are also fixed in the inner walls of the tank. The tank is covered and is provided with manhole, ventilation pipe and lightning conductors to serve as a protection against lightning.

These tanks can be built for storing large quantity of water. The various shapes depend on the designing engineer. Steel tanks should to inspected regularly and cleaned and painted every 4 or 5 years with aluminium paints etc.

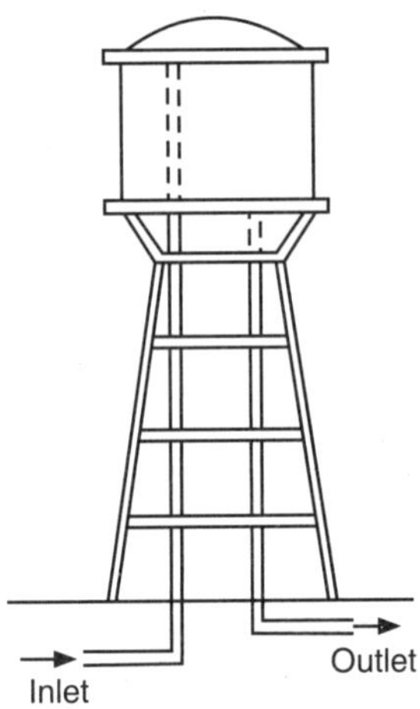

Fig. 6.28.

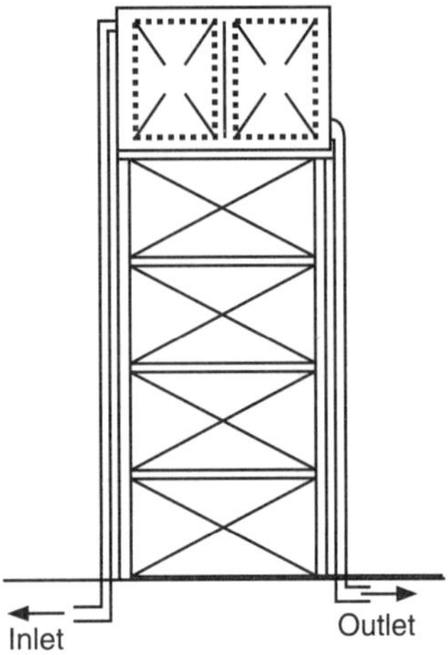

Fig. 6.29.

6.18. ADVANTAGES AND DISADVANTAGES OF RESERVOIRS

Elevated Reservoirs	Surface Reservoirs
(*i*) These are expensive e.g., cost of storage for (4.3 m litres) tank to be four times that of a surface reservoir.	(*i*) These are economical.
(*ii*) These give reduction of pump capacity and the pumping cost.	(*ii*) This does not happen here.
(*iii*) These give reduction of maximum pressures required at the pump discharge and reduction of peak demand on the pumps.	(*iii*) This does not happen here.
(*iv*) For large structures bearing power or foundation should be determined by tests.	(*iv*) Bearing capacity tests are not required here. Only permeability conditions are determined.

(Contd...)

(*v*) Unequal settlement causes severe strains on the various members of elevated tank and may cause serious leakage.	(*v*) This does not happen here.
(*vi*) Design of foundation etc., is complicated.	(*vi*) Design is simple as compared to elevated tank.
(*vii*) Lighting conductors are fixed to serve as a protection against lighting.	(*vii*) These are not essential here.
(*viii*) More maintenance is required for steel built elevated tanks.	(*viii*) Maintenance is less in case of surface reservoirs.

6.19. BOOSTING

A supply system is satisfactory only when it provides good pressures in the various pipes under actual or expected working conditions. Very high pressures are also to be avoided as they mean unnecessary expenditure for safety and maintenance. In a city there may be certain areas where foot pressure is not maintained, specially when some development of the city has taken place and pipes have been added in several places to serve new areas. The capacity of the mains also falls short of requirements. Booster pumps are installed at different points as required and they serve the following purposes:

(*a*) They allow water mains to carry more discharge from the source to the treatment plant.

(*b*) Pressures in the entire distribution system where consumption has gone up, are improved.

(*c*) They also allow distribution pipes to carry more discharge to meet increased demand of water.

(*d*) Where boosting is restored to, water mains can be made of smaller size, thereby reducing cost.

Boosting increases the pressures upto 50% or more. It consists of direct pumping into pipes by drawing water under pressure.

MODEL QUESTIONS

1. (*i*) Minimum pressure required for four storeyed building is _____ kg/cm^2.

(ii) At the ridges _____ valve is used.
(Air valve/Safety valve/Reflux valve)

(*iii*) At the root of the suction pipe _____ valve is used.
(Foot valve/Float valve/Air valve)

(*iv*) Elevated reservoirs are _____ as compared to surface reservoirs.
(Expensive/Economical)

(*v*) When the use of stand pipes becomes impracticable and uneconomical _____ Tanks are constructed. (Underground/ Elevated)

(*vi*) Cities which are irregularly developed _____ Distribution system is used.
(Dead End/Radial/Ring)

2. Give a plan of locality, propose a suitable distribution layout.
3. What do you mean by distribution system? Discuss the different types of the same.
4. (*a*) Name the various types of reservoirs used in distributing water. Describe any two of them with the help of neat diagrams.
   (*b*) How would you determine the TOTAL capacity of a service reservoir.
5. What are the advantages and disadvantages of intermittent and constant system of supply?
6. What is service reservoir? Give its importance in a distribution system. Draw a neat sketch of an elevated tank and show on all its component parts and appurtenances.
7. Discuss the factors that govern the layout of a distribution system in a city.
8. What is the function of a stand-pipe in a distribution system?
9. What is the function of hydrants in the distribution system? With the help of neat sketches describe hydrants.
10. Explain with line sketches in water supply distribution:
    (*i*) Radial system (*ii*) Ring system
11. Write short note on Air-relief valve.
12. Explain with tine sketch of water supply distribution:
    (*i*) Grid iron system (*ii*) Dead end system
13. Describe the functions of a service reservoir. Sketch on overhead RCC Service Reservoir, label the various components.
14. Sketch the radial system of water supply distribution.
15. What are the functions of a service reservoir? Name the various types of Sketch an overhead RCC service reservoir.
16. (*a*) Write a note on sluice valve. Explain its function on water supply lines.
    (*b*) Write notes on the following:
    (*i*) Fire Hydrants (*ii*) Sluice valve
    (*iii*) Reflux valve
17. Differentiate between air-relief valve and pressure relief valve.
18. Why and where are valves located? Explain their working.
19. Discuss in detail the various requirements of distribution system of water supply to a town.
20. Explain in detail the various types of layout of distribution system.
21. What are the different ways used for the conveyance of water from a source to purification works situated at a great distance. Discuss the factors that govern the adoption of one-type or the other.
22. (*i*) For the distribution system in a building _____ pipe is generally used. (Galvanized iron/Cast iron steel)
    (*ii*) The minimum size of cast iron pipe used in the distribution is _____ cm.
    (*iii*) For rising main _____ pipe is used. (Galvanished Iron/Steel/Cast iron)
    (*iv*) Rubber ring joints are _____ than lead joints. (Costlier/economical)
    (*v*) For elevated reservoir _____ joint is used. (Flange/Spigot and Socket)
    (*vi*) _________ joints are used on pipes exposed to considerable difference of temperatures. (Flexible/Expansion)
23. What are different kinds of pipes available for use in water supply system? Discuss mertis of each.

24. Discuss the comparative merits and demerits of steel pipes, cast iron pipes and concrete pipes. Describe a suitable joint for a cast iron pipe.

25. Describe a joint suitable for 125 cm cost iron pipe used for conveyance of water.

26. Under the following places, what type of pipes you would recommend? Also give reasons of your recommendations:

(*a*) Rising mains from pump of the over-head tank.

(*b*) A long gravity main type bringing water to the treatment point from the source.

(*c*) Distribution pipes for carrying treated water from the treatment plant to the city.

(*d*) Distribution pipes from which connections are to be made to individual houses.

(*e*) Pipes used for distributing water inside the building.

(*f*) Pipes for carrying acidity and alkalinity impurities.

(*g*) Pipes for carrying hot water.

27. (*i*) Describe common types of water supply pipes.

(*ii*) Sketch either a flanged joint or a socket and spigot joint or a socket and spigot joint.

28. Write short not on joints in CI pipes.

29. Of what materials are the pipes meant to carrying drinking water generally made of? Please name pipes material-wise, giving advantages/disadvantages in each case very briefly.

30. Draw a sketch of a flexible joint of AC pipes.

31. Explain in brief various types of pipes used in water works system and describe briefly their relative advantages and disadvantages.

32. Write short note on PVC pipes.

33. Write short note on "The necessity of wrought iron main in water supply".

34. Write short notes on the following:

(*i*) Air valves (*ii*) Reflux valves

(*iii*) Sluice valves (*iv*) Check valves.

35. What are the advantages and disadvantages of intermittent and constant system of supply?

36. Discuss the factors that govern the layout of a distribution system in a city.

37. Describe the functions of a service reservoir. Sketch on overhead RCC service reservoir, lable the various components.

38. Sketch the radial system of water supply distribution.

39. Explain ring and radial system of distribution of water supply scheme.

40. Describe systems of water supply (Intermittent and continuous), giving advantages and disadvantages of each.

41. (*a*) Given the merits and demerits of CI, pipes as used for water distribution. Explain with the help of a neat sketch joining of CI pipe with proper joint.

(*b*) What are the requirements of a good water distribution system?

42. Differentiate between the following:

(*i*) Ring system and grid system

(*ii*) Flanged joint and spigot and socket joint

(*iii*) Reflux valve and air valve

(*iv*) Check valve and slaves valve

(*v*) Intermittent and continuous water supply system.

❑❑❑

CHAPTER 7

LAYING OUT PIPES

Learning Objectives

- General
- Pipe Laying
- Disinfection of Pipe Lines Before Use

7.1. GENERAL

Before taking up any pipe laying work, a detailed map of the city is prepared. All important roads, streets, lanes are marked clearly on it. Position of existing pipe lines, sewer lines, gas pipes, telephone and electrical conduits are shown on the map. The length and size of the proposed pipe line should also be shown on the map. Proposed positions of valves, service connections, fire hydrants etc., should also be anticipated, so that provisions for their installation may be made during the laying of pipe line.

7.2. PIPE LAYING

Pipe laying operation includes the following steps:

(*a*) Setting out alignment of pipes

(*b*) Excavation of trenches for laying of pipes

(*c*) Bottoming up of the excavated trench

(*d*) Lowering of pipes into the trench.

(*e*) Laying of pipes

(*f*) Jointing and anchoring of pipes

(*g*) Testing of pipes lines

(*h*) Back filling

7.2.1. Setting Out Alignment of Pipes

After preparation of the detailed maps and after getting the sanction, the work on the pipe line is started by making the alignment of the pipe line on the ground. The centre line is marked by driving stakes at 30 m interval on the straight line. For the curved alignment, the spacing of stakes is reduced to 5 m to 15 m depending upon the sharpness of the curve. If at the alignment there is a pukka road and driving of stakes is not possible, then the distance of centre lines of pipe may be marked from the curb line.

7.2.2 Excavation of Trenches for Laying of Pipes

After fixing the centre line of the pipe on the ground, the excavation of trenches is started. The width of the trench should be sufficient enough to allow the pipes to be laid and jointed properly. The depth of the trench should be sufficient enough to give adequate protection to the pipes against impact of traffic and other factors. Width is usually kept 300 mm to 450 mm more than the outer diameter of the pipe. Depth of the trench should be such as to give a ground cover of about 900 mm from the top of the barrel of the pipe. At every joint the depth of excavation should be little more than the normal depth of the trench. It should be kept in mind during excavation that full length of pipe remains supported on soil but the joints should remain over hanging. Excavations may be done by manual labour or by excavation machines.

Precautions to be taken during Excavation

(*i*) The excavated soil from the trench is deposited only on one side and a little away from the trench.

(*ii*) Excavated soil is deposited in one side and pipe lengths are to be stacked on other side of trench.

(*iii*) Where the excavation is done below the ground water table, arrangements should be made for pumping of water.

(*iv*) In case of excavations in loose soils where the depth of excavation is large, the sides of the trenches should be prevented from caving in by timbering.

(*v*) If the trenches are wet due to sub soil water, only small length of trench should be dug and finished with pipe laying.

7.2.3 Bottoming up of the Excavated Trench

After excavation, the bottom of the trench should be prepared carefully, so that the barrels of the pipe, when laid, are well bedded for their whole length on a firm surface and are true to line and gradient. If trench bottom is very soft and unreliable, cement concrete bedding or cement concrete block supports may be provided for the pipes to rest upon. If this is not done, the line may deflect later on and may caused damage to the pipe joints.

Several methods of bedding the pipe in trench bottom are used, and they are shown in the following diagrams :

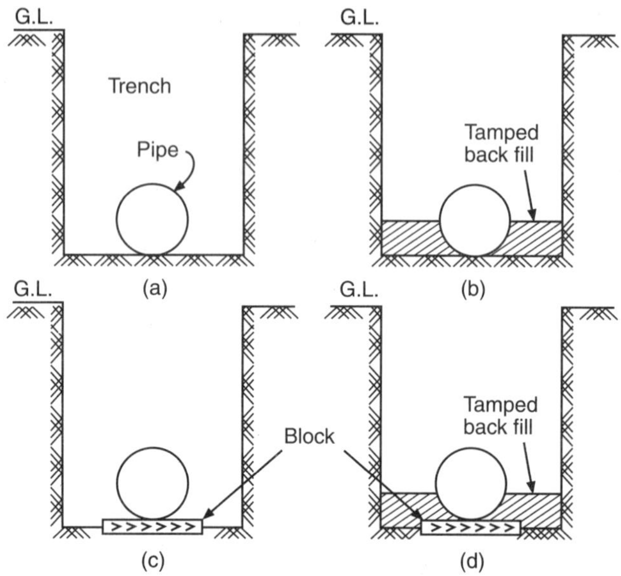

Fig. 7.1. Type of trench bottoms

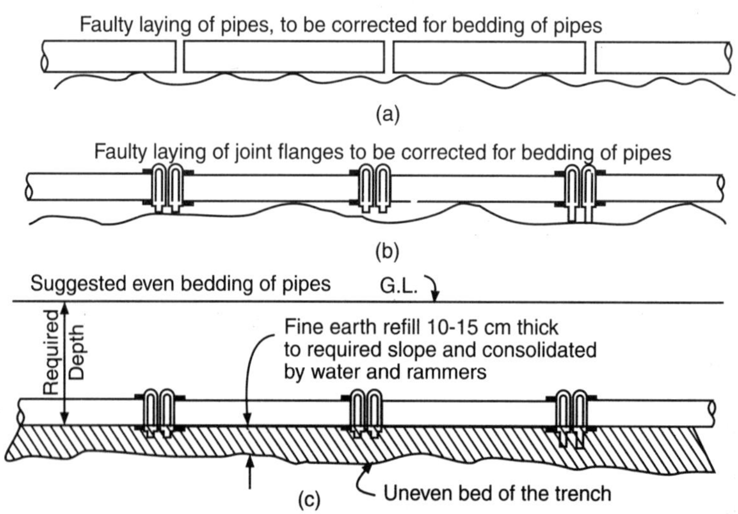

Fig. 7.2. (a) and (b) Faulty pipe laying and (c) Suggested even bedding of pipes

The use of blocks as in type *c* and *d*, has the advantage of allowing easy levelling and alignment, but it produces greater crushing loads than type *a* and *b*. When blocks are used they should be placed on undisturbed earth and set in slots so that they project about 20 mm above the trench bottom. Back fill should be carefully tamped around and under the pipe, as in *d*, so that the block carry only a small part of the load. Type *c* and *d* are commonly recommended. Refer Fig. 7.2 (*a*), (*b*) and (*c*) for faulty laying of pipes and suggested even bedding of pipes.

7.2.4 Lowering of Pipe into the Trench

Pipes are generally staked on one side of the trench after transporting these to the site. Pipes should be gently lowered into the trench to avoid any damage to the pipes. If pipes are heavy these should be lowered in the trench by using pulley blocks fixed on derricks, small size pipes are lowered manually. Before lowering any pipe in the trench it should be cleaned of all foreign matter. They should be throughly brushed out. The pipe should be examined for damage to its coating, sheathing or wrapping and repaired if necessary.

7.2.5 Laying of Pipes

Pipes are generally laid with a flat slope parallel to the hydraulic gradient to avoid any air locking trouble. Where there is slope, pipe laying should be done in an uphill direction to facilitate joint making.

7.2.6 Jointing and Anchoring of Pipes

Jointing of different types of pipes are already discussed in the previous chapter.

Thrust blocks of concrete should be formed at all bends, except in case of small pipes under low pressure, to transmit the hydraulic thrust and distribute it over a wide area of the ground. Where the hydraulic thrust is in upward direction as in case of pipes on sloping grounds, anchor blocks of sufficient weight should be provided to which the pipes should be secured with steel straps. Table 7.1 shows the displacing forces in the mains due to end and side thrust at bends.

Table 7.1. End thrust and radial thrust on bends in mains calculated for a pressure of 1 N/mm^2

Main diameter	End thrust	Radial thrust on bends at the following angles			
		90°	45°	22½	11¼
mm	kN	kN	kN	kN	kN
50	1.96	2.78	1.50	0.77	0.36
80	5.03	7.11	3.85	1.96	0.91
100	7.85	11.11	6.01	3.06	1.42
150	17.67	24.99	13.53	36.90	3.21
225	39.76	56.23	30.43	15.51	7.21
300	70.67	99.94	54.09	27.57	12.82

7.2.7 Testing of Pipe Lines

After a new pipe line has been laid and jointed, it shall be subjected to the following two tests:

(*a*) Pressure test at a pressure of atleast double the maximum working pressure.

(*b*) Leakage test at a pressure to be specified by authority. This test is to be conducted after the satisfactory completion of the pressure test.

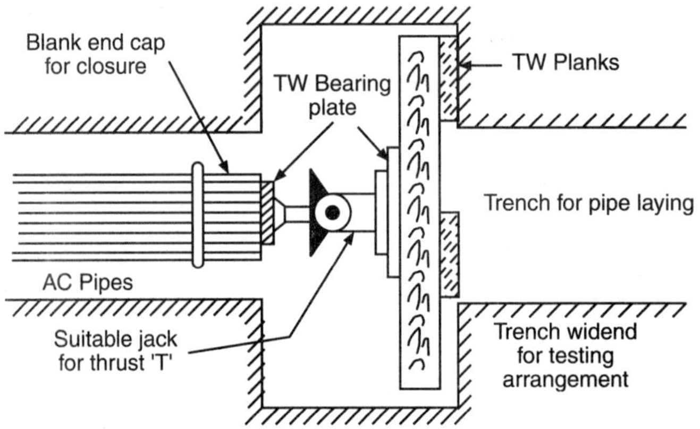

Fig. 7.3. Testing of pipeline

(*a*) Pressure Test: The step by step procedure adopted for pressure testing of pipes is described below:

(*i*) The pipe line is tested from section to section. At a time only one section lying between two sluice values is taken up for testing.

(*ii*) First the downstream sluice valve of the section is closed and water is admitted in the section through the upstream sluice valve. During filling air valve is properly operated to remove all air from the pipe.

(*iii*) Then the upstream valve of the section is closed to completely isolate the section from the rest of the pipe line.

(*iv*) Pressure gauges are then fitted along the pipe length of the section at suitable interval (generally 1000 mm so) on the crown through holes left for this purpose.

(*v*) The pipe section is then connected to the delivery side of a pump through a small by-pass valve and the pump is started to increase the pressure in the pipe. The operation is continued till the pressure inside the pipe reaches a pressure at least double of the maximum working pressure.

(*vi*) The by-pass valve is then closed and the pump is discontinued.

(*vii*) The pipe is kept as it is for 24 hrs and inspected for any fall of pressure. This completes the pressure testing of pipes.

(*b*) Leakage Test

After successfully completing the pressure test, the leakage test is carried out. Leakage test is to test maximum allowable leakage which is determined by the formula:

$$Q = \frac{ND\sqrt{P}}{3.3}$$

where Q = allowable leakage in cm^3/hr

N = number of joints in the length of pipe line

D = diameter in mm

P = the average test pressure during the leakage test in kg/cm^2.

7.2.8 Back Filling

After testing the pipe line, the trench should be refilled with the excavated material.

The process of placing the excavated soil into the trench all around the pipes is called back filling.

(*a*) For the purpose of back-filling the depth of the trench shall be considered as divided into the following three zones from the bottom of the trench to its top:

Zone A: Form the bottom of the trench to the level of the centre line of the pipe.

Zone B: From the level of the centre line of the pipe to a level 30 cm above the top of the pipe, and

Zone C: From a level 30 cm above the top of the pipe to the top of the trench.

(*b*) Back filling in Zone A shall be done by hand with sand, fine gravel or other approved material placed in layers of 8 cm and compacted by tamping. The back filling material shall be deposited in the trench for its full width of each side of the pipe, fitting and appurtenances simultaneously.

Back filling in Zone B shall be done by hand or approved mechanical methods, special care being taken to avoid injuring or moving the pipe. The type of back fill material to be used and the method of placing and consolidating shall be prescribed by the authority to suit individual locations.

Back filling in Zone C shall be done by hand or approved mechanical methods. Type of back fill material and method of filling shall be as prescribed by the authority.

7.3. DISINFECTION OF PIPE LINES BEFORE USE

After testing and correcting the defects of the pipe line, it is ready for transporting water. When the pipes are carrying treated water, they must be disinfected before use. They are disinfected by keeping them, full with water and adding chlorine in such an amount so as to maintain a residue of 50 ppm. This residue is maintained for 12 hrs and the pipe is emptied and flushed with fresh treated water.

MODEL QUESTIONS

1. (*i*) Minimum cover for a water supply pipe line is ______ .
   (*ii*) Water supply pipe line should be tested for ______ pressure.
   (*iii*) Top site rails follow the grade of pipe line. (Yes/No).
   (*iv*) ______ in water supplies will stain Household fixtures.
   (Copper and Iron/ Lead or Zinc)
   (*v*) ______ dissolved may cause water to be toxic.
   (Copper and Iron/ Lead or Zinc)
2. What problems may arise during the operation of excavation, lowering and laying the pipe and how will you overcome them?
3. Describe the laying of 22.5 cm and CI main with socket and spigot joint.
4. How will you conduct a leakage test?
5. Write short notes on the following:
   (*i*) Tuberculation in a pipe-line (*ii*) Cathodic protection
   (*iii*) Back filling
6. Explain the precautions to be taken during excavation and till the back filling is complete.
7. Explain the following with neat sketches:
   (*a*) Timbering of pipe trenches. (*b*) Shoring up of buildings.
8. What precautions are to be taken while laying pipe lines for drinking water supplies?
9. Write a brief note on testing of water supply lines.
10. Write short notes on the following:
    (*i*) Tuberculation in a pipe-line (*ii*) Cathodic protection
11. What precautions are to be taken while laying pipe lines for drinking water supplies?
12. Write a brief note on testing of water supply lines.
13. Mention the method to calculate the most economical size of a rising main. List the appurtenances necessary to be installed on a rising main between pumps and overhead tank. Indicate their relative positions on a line diagram.
14. Describe the procedure for laying and jointing CI pipes having S/s ends; giving the precautions required in the laying of these pipes.
15. Describe how you will lay and joint a 40 cm CI main in a poor soil.
16. What points should be kept in mind while designing a pipe line.
17. Write short note on "Back filling of Trenches".
18. What precautions are taken for traffic control white laying pipe lines.
19. Write short notes on the following drawing neat sketches where necessary:
    (*i*) Shoring of Buildings. (*ii*) Special foundations in poor soil
    (*iii*) Pipe corrosion (*iv*) Corrosion control
    (*v*) Testing of pipe lines (*vi*) Alignment of trenches

□□□

CHAPTER

8

BUILDING WATER SUPPLY

Learning Objectives

- General
- Terms used in Water Supply System
- The House Water Connection
- Storage of Water
- Essential for Building Water Supply
- Layout of Water Supply Pipes for a Building (Single and Multistorey)
- Water Supply Fixtures and Its Maintenance
- Estimation of Pipe Size
- Hot Water Supply

8.1 GENERAL

Water is taken into the building through a service pipe connected with the main pipe. When the public supply pressure is inadequate, pumps may be installed to supply water to higher floors. During the construction of the building, all principal water risers and municipal drain pipes are installed. G.I. supply mains are used upto 7.5 cm pipe diameter and when higher diameter of pipes are required, cast iron pipes may be used. All water pipers must be capable of meeting the peak demand in the house. Water can be supplied in buildings in two patterns.

(*a*) Upfeed system and

(*b*) Downfeed system.

Down feed system is more popular in India.

Upfeed System: When the pressure in the municipal supply is such that water can reach upto the top floor of a multistorey building, then water is supplied to all floors of the buildings. This is called upfeed system.

Downfeed System: In this system, water is pumped from a sump to a overhead tank then water is supplied to all the floors of the building by gravity.

8.2. TERMS USED IN WATER SUPPLY SYSTEM

(*a*) **Air Gap:** It is the vertical distance from the mouth of a water supply faucet to the rim level of the fixture.

(*b*) **Available Head:** It is the pressure of water available from a water main at the plinth level of the premises.

(*c*) **Residual Head:** It is the pressure available at the tail end of the distribution system.

(*d*) **Water Main:** It is a water supply pipe line laid and maintained by the local authority (like P.H.E. Deptt.) for the purpose of giving general supply of water.

(*e*) **Service Pipe:** It is the pipe extending out of the water main and used for feeding water from mains to the building.

(*f*) **Stop Cock:** Stop cock is a control valve used in a pipe line for controlling the flow of water.

(*g*) **Stop Tap:** Stop tap includes stop-cock, stop valve or any other device for stopping the flow of water in a line or system of pipe at well.

(*h*) **Direct Tap:** A tap which is connected to a supply pipe and is subject to pressure from the water main.

(*i*) **Downtake Tap:** A tap which is connected to a system of piping not to subject to water pressure from the water main.

(*j*) **Supply Pipe:** It is the pipe which extends from stop cock to the entrance of the storage tank and is under direct pressure from the water main.

(*k*) **Communication Pipe:** The part of the service pipe extending from the water main upto and including the stop cock which is under the control of the administrative authority.

(*l*) **Consumer's Pipe:** It is the portion of the service pipe used for supplying water to the building and is not the property of administrative authority.

(*m*) **Cross-connection:** A connection between two normally independent pipelines which permits flow from either pipeline into the other.

(*n*) **Back Flow:** It is the flowing back of water or other liquids into the distributing pipes of a potable supply of water from any source other than its internal source.

(*o*) **Back Siphonage:** It is the flowing back of used, contaminated or polluted water from a plumbing fixture or vessel into a water supply due to a reduced pressure in such pipe.

(*p*) **Plumbing:** Plumbing is the installation of the foregoing pipes, fixtures and other apparatus inside a building for bringing the water supply and removing the liquid and waterborne wastes.

(*q*) **Plumbing system:** The entire system of pipings, fixtures, pipelines etc., for providing water supply or drainage to premises is called plumbing system. It includes the water supply and distribution pipes, plumbing fittings and traps, vent pipes and anti-siphonage pipes, building drains and sewers including the connections etc.

8.3. THE HOUSE WATER CONNECTION

To provide water supply in a building or any premises, the first step is to get a water connection from the Muncipal water main. Fig. 8.1. shows a typical water connection connecting service pipes with the Muncipal water main. The water connections comprises of the following:

(*i*) Ferrule (*ii*) Goose neck
(*iii*) Service pipe (*iv*) Stop cock
(*v*) Water meter.

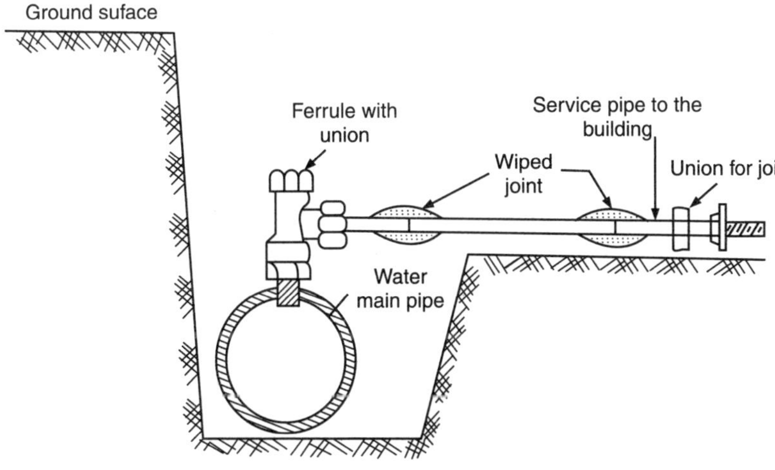

(a) Service pipe connection from water main

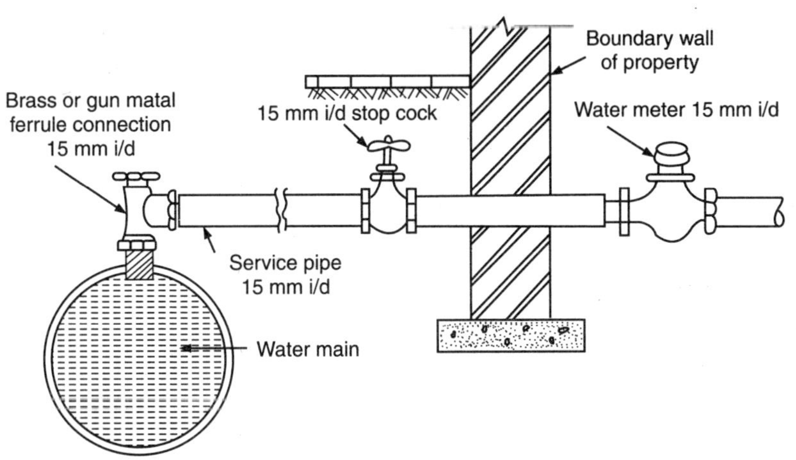

(b) Service pipe connection from water main

Fig. 8.1.

(*i*) **Ferrule:** It is a right angle sleeve made of brass or gun metal and is jointed to an opening drilled in the water main, to which it is screwed down with a plug. It is usually of size varying from 10 mm to 50 mm bore. A tee branch connection off the water main is used if the connections required are more than 50 mm bore.

(*ii*) **Goose Neck:** It is a small sized curved pipe of a flexible material, usually lead and about 750 mm in length forming a flexible connection between the water main and service pipe.

(*iii*) **Service Pipe:** Service pipe is usually of G.I. pipe of size less than 50 mm bore. It should be laid underground in a trench in which no sewer or drainage pipe is laid. The service pipe which supplies water to the building from the water main is connected to the main through the ferrule.

(*iv*) **Stop Cock:** It is provided before the water meter in the service line. Stop cock is housed in a suitable masonary chamber with a removable cover and is fixed in the street close to the boundary wall in an accessible position. It should be of size sufficient to pass the required discharge through the service pipe. Sometimes it is placed just before the water meter for keeping both of them in one chamber.

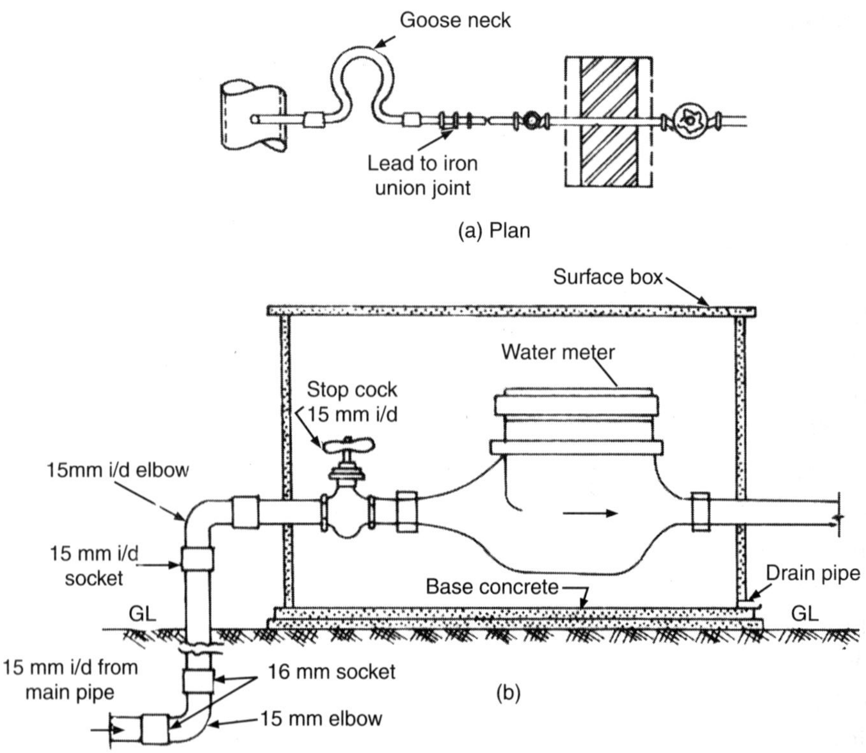

Fig. 8.2. Water meter fixing

(*v*) **Water Meter:** Water meter is used to measure the quantity of water consumed. For domestic water supply, water meters are fitted into the service pipe with unions. The union enables the meter to be changed whenever necessary. It is usual to fix the water meter underground, housed in a suitable brick masonry chamber and cover it with any iron surface box.

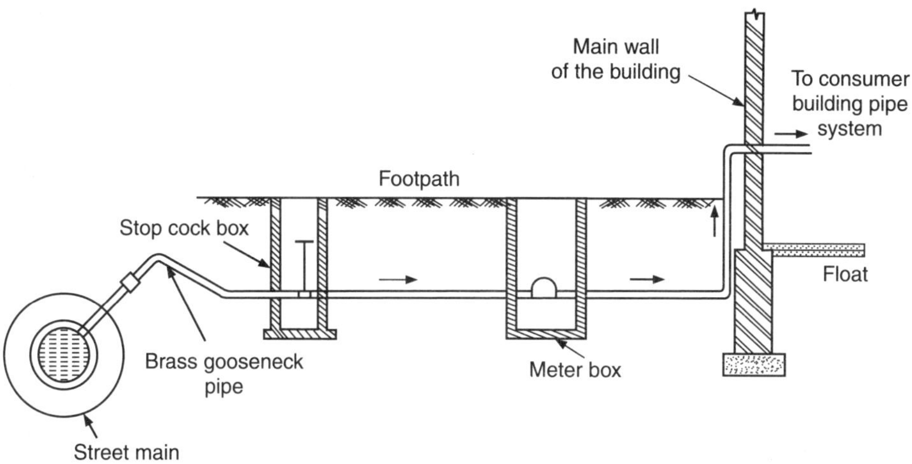

Fig. 8.3. Service connection

8.4. STORAGE OF WATER

1. Purposes of providing storage: In a building, provision is required to be made for storage of water for one or more of the following reasons:

(*a*) To provide against interruptions of the supply caused by repairs to main etc;

(*b*) To reduce the maximum rate of demand on the mains; and

(*c*) To tide over periods of intermittent supply

2. Storage Capacities: The quantity of water to be stored shall be calculated taking into account the following factors:

(*a*) House of supply at sufficiently high pressure to fill up the overhead storage tanks;

(*b*) Frequency of replenishment of overhead tanks, during the 24 hours;

(*c*) Rate and regularity of supply; and

(*d*) Consequences of exhausting storage particularly in case of public buildings like hospitals.

If the water supply is intermittent and the hours of supply are irregular, it is desirable to have a minimum storage of half a day's supply.

3. Types of Storage Tanks: These are small tanks installed above the roof level of the buildings and intended to serve the requirements of water storage of the buildings.

Construction: A storage tank is made either of cast iron or galvanized mild steel pressed plates usually 1.2 m (120 cm) square, jointed together and made water tight by the use of putty. It may also be constructed of masonry or reinforced cement concrete when the size is to large. Other accessories of a storage tank are:

(*i*) A close fitted light, mosquito proof cover with a locking arrangement.

(*ii*) Equilibrium ball valve with float, provided near the inlet to the tank and controlling the inflow of water. The arrangement works automatically admitting in water when level in the tank falls and cutting of supply when water level reaches the full supply level (FSL).

(*iii*) Overflow pipe with its lips projecting about 20 mm above the FSL. The overflow pipe is intended to function in case of failure of the ball valve to function properly. The pipe discharges its flow either through the rainwater downtake pipe or through gully to drain of sewer.

(*iv*) Outlet pipe usually fixed at the end opposite the inlet to avoid any stagnation of water. The pipe projects about (50 mm) above the bottom of the tank.

(*v*) Stop valve provided at the outlet pipe, prevents the necessity of emptying the tank in order to enable any repairs to be carried out to down-take pipe etc.

(*vi*) Scour or drain pipe enables the tank to be periodically cleaned out. For this purpose, the bed of the tank is kept sloping towards the pipe.

Position: The storage tank is usually supported on the roof-slab of the building or on separate bearers so as to distribute the load uniformly. The tank should be located in an easily accessible positions and should admit of proper inspection and cleaning. (See Fig. 8.4)

Capacity: This is dependent on such factors as:

(*i*) Hours of supply at sufficiently high pressure to fill up the over head storage tanks.

(*ii*) Frequency with which these tanks can be replenished during the 24 hours and

(*iii*) Rate and regularity of supply.

It is desirable to have a minimum storage of half a day's supply where water supply is intermittent and hours of supply are irregular.

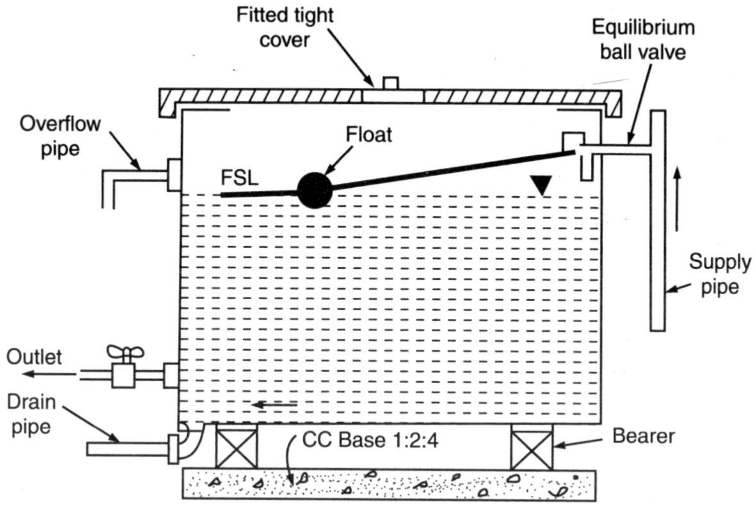

Fig. 8.4. Storage tank

(*b*) **Domestic Storage Tanks:** In case of multi-storeyed buildings, it is usual for the lower storeys (i.e., first, second or even third floor) to be fed from water supply directly from the main, while the higher storeys (fourth floor and above) where direct supply is not likely to be easily available, are connected and fed by the domestic storage tank with storage adequate for one day's supply. The domestic storage is based on the number of downtake fittings such as taps, showers, bath tubs etc., that may be fed from the storage tanks and as given below:

Downtake taps ...(70 litres) each
Showers ...(135 lits) each
Bath tubs ...(200 lits) each

The domestic storage capacities for different classes of buildings are given in Table 8.1 below:

Table. 8.1. Domestic storage capacities

S.No.	Storage	Remarks
	For premises occupied as tenements with common conveniences.	
(*i*)	Ground floor – Nil	Provided no down-take fittings are installed.
(*ii*)	1st, 2nd, 3rd, 4th and upper floors – 500 litres per tenement	
	For premises occupied as flats or blocks.	
(*i*)	Ground floor – Nil	Provided no down-take fitting are installed.
(*ii*)	1st, 2nd, 3rd, 4th and upper floors – 800 litres per treatment	

(*c*) **Flushing Storage Tanks:** These are required to supply water to all flushing cisterns through downtake pipes. The supply pipes are not supposed to feed these cisterns directly. Flushing storage is, therefore, usually based on the number of water closet (WC) seats in case of residential buildings and on the number of WC seats and urinals in the case of public buildings. The flushing storage capacities for different classes of buildings are shown in the Table 8.2.

Table 8.2. Flushing storage capacities

S.No.	Classification of buildings	Storage capacities
(*i*)	For tenements having common conveniences.	900 lit. net per WC seat.
(*ii*)	For residential premises other than tenements having common conveniences.	270 lit. net for each WC seat and (180 lit.) for each additional seat in the same flat.
(*iii*)	For factories and workshops.	900 lit. per W.C. seat and (180 lit. per urinal seat)
(*iv*)	For cinemas, public assembly halls etc.	900 lit. per W.C. seat and (360 lit.) per urinal seat.

8.5. ESSENTIAL FOR BUILDING WATER SUPPLY

(*i*) All pipe works shall be so designed, laid or fixed, and maintained as to be and to remain completely watertight, thereby avoiding wastage of water, damage to property and the risk of contamination of the water conveyed.

(*ii*) There should be no risk of contamination of water supply for domestic purposes. The following points should be kept in mind to ensure this.

- (*a*) There should not be any cross connection anywhere between a pipe containing potable water and pipe containing used or polluted water.
- (*b*) The design of pipe work shall be such that there should be no back flow from any cistern of appliance towards the source of supply. Minimum 15 cm air gap between the outlet end of the supply pipe discharging water into the cistern and its flood level rim will prevent any occurrence of backflow.
- (*c*) The water supply and the drainage pipe should not be laid very close to each other.

(*iii*) In designing and planning the layout of the pipe work due attention should be given to the maximum rate of discharge required, economy in labour and materials protection against damage and corrosion, protection from frost if required, and avoidance of air-locks, noise etc.

(*iv*) Underground pipe work (as service pipe) should have a 60 cm earth cover. When pipe work is laid above ground and in exposed situations it should run clear of the wall with 2 cm clearance and when crossing wall or floor it should be contained in suitable sleeves for entire length to avoid any damage.

(*v*) To reduce frictional losses, piping shall be as smooth as possible inside. Method of jointing shall be such as to avoid internal roughness and projection at the joint, whether of the jointing materials or otherwise.

(*iv*) Water supply pipes should carry water inside building under sufficient pressure in the water main. In case of multi-storeyed buildings where available pressure are not sufficient, and it is necessary to pump water, separate storage tanks are used for providing necessary suction-lift. The booster pump should not be allowed on the service pipe as its suction lift decrease pressure of water supply to the adjoining buildings.

(*vii*) Change in diameter and in direction should be gradual rather than abrupt to avoid under-loss of head. No bend or curve in piping shall be made so as to alter the cross-section.

(*viii*) No boiler for generating steam or closed boilers or any machinery shall be supplied direct from a service or supply pipe. Every such boiler or machinery shall be supplied from a feed cistern.

8.6. LAYOUT OF WATER SUPPLY PIPES FOR A BUILDING (SINGLE AND MULTISTORYEYED)

(*i*) The lay-out of water supply pipes should be simple and direct as far as possible. This will ensure economy both in operation and maintenance of pipe-work. The direct lay-out would reduce lengths of pipings and connections required and thus effect saving in cost of materials.

(*ii*) The communication pipe between the water main and the stop cock at the boundary of the premises shall be laid by the administrative authority.

(*iii*) The pipes should be laid out as straight as possible. This is to minimize friction losses at bends etc.

(*iv*) Every consumers pipe shall be, as practicable; laid so as to follow the shortest route between the water main and the consumer's premises.

(*v*) The lay-out should be such as to admit easy inspection, replacement and repair. Where it is not desired to allow pipes run parallel to the external walls in an exposed condition, they may be conveniently grouped and passed through recesses made in the walls or through vertical shafts provided in accessible position.

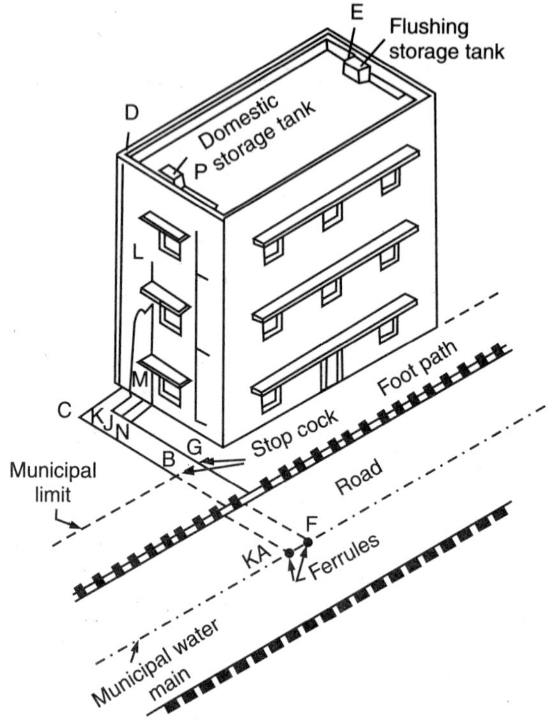

Fig. 8.5. Multistorey building showing water supply arrangement

(*vi*) As far as possible, burying of pipes in walls or floors should be avoided. Where unavoidable piping may be buried for short distance and to facilitate removal of pipe, casing floor boards covering piping shall be fixed with screws of bolts.

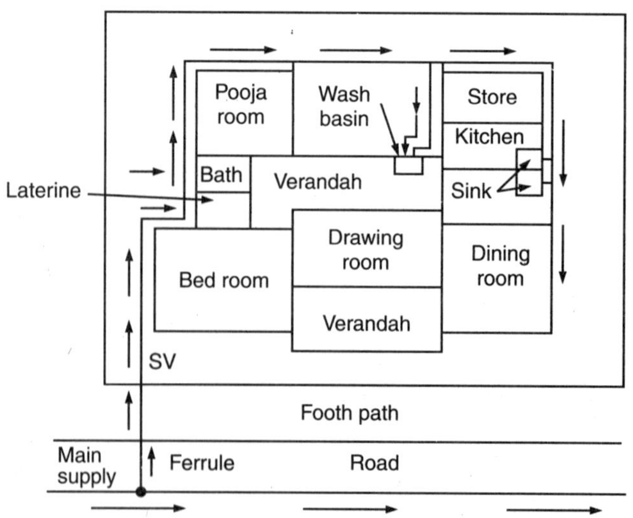

Fig. 8.6. Plan of water supply arrangement

(*vii*) The layout should be free from the occurrence of air pockets, noise transmission etc. This can be achieved by avoiding upward bends and downwards dips.

(*viii*) The rising pipe to the storage cistern or feed cistern (in case of multistoreyed building) shall be taken as directly as possible to the cistern and shall be fixed away from windows or ventilators.

(*ix*) Piping shall be so located that it is not unduly exposed to accidental damage, and shall be fixed in such positions as to facilitate cleaning and avoid accumulation of dirt.

(*x*) When it is necessary for a pipe to pass through a wall or floor, a sleeve shall be fixed there in for reception of the pipe and to allow freedom for expansion and contraction and other movement. The layout of pipes in wooden floor shall, where possible to parallel with the joints.

8.7. WATER SUPPLY FIXTURES AND ITS MAINTENANCE (Refer Fig. 8.7)

The various fixtures used in internal water supply are:

(*i*) Bib cock
(*ii*) Stop cock
(*iii*) Ball valve
(*iv*) Non-return valve
(*v*) Brass ferrule (in mains only)
(*vi*) Lead connection pipe
(*vii*) Union
(*viii*) Shower and Rose
(*ix*) GI or MS storage tank in buildings

The various fittings require in a tank are:

(*a*) Scour pipe
(*b*) Overflow pipe
(*c*) Mosquito proof-coupling
(*d*) Ball valve (floating type)
(*e*) Cover
(*f*) Pads for inlets and outlets
(*g*) GI pipe
(*h*) GI pipe fittings

GI pipe fittings consists of:

(*i*) Clamp
(*ii*) Tee
(*iii*) GI back nuts
(*iv*) GI bends
(*v*) GI crosses
(*vi*) GI elbows
(*vii*) GI nipples
(*viii*) GI plugs
(*ix*) GI socket
(*x*) GI reducer
(*xi*) GI long screw

The maintenance of these fittings consists of:

(*a*) Attending to leakage
(*b*) Cleaning
(*c*) Disinfecting
(*d*) Removal of blockages
(*e*) Repainting

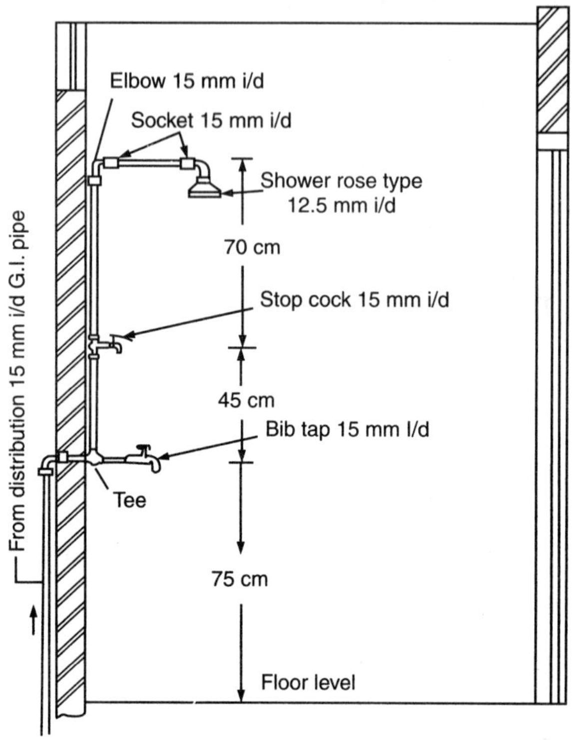

Fig. 8.7. Water supply fixtures of bathroom

8.8. ESTIMATION OF PIPE SIZE

The required sizes for various water supply pipes to supply water to various fixtures in the building can be found as follows:

(a) Rate of flow of each fixture is the rate of demand load of each fixture in terms of fixture units and is calculated from the Table 8.3.

(*b*) Sum up the total number of fixture units for the entire building and find out the total probable peak demand from the probability peak demand graph or by calculating effective fixture unit from Table 8.4 and multiplying this unit with the unit rate of flow (28.316 lpm)

(*c*) Compute the head loss from the supply point to the highest fixture in the building.

(*d*) Compute the development length of piping from the supply point to the highest fixture outlet (from code).

(*e*) Find the pressure factor (head available to overcome fricition for a developed length of 100 mm with reference to the highest fixture) for 100 m of developed length.

(*f*) Compute the diameter of service main to produce the total probable demand rate of the building at the required pressure at the highest outlet from standard graph.

Generally the fixture supply pipes will not be less than the following sizes:

Fixture	Minimum pipe size (mm)
Bath	12.7
Water closet flush tank type	8.5
Commercial kitchen sink	19.0
Water closet flush valve type	25.0

Table 8.3. Fixture unit values for sanitary fixtures

S.No.	Type of Fixture	Fixture unit value
1.	Flushing tank	1
2.	Ablution tap	1
3.	Wash basin	1
4.	Shower	2
5.	Kitchen tap	2
6.	Kitchen sink	2
7.	Sink cum draining board	3
8.	Bath tub	3
9.	Urinal	4
10.	Water closet (flush tank operated)	4
11.	Water closet (flush valve operated)	8

Table. 8.4. Probable demand load on water pipes in buildings

No. of fixture units	Probable demand load (l.p.m.)	
	System with flush tanks	System with flush valves
20	56.70	133.10
40	93.40	178.40
60	121.70	209.50
80	144.40	235.00
100	161.40	257.70
120	181.20	277.50
140	201.00	294.50
160	215.20	311.50
180	232.20	328.50
200	243.50	348.30
220	260.50	359.60
240	272.00	371.00

(Contd...)

300	322.80	416.20
400	396.40	481.40
500	472.90	538.00
600	549.30	597.50
700	606.00	651.30
800	682.40	693.70
900	739.00	739.00

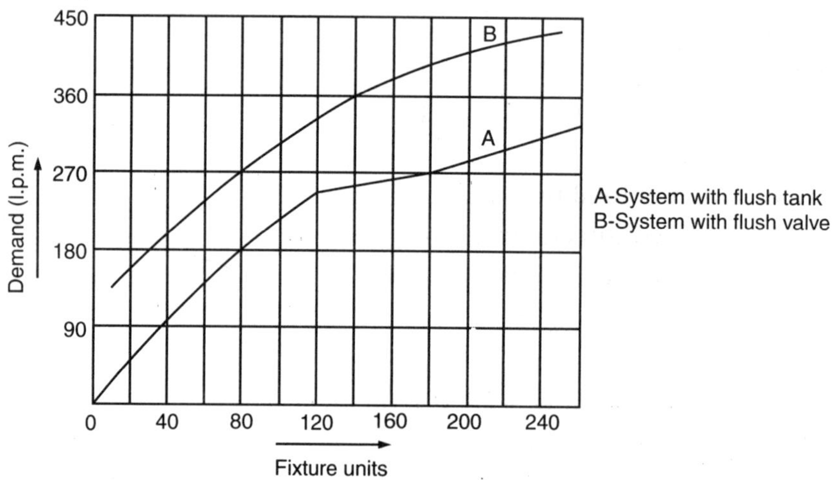

Fig. 8.8. Probability demand graph

The following example illustrates the use of the above tables and chart.

EXAMPLE. 8.1. A two-storeyed residential building has the following sanitary fixtures:

Two toilet rooms each having one tap, one shower, one wash basin, one water closet (Flush tank operated) and ablution tap.

One kitchen having a sink and kitchen tap.

The highest fixture, i.e., the shower in the toilet room on the first floor is located at 5.5 m above the service pipe where the available head of water is 12 m. The developed length of the supply pipe upto the level of the highest fixture is 12 m.

Calculate:

(*a*) the maximum probable demand of water supply

(*b*) the required size of the supply pipe.

Solution: (i) Calculation of fixture units (using Table 8.3)

Toilet room

Type of fixtures	No. of fixture	Fixture units
Tap	2	4
Shower	2	4
Wash basin	2	2
Water closet (flush tank type)	2	8
Ablution tap	2	2
Kitchen		
Sink	1	2
Tap	1	2

Total no. of fixture units = 24

(*ii*) Calculation of total probable peak demand.

It can be obtained directly from the graph shown in Fig. 8.8.

or it can be calculated like the way shown below:

$$\text{Effective fixture units} = 2 + \frac{1.3}{20} \times 4 = 2.26$$

$$\text{1 unit rate of flow} = 28.316 \text{ lpm}$$

$$\therefore \text{ Maximum probable demand} = 2.26 \times 28.316 = 64 \text{ lpm}$$

$$= \frac{64 \times 60 \times 24}{1000} = 92.16 \text{ kld}$$

(*iii*) Calculation of loss of head in the developed length of the pipe

Assuming frictional loss of 1.5 m

The operating head for the highest fixture = 12 – 5.5 – 1.5 = 5 m

$$\text{Developed length} = 12 \text{ m}$$

$$\therefore \quad \text{Loss of head} = \frac{5}{12} = 0.42 \text{ m/m} = 42 \text{ m/100 m}$$

(*iv*) Calculation of size of pipe by using the standard graph

For discharge = 92.16 kld

Slope = 42 m/100 m

From the graph

Size of pipe = 25 mm **Ans.**

8.9. HOT WATER SUPPLY

8.9.1. Design Consideration of Installations

The design of hot water supply system and its appliances shall be based on the temperatures at which water is normally required for the various uses, like

Sink 60°C
Hot bath 41°C
Warm bath 37°C
Tepid bath 29.5°C

Storage Capacity: The hot water is generally taken from the geyser. The size of the geyser is governed by the maximum short time demand of the domestic premises. For a house with bath tub, 50–75 litres capacity is enough whereas for a shower or a tap 25 litres capacity is generally used.

Rate of Flow: With storage type installation like geyser the recommended minimum rate of flow for different types of fixtures is given in Table 8.5.

Table 8.5. Rate of hot water flow

S.No.	Fixture	Rate of flow in litre/min
1.	Bath tub	22.5
2.	Kitchen sink	18.0
3.	Wash basin	7.0
4.	Shower	7.0

8.9.2 Cold Water Supply to Storage Water Heaters

(*a*) The feed pipe connecting cold water tank with the water heater shall not be less than 20 mm bore and it shall leave the cold water tank at a point not less than 50 mm above the bottom of the tank and shall connect into the hot water heater near its bottom.

(*b*) In the case of multistoreyed building a common cold water pipe may be used for feeding to water heaters, but each water heater shall be provided with check valve.

(*c*) Care shall be taken in installing the piping to prevent air locks in the piping and negative pressure in the hot water heater.

8.9.3 Hot Water Piping

1. **Vent Pipe :** Each pressure type hot water heater shall be provided with a vent pipe of not less than 20 mm bore. The vent pipe shall rise above the water line of the cold water tank by at least 150 mm plus 10 mm for every 300 mm height of the water line above the bottom of the heater.

   The vent pipe shall be connected to the highest point of the heater and shall not project downward.

   At no point the vent pipe after leaving the vessel shall dip below the level of its connection.

   The vent pipe shall not be provided with any valve or check valve.

2. The design of hot water taps shall be such that its would cause the minimum friction. Oversized tap may be provided such as a 20 mm tap on a 15 mm pipe.
3. The distributing system shall be so designed that the time lag between opening of the draw off taps and discharge of hot water is reduced.
4. The pressure in the cold and hot water system shall be equal when the mixing of hot and cold water is done by mixing fitting.

Table 8.6. Maximum permissible lengths of hot water draw-off pipes

S.No.	Largest internal diameter of pipe	Length in m
1.	Not exceeding 20 mm	12
2.	Exceeding 20 mm but not exceeding 25 mm	7.5
3.	Exceeding 25 mm	3.0

8.9.4 Type of Hot Water Heaters

The different types of water heaters used for preparation of hot water are:

1. Electric Storage Heaters

(*i*) Non-pressure or open outlet type
(*ii*) Pressure type
(*iii*) Cistern type
(*iv*) Dual heater type

2. Gas Water Heaters

(*i*) Instantaneous type
(*ii*) Storage type

MODEL QUESTIONS

1. Distinguish between:
(*i*) Elbow and bend
(*ii*) Socket and union socket.

2. What is the use of the following
(*i*) Goose neck
(*ii*) Ferrule
(*iii*) Nipple
(*iv*) Stop-cock
(*v*) Service pipe
(*vi*) Bib cock

3. Draw a sketch of the service connection from the street main to a residential building and state the function of each fitting shown.

4. Why is it necessary to store water in case of residential buildings. Sketch a storage tank for domestic use. Name the various parts and explain their function. Indicated the various connections made. What factor determines the storage capacity of such a tank?

5. Draw a neat diagram showing water bath connections.
6. (*a*) Prepare a list of materials required for bath room fittings and, wash basin. (connection in a residential buildings)

   (*b*) Fill in the blanks:

   (*i*) Shower is fitted _____ from the floor level.

   (*ii*) Rate of flow in case of shower is _____ litres per minute.

   (*iii*) Rate of flow in case of wash basin and kitchen sink is _____ litres/ min. respectively.

   (*iv*) Capacity of bowl of wash basin is about _____ litres.

   (*v*) Wash basin when fixed has a height of about _____ cm above the floor level.
7. Sketch the arrangement to be made for taking water supply connection for a house from the water supply street main.
8. Write short note on "Tapping of Water-main.
9. Draw a neat sketch showing various fixtures in water supply, starting from water main upto water meter. Show clearly how the water is tapped from the water main, naming the various parts.
10. Why and where storage tanks are provided.
11. Draw a neat sketch showing water supply fixtures and their installation from water-main to water meter. Name the fixtures on the sketch.
12. Sketch arrangement to be made for taking water supply connection for a house from the water supply street mains.
13. Explain the important points for general layout of water supply arrangements for a single and multistoryed building.

❑❑❑

CHAPTER 9

WATER POLLUTION CONTROL AND MANAGEMENT

Learning Objectives

- Pollution
- Types of Water Pollution
- Effects of Water Pollution
- Preventive Measures
- Remedial Measures to Control Fresh Water Pollution
- Water Pollution Legislation
- Water Management for Resh a Ping; Local Water Balance use and Conservation of Water Resources

9.1 POLLUTION

In recent times the term "pollution" has become very popular and every body seems to be talking about it irrespective of the fact whether one knows its meaning or not. The Oxford English Dictionary defines the term "Pollution" as under:

Pollution: Contamination or defilement of environment or destruction of the purity and sanctity.

Pollution can also be defined as the direct or indirect change of alteration of the physical, chemical, biological, thermal and radioactivities of any part of the environment in such a way as to create a danger/hazard or potential hazard to the health safety or welfare of any living species.

9.1.1 Water Pollution

It may be defined as the adverse change in the condition and composition of water to such an extent that it becomes unsuitable for the purpose for which it is intended to be used.

Water pollution in simple words can be expressed as a condition created by the presence of harmful or objectionable mate-rial in water.

9.1.2 Sources of Water Pollution

There are numerous sources of water pollution. The important ones are classified as:

1. Natural sources
2. Artificial sources

1. Natural Sources: These are the sources of water pollution over which human being have no check or control. The examples of natural sources of water pollution are.

(*i*) Aerial contaminants *e.g.*, Gases, dust (mixed in water body through the agencies like rains, snow, hails etc.

(*ii*) Surface contaminants *e.g.*, Soluble salts leached from the earth by surface stream and rivers etc. Mixing of animal excretes, decayed dead organic matters as in lakes, ponds etc.

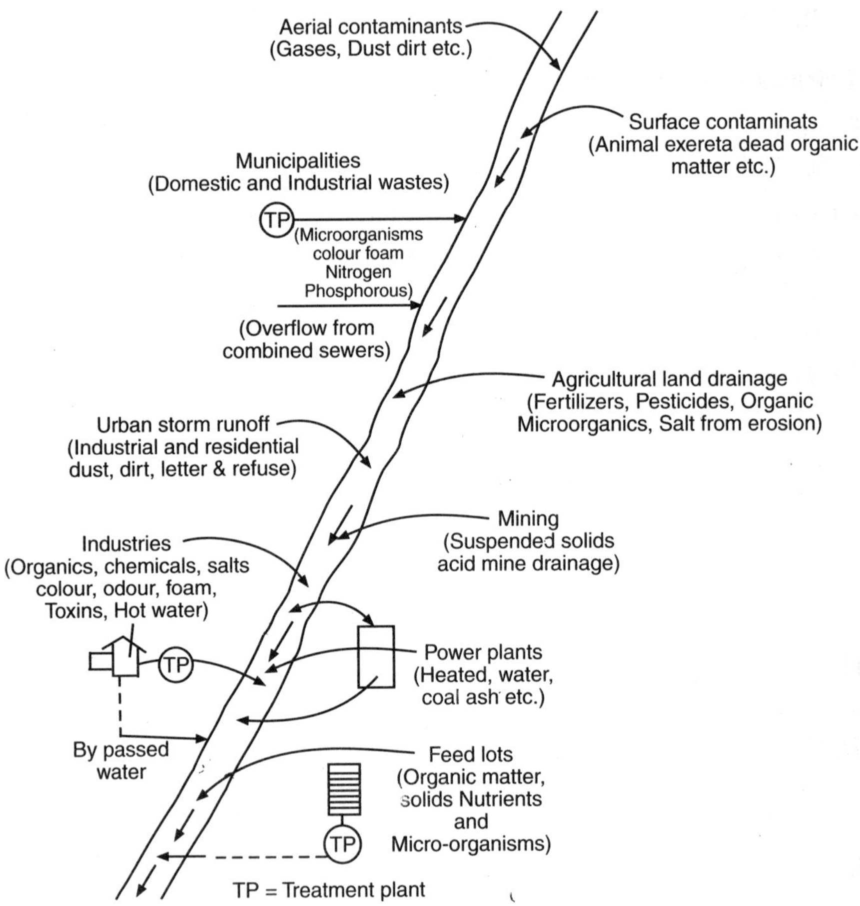

Fig. 9.1. Principal sourcé of water pollution

2. Artificial Sources: These pollutants are man-made and are classified as

(*i*) Domestic wastes and sewage

(*ii*) Domestic detergents

(*iii*) Agricultural wastes

(*iv*) Drainage from irrigated land

(*v*) Fuel gases

(*vi*) Petrochemicals, oils and greases

(*vii*) Excess of water purifies such as cougulation, disinfectants etc.

(*viii*) Pollution of underground water by the substances leached from refuse and spoil heaps.

9.2. TYPES OF WATER POLLUTION

Depending upon the contaminants water pollution can be of the following three types:

1. Physical pollution
2. Chemical pollution and
3. Biological pollution

9.2.1. Physical Pollution

The physical pollution of water is caused due to physical pollutants. It can be of the following categories.

(*i*) **Colour:** The water in its natural form is obtained from various sources such as rain, snow, surface and subsurface sources.

The natural colour and brightness of colour changes due to Municipal and industrial effluents mixing with it.

(*ii*) **Taste and Odour:** Disagreeable taste and odour of water may be due to the presence of:

(*a*) Dissolved gases *e.g.*, oxygen, carbon dioxide, methane, hydrogen sulphide etc.

(*b*) Mineral substances *e.g.*, chlorides, sulphates, carbonates of various substances

(*c*) Micro-organism

The taste and odour of water may also originate in the treatment process, conveyance and distribution etc.

(*iii*) **Temperature or thermal pollution:** Hot effluents from industries and hot air may also pollute the water rendering it unsuitable for drinking and other uses.

(*iv*) **Foam Formation:** Industrial effluent may contain soaps, detergents etc. as from paper and cloth industries. The detergent and other sub-stances having a lower surface tension then water produce foam as such this water cannot be used for drinking, cooking etc.

(*v*) **Suspended Impurities:** Undissolved suspended impurities cause turbidity in water, a cause of water pollution. Turbid water is not liked for drinking and cooking.

9.2.2. Chemical Pollution

Water containing chemical (organic/inorganic) substances is also polluted and hence can be of the following type of pollution :

(*a*) **Organic Pollution:** Organic chemicals and substances such as carbo-hydrates, protiens, fatty acids, detergents, fats, pesticides etc. are known as organic pollutants. The water containing such pollutants is left with very small quantities of dissolved oxygen and hence is rendered unfit for the survived of aquatic life.

The main sources of chemical pollutants are sugar mills, oil expellers, oil refineries, Alcohol industries and distilleries.

(*b*) **Inorganic Pollutants:** Inorganic substances such as acids, alkalies and salts are known as inorganic pollutants. The water containing such pollutants affects the pH value of water to a great extent rendering it unfit for human consumption.

The inorganic pollutants may also cause water to become 'hard' and even poisons. The main inorganic pollutants of water are sulphuric acid, soda ash, chlorides, sulphates, carbonates, nitrates cyanides, sulfides, mercury, lead, cadmium. Inorganic pesticides and fertilisers.

9.2.3. Biological Pollution

The presence of bio-organic such as bacterias, viruses, algae. Fungi, parasites etc., in water, is also a source of pollution. Bio-pollutants are added from domestic sewage, agro-wastes and industrial wastes. These pollutants cause a number of diseases such as hepatitis, cholera, typhoid etc.

9.3. EFFECTS OF WATER POLLUTION

Polluted water in any form is highly objectionable and damaging for many reasons. Ill effects of water pollution are described in brief.

(*i*) **Effects on Human health and safety:** The polluted water containing pathogenic bacterias from domestic sewage causes a number of diseases named above. Toxic and poisonous substances are a potential hazard to the human health, aquatic life and their safety.

(*ii*) **Effects on Aquatic and other life:** Polluted waters play a havoc on fish and other organic organism naturally inhibiting water bodies. Suspended solids such as silt and coal in water create a number of problems for aquatic life.

(*iii*) **Damaging effect on property:** Acid rains and corrosive water contaminated with acids, alkalies etc., attacks the metal, stone and wood use in building construction. Boilers scale is one such problem created by using hard water for generating steam.

(*iv*) **Economic Losses:** Polluted water affects the tourism potential and allied services of a particular place resulting into economic losses to public and the state alike.

9.4. PREVENTIVE MEASURES

As can be seen from the ill effects of water pollution it becomes the moral duty of every one of us to minimise the pollution by taking various preventive measures described below, broadly various preventive measures can be classified as:

(*i*) Measures at the source of water

(*ii*) Measures during treatment, conveyance, distribution and consumption

(*iii*) Measure at the discharge point

(*a*) Preventive measures at the source of water: The following controls can be exercised at the source of water to check pollution:

(*i*) The source of water should be away from the municipal out fall and discharge points of mines, industries and power plants.

Water in-take works, surface sources such as rivers, streams should be located on the upstream of the discharge point if any.

(*ii*) The degree of pollution and extent of harmful substances in the water source should be get tested in the laboratory before finally selecting the source.

(*iii*) Extensive and intensive studies should be done to ensure that there is not going to be any danger of source being polluted in the near future.

(*b*) Preventive Measures during treatment, conveyance, distribution and consumption: It has been observed that most of the water pollution is contributed during its conveyance and consumption hence the following points should be considered to check and control the pollution.

(*i*) No chemicals, toxins or other hazardous substances should be left in water as residuals during its treatment.

(*ii*) Water should be pumped through corrosionless pipes. All valves, taps and joints should be made of non-toxic materials. As far as possible leads joints and lead water seals should be avoided.

(*iii*) Water distribution mains should be laid away from sewer lines so as to avoid accidental mixing of sewage in the public water supply, (iv) Drinking water and water for lawns should be distributed in separate pipes.

(*c*) Preventive measures at discharge points: The following important measures can be exercised to control water pollution of the waste water discharge points.

(*i*) All waste water from municipalities and Industries should invariably be treated so as not to pose any threat to water bodies receiving discharge.

(*ii*) Sewerage system of the city should be designed and laid out in such a way that the out fall sewer is located on the down stream side of the locality in case the waste water is discharged into a river/stream.

(*iii*) Enough space should be available between the discharge point and the intake on the down stream of the river so that self-purification of rivers could take place effectively.

(*iv*) Toxic wastes should not be discharged in standing water bodies such as pond, lakes and pools etc.

9.5. REMEDIAL MEASURES TO CONTROL FRESH WATER POLLUTION

Depending upon the sources fresh water pollution can be controlled by taking the following remedial measures.

1. Wells and Tube wells should be located away the potential hazards of a nearby waste water unit. Deep wells should be preferred to open wells.
2. Reduction of wastes entering into the source of water.
3. Realization, recycling, renovation of waste matter and recharge of ground water with safe water.
4. Strict compliance of local effluent standards, effective monitoring and control of water pollutants.

9.6. WATER POLLUTION LEGISLATION

Water pollution has become a universal problem essentially due to the increase in population and industrial growth. The water pollution problem in India is assuming alarming proportions and is worsening every day.

It has been observed that in India about 70 per cent of the surface water pollution is caused from domestic sewage and the rest from other sources. Voluntary measures to check pollution of water are neither sufficient nor strictly followed; therefore the Govt. of India had to specify certain laws to control water pollution.

These specified laws have been formed to be implemented at state and central level.

To formulate such laws Indian Parliament in 1974 passed a "Water pollution Act". Under article 252 of the constitution to check water pollution. This act makes provisions for the following:

1. Constitution of central and state water pollution control boards.
2. Terms and conditions of selection and service of the members of these boards.
3. Meeting of Board, disqualification and vacation of seats by its members.
4. Functions and powers of the water pollution control boards.
5. Setting up of water testing laboratories, penalties and prosecution of industrial units/municipalities found flouting and norms of the Board.
6. Provision of funds and infrastructure by state and centre.

The aforesaid act has since been modified and amended in 1975, 1977 and 1978.

We now have a central and water Pollution Control Board almost in every state of the country.

9.7. WATER MANAGEMENT FOR RESHAPING LOCAL WATER BALANCE, USE AND CONSERVATION OF WATER RESOURCES

On the planet earth water occupies about 3/4 of the surface of the globe in the form of oceans, seas, lakes, rivers etc. The other form of water available on the earth is subsurface water, water in vapour form and as snow on the earth is subsurface water, water in vapour form and as snow on the earth surface. The whole water system follows what is known as "Water Cycle". Inspite of gigantic quantities of water available only a fragment of it is only suitable for human use. Hence there is great need to properly manage the very little water available to us. Water management deals with the optimum and economical use of water.

In reshaping local water balance concerted efforts have to be made as potable water is still a scarce commodity in large parts of the world population. The problem of provision of drinking water is enlighted each year by the UNESCO and other agencies of the United Nations.

A few measures suggested in reshaping the local water balance are:

1. Improving the characteristics of the local catchment areas in order to get enhanced quantities of water in rivers and streams etc.
2. Improving and adopting new techniques of farming such as sprinkler and drip irrigation system to economical irrigation water.
3. Checking and making use of flood waters
4. Construction of small bunds, dams and reservoirs
5. Enforcement of laws to stop deforestation and afforestation
6. Making effective and economic uses of water
7. Effective enforcement of water pollution legislation.

Use and Conservation of Water Resources. All humans will suffer acute scarcity of water if due importance is not given to optimum use and conservation of water resources. Some of the helpful measures in this direction are:

1. Judicious and environment friendly planning and construction of multipurpose projects.
2. Economic use and conservation of water resources
3. Development and adoption of industrial processes using less water
4. Adoping agri-techniques using less water e.g., preferring drip and sprinkler irrigation to flooding the fields method.
5. Prevention of dumping wastes and sewage into natural water sources.

MODEL QUESTIONS

1. Define pollution. Also explain water pollution and its sources.
2. Draw a neat diagram showing principal sources of water pollution.
3. Explain in detail the following :
   (*a*) Physical Pollution (*b*) Chemical Pollution (*c*) Biological Pollution.
4. Explain briefly effects of water pollution. Also give Remedial measures to control fresh water pollution.
5. Explain briefly the following:
   (*i*) Water pollution legislation
   (*ii*) Preventive measures to minimise pollution
   (*iii*) Conservation of water resources
6. Enlist the measures to be taken in reshaping the local water balance.

❑❑❑

OBJECTIVE TYPE QUESTIONS

WITH ANSWERS

IN

WATER SUPPLY ENGINEERING

MULTIPLE CHOICE OBJECTIVE TYPE QUESTIONS

1. Most common surface water supply source is
(*a*) sea (*b*) river (*c*) pond (*d*) lake

2. Which of the following gives underground water
(*a*) spring (*b*) river (*c*) lake (*d*) reservoir

3. All water that occurs below the surface of the earth is called
(*a*) groundwater (*b*) underground water
(*c*) sub-surface water (*d*) all the above are correct

4. The sub-surface water obtained under pressure is generally known as
(*a*) open well (*b*) tube well
(*c*) infiltration well (*d*) artesian well

5. Shallow horizontal tunnels constructed along the river on its bank to intercept ground water are called
(*a*) canals (*b*) infiltration galleries
(*c*) spring (*d*) lakes

6. The vertical wells sink along the blanks of a river to draw ground water in dry season are called
(*a*) open walls (*b*) tube wells
(*c*) artesion wells (*d*) infiltration wells

7. A pipe sunk into the ground to tap the underground water is called
(*a*) open well (*b*) tube well
(*c*) artesian well (*d*) infilteration well

8. Ground water from artesian wells
(*a*) contains no suspended materials
(*b*) contains dissolved salts
(*c*) requires less treatment
(*d*) all the above are correct

9. The wells used by small town for public water supply are
(*a*) dug well (*b*) driven well (*c*) tube well (*d*) hand pump

10. Dug wells are preferred when they have to be used up to a depth of
(*a*) 10 metres (*b*) 20 metres
(*c*) 50 metres (*d*) 100 metres

11. Generally the maximum discharge from dugwell ranges in
(*a*) 10 cu. m/hours (*b*) 15 cu. m/hour
(*c*) 20 cu. m/hour (*d*) 50 cu. m/hour

12. The maximum yield from a tube well is generally about
(*a*) 50 litres/sec. (*b*) 30 litres/sec.
(*c*) 20 litres/sec. (*d*) 10 litres/sec.

13. The suitable method for boring in hard rock, hard soil and boulder region is
(*a*) precussion method (*b*) rotary method
(*c*) core drilling (*d*) jetting method

14. The specific retention least in
(*a*) coarse gravel (*b*) sand
(*c*) clay (*d*) silt

15. To determine the velocity of flow of ground water, the most commonly used non-empirical formula is
(*a*) Darcy's formula (*b*) Slichter's formula
(*c*) Hazen's formula (*d*) lacy's formula.

16. The location of a well is considered to be good, if it suck into
(*a*) coarse gravel (*b*) silt
(*c*) sand (*d*) clay

17. Water to the consumers may be supplied from
(*a*) rainy wells sunk to the water level.
(*b*) infiltration well dug out on the banks of rivers.
(*c*) infiltration galleries connected to sump well.

18. The most widely used tube well in India, is
(*a*) strainer well (*b*) cavity well
(*c*) slotted well (*d*) perforated well

19. For insignificant effect on the water table as a thumb rule, the tube wells should be spaced as one in every
(*a*) 0.5 sq km. (*b*) 1.0 sq km
(*c*) 1.5 sq km. (*d*) 2.0 sq km.

20. "Shrouding" is provided in case of
(*a*) slotted type wells (*b*) strainer wells
(*c*) cavity wells (*d*) all the above

21. Strainer type tube wells are considered unsuitable for
(*a*) coarse gravel (*b*) fine sand strata
(*c*) clean gravel (*d*) all the above.

22. Per capita water demand is calculated in litres
(*a*) per person per day (*b*) per person per month.
(*c*) fire and public-use demand(*d*) none of the above.

23. Water demand of a city includes
(*a*) domestic water demand.
(*b*) commercial and industrial demand.
(*c*) fire and public use demand.
(*d*) all the above.

24. Under normal conditions, the average domestic consumption in India per person per day in litres is
(*a*) 105 (*b*) 135 (*c*) 180 (*d*) 215

25. The total domestic consumption in a city as compared to total demand of city is

(*a*) 20% (*b*) 30%
(*c*) 60% (*d*) 75%.

26. Industrial and commercial water demand in a city as compared to total demand of the city is

(*a*) 10 to 15% (*b*) 15 to 20%
(*c*) 20 to 25% (*d*) 30 to 40%

27. As per norms, 45 litres of water per person per day is provided in case of

(*a*) hotels (*b*) hospitals
(*c*) office buildings (*d*) public places

28. In a city having population approximately as 8 lakhs, the per capita demand of water will be

(*a*) 200 to 250 litres (*b*) 180 to 200 litres
(*c*) 160 to 180 litres (*d*) 100 to 120 litres

29. Per capita consumption will be higher if

(*a*) pressure in distribution system will be more
(*b*) quality of water will be good
(*c*) the living standard of people is higher
(*d*) all the above are correct

30. Per capita consumption is highest in

(*a*) schools (*b*) hospitals
(*c*) offices (*d*) factories

31. The fire demand of a city is generally calculated by

(*a*) Under writers formula (*b*) Freeman formula
(*c*) Kuichling's formula (*d*) All the above

32. If P is the population of place in thousands then the fire demand of water in liters per minute, according to Kuichling formula, is given by

(*a*) $5663\sqrt{p}$ (*b*) $3182\sqrt{p}$
(*c*) $4640\sqrt{p}$ (*d*) None of these

33. In India, the storage of water for fire on per capita basis is generally taken as

(*a*) 2 litres (*b*) 3 litres (*c*) 5 litres (*d*) 10 litres

34. Water demand for public use as compared to total demand, kept in design of water works for a city is

(*a*) 5% (*b*) 10% (*c*) 20% (*d*) 25%

35. To compensate for losses, thefts and wastage of water, an allowance provided is

(*a*) 5% (*b*) 10% (*c*) 15% (*d*) 25%

36. The per capita consumption of a locality is affected by
(*a*) climatic condition (*b*) quality of water
(*c*) distribution pressure (*d*) all the above

37. Maximum consumption of water is noticed in
(*a*) paper factory (*b*) steel plant
(*c*) automobile industry (*d*) foundary

38. Per capital consumption decreases when
(*a*) use of metering system is adopted
(*b*) hot climate persists
(*c*) good quality water is supplied
(*d*) All the above

39. The ratio of the maximum daily consumption to the average daily demands is
(*a*) 1.2 (*b*) 1.5 (*c*) 1.8 (*d*) 2.7

40. The ratio of the maximum hourly consumption and average hourly consumption of the maximum day is
(*a*) 1.2 (*b*) 1.5 (*c*) 1.8 (*d*) 2.7

41. If the average daily consumption of a city is $10^5 m^3$, the maximum daily consumption on peak hourly demand will be
(*a*) $1.2 \times 10^5 m^3$ (*b*) $1.5 \times 10^5 m^3$
(*c*) $1.8 \times 10^5 m^3$ (*d*) $2.7 \times 10^5 m^3$

42. The distribution mains in water supply system are designed for
(*a*) maximum daily demand
(*b*) peak hourly demand
(*c*) average daily demand
(*d*) maximum hourly demand on maximum consumption day

43. Factor affecting, prediction of future population of a city is
(*a*) births (*b*) deaths
(*c*) migrations (*d*) All the above

44. As compared to geometric increase method of forecasting population, arithmetical increase method gives
(*a*) higher value (*b*) lesser value
(*c*) same value (*d*) more accurate value

45. The accurate method for predicting population for a sound and rapidly increasing city is
(*a*) arithmetical increase method
(*b*) geometrical increase method
(*c*) incremental increase method
(*d*) graphical method.

46. Water supply scheme is usually designed for a line of
(*a*) 100 years (*b*) 50 years
(*c*) 25 years (*d*) 10 years

47. Which of the following sources gives water relatively free from impurities?
(*a*) River (*b*) Well
(*c*) Lake (*c*) Spring along the hill slope

49. The water of a river has an important property called
(*a*) consists of undesirable substances rendering it unfit for drinking and domestic use
(*b*) is wholesome
(*c*) contains pathogens
(*d*) none of the above

49. The water of a river has an important property called
(*a*) turbidity (*b*) self-purification
(*c*) permeability (*d*) infilteration capacity

50. Ground water is generally free from
(*a*) suspended impurities (*b*) dissolved impurities
(*c*) both (*a*) and (*b*) above (*d*) none of the above

51. The water which is not chemically pure but does not contain anything harmful to human health is known as
(*a*) pure water (*b*) wholesome water
(*c*) polluted water (*d*) none of the above

52. The impurities caused by dispersion of solid particles that are large enough to be removed by filtration and settling, are known as
(*a*) suspended impurities (*b*) colloidal impurities
(*c*) dissolved impurities (*d*) none of the above

53. Suspended impurities consists of
(*a*) iron (*b*) chlorine
(*c*) bacteria (*d*) all the above

54. Dissolved impurities consist of
(*a*) bacteria (*b*) iron (*c*) silt (*d*) fungi

55. In water analysis, the test carried out is
(*a*) physical test (*b*) chemical test
(*c*) biological test (*d*) all the above

56. The turbidity of water is caused due to
(*a*) clay
(*b*) slit
(*c*) finely divided organic material
(*d*) all the above

57. Dissolved impurity in water is
(*a*) silt (*b*) clay
(*c*) calcium carbonate (*d*) algae

58. Turbidity
   (*a*) is a measure of the resistance of water to the passage of light through it
   (*b*) is expressed in parts per million or miligrams per liter
   (*c*) produced by one miligram of silica in one liter of water is the unit of turbidity
   (*d*) all the above

59. Turbidity is measured on
   (*a*) standard cobalt scale (*b*) standard silica scale
   (*c*) standard platinum scale (*d*) none of the above

60. The turbidimeter used for the measurement of turbidity from 0-1000 ppm directly is called
   (*a*) Jackson's turbidimeter
   (*b*) Baylis turbidimeter
   (*c*) Helligc turbidimeter

61. The range of Hellige turbidimeter is
   (*a*) 0-2 ppm (*b*) 0-50 ppm
   (*c*) 0-500 ppm (*d*) 0-1000 ppm

62. The range for Baylis turbidimeter is
   (*a*) 0-2 ppm (*b*) 0-50 ppm (*c*) 0-500 ppm (*d*) 0-1000 ppm

63. Hard water may contain
   (*a*) calcium carbonate (*b*) magnesium sulphate
   (*c*) magnesium bicarbonate (*d*) any of the above

64. Permanent hardness of water can be removed by
   (*a*) lime soda process (*b*) base-exchange process
   (*c*) de-mineralisation process (*d*) all the above are correct

65. Maximum permissible turbidity for drinking water, based on silical scale is
   (*a*) 5 ppm (*b*) 10 ppm (*c*) 20 ppm (*d*) 50 ppm

66. The most desirable temperature of water for public supply is
   (*a*) between 4.4°C to 10°C (*b*) between 10°C to 15°C
   (*c*) 15°C to 2°C (*d*) above 26°C

67. The maximum permissible colour for domestic supplies based on cobalt scale is
   (*a*) 5 ppm (*b*) 10 ppm (*c*) 20 ppm (*d*) 50 ppm

68. The colour of water is measured on
   (*a*) turbidity scale (*b*) silical scale
   (*c*) platinum cobalt scale (*d*) threshold scale

69. Taste and odour in the water are caused due to presence of
   (*a*) living algae (*b*) decaying organic matter
   (*c*) phenolic substances (*d*) all the above

70. Odour is measured in terms of
(*a*) Silical scale (*b*) cobalt scale
(*c*) nickel scale (*d*) threshold odour

71. The odour of water can be determined by
(*a*) thermometer (*b*) osmoscope
(*c*) Jackson's turbidimeter (*d*) none of the above

72. The permissible odour in domestic water should be
(*a*) no odour (*b*) no objectionable odour
(*c*) scented odour (*d*) 15 ppm

73. For public water supply threshold odour number should be
(*a*) one (*b*) between 1 and 3
(*c*) three (*d*) more than 3

74. The total amount of dissolved salts present in water can be easily estimated by measuring the
(*a*) temperature of water
(*b*) turbidity of water
(*c*) specific conductivity of water
(*d*) any of the above

75. Non-carbonate hardness of water is mainly due to
(*a*) silicates (*b*) aldehydes
(*c*) nitrates and sulphates (*d*) sulphates and chlorides

76. Maximum acidity will occur at pH value
(*a*) 0 (*b*) 2 (*c*) 7 (*d*) 14

77. pH value representing a stronger acid is
(*a*) 1 (*b*) 5 (*c*) 7 (*d*) 10

78. The permissible pH value for public supply water is between
(*a*) 4.5 to 5.5 (*b*) 5.5 to 6.5
(*c*) 6.5 to 8.5 (*d*) 8.5to 10.5

79. The most common cause of acidity in water is
(*a*) hydrogen (*b*) oxygen
(*c*) carbon dioxide (*d*) all the above

80. Alkalinity in water is caused by all quoted below
(*a*) Sodium carbonate (*b*) Potassium carbonate
(*c*) Potassium hydroxide
(*d*) Calcium and magnesium carbonates

81. The hard water has disadvantage of
(*a*) greater soap consumption (*b*) scalling of boilers
(*c*) making food tasteless (*d*) all the above

82. pH value of fresh sewage is usually
(*a*) zero (*b*) 1
(*c*) any where between 7 and 14
(*d*) any where between 1 and 7

83. A moderately soft or slightly hard water will have hardnes in the range
(*a*) 100–150 mg/litre of $CaCO_3$
(*b*) 150–250 mg/litre of $CaCO_3$
(*c*) 250–400 mg/litre of $CaCO_3$
(*d*) 400–1000 mg/litre of $CaCO_3$

84. Hardness of water is expressed in term of
(*a*) percentage
(*b*) fraction of hardness causing solids to total solids
(*c*) parts per million or mg/litre of calcium carbonate

85. For removal of temporary hardness of water
(*a*) water is filtered (*b*) water is boiled
(*c*) lime is added (*d*) chlorination is done

86. Which of the following process can effectively remove permanent hardness in water?
(*a*) Zeolite process (*b*) Boiling process
(*c*) Sedimentation process (*d*) Lime process

87. One degree of hardness is equivalent to
(*a*) 1.425 mg/litre (*b*) 14.25 mg/litre
(*c*) 142.5 mg/litre (*d*) 1425 mg/litre

88. Zeolite is
(*a*) a naturally occurring salt (*b*) hydrated silica
(*c*) hydrated alumino-silicate (*d*) dehydrated calcium silicate

89. Soft waters
(*a*) have corrosive action on metals
(*b*) dissolve lead
(*c*) corrode iron
(*d*) all the above

90. The maximum permissible total solid content in water for domestic purposes should not exceed
(*a*) nil (*b*) 400 ppm
(*c*) 500 ppm (*d*) 1000 ppm

91. For public water supply, hard water is not used because
(*a*) it consumes more soap
(*b*) it is turbid
(*c*) it possesses bad taste and odour
(*d*) all the above

92. Hardness is generally defined as the calcium carbonate equivalent of
(*a*) calcium ions present in water
(*b*) magnesium ions present in water
(*c*) both calcium and magnesium ions present in water
(*d*) none of the above

93. If the harness of a sample of water expressed in ppm of calcium carbonate is 50, the sample will be
(*a*) hard water (*b*) soft water
(*c*) moderately hard water (*d*) none of the above

94. Hardness limit for public supplies water ranges between
(*a*) 50 to 75 ppm (*b*) 75 to 115 ppm
(*c*) 115 to 200 ppm (*d*) 200 to 300 ppm

95. Chlorides impart
(*a*) salty taste to water (*b*) sweet taste to water
(*c*) irritating taste to water (*d*) none of the above

96. In case of pubic water supplies the permissible upper limit of chloride content is
(*a*) 100 ppm (*b*) 150 ppm (*c*) 200 ppm (*d*) 250 ppm

97. The presence of high quality of chloride in a river or stream waters indicate
(*a*) pollution of water due to sewage on industrial wastes
(*b*) stage of decomposition of organic matter
(*c*) hardness of water
(*d*) all of the above

98. Chlorides are normaly present in water in the form of
(*a*) zinc chloride (*b*) ferric chloride
(*c*) mercuric chloride (*d*) sodium chloride

99. Presence of free ammonia in water indicateds the presence of
(*a*) undercomposed organic matter
(*b*) partly decomposed organic matter
(*c*) full decomposed organic matter
(*d*) none of the above

100. The permissible amount of nitrites present in potable water is
(*a*) nil (*b*) 5 ppm (*c*) 10 ppm (*d*) 20 ppm

101. The permissible limit of nitrate content in potable water is
(*a*) 25 ppm (*b*) 45 ppm (*c*) 100 ppm (*d*) 150 ppm

102. Natural mineral contaminant in water is
(*a*) calcium (*b*) flourine (*c*) iron (*d*) sodium

103. Mottling of teeth is associated with the presence of
(*a*) chlorides in water (*b*) flourides in water
(*c*) calcium in water (*d*) sulphur in water

104. The phenolic compound in public water supply is limited to
(*a*) 0.1 ppm (*b*) 0.01 ppm (*c*) 0.001 ppm (*d*) 0.0001 ppm

105. Flourides are considered beneficial in water when their concentration in milligrams per litre is below
(*a*) 100 (*b*) 50 (*c*) 10 (*d*) 1

106. The maximum allowable limit for flouride in drinking water is
(*a*) 0.5 mg/litre (*b*) 1.0 mg/litre
(*c*) 1.5 mg/litre (*d*) 5.0 mg/litre

107. Which metal if present in water causes poisoning ?
(*a*) Mercury (*b*) Lead
(*c*) Arsenic (*d*) All the above

108. Presence of which material is not considered dangerous in drinking water?
(*a*) Lead (*b*) Zinc (*c*) Calcium (*d*) Copper

109. Lead poisoning occurs when the lead content in water exceeds
(*a*) 0.05 ppm (*b*) 0.1 ppm (*c*) 0.15 ppm (*d*) 0.5 ppm

110. The BOD (biochemical oxygen demand) of raw water indicated
(*a*) the extent of organic matter present
(*b*) the amount of treatment required for purifying
(*c*) both (*a*) and (*b*) above
(*d*) none of the above

111. BOD of safe drink water should be
(*a*) nil (*b*) 10 ppm (*c*) 20 ppm (*d*) 40 ppm

112. The standard BOD at 20 symbol C is taken for the consumption in
(*a*) two days (*b*) three days (*c*) four days (*d*) five days

113. The dissolved oxygen level in natural unpolluted water at normal temperature is found of the order of
(*a*) 1 mg/lit. (*b*) 10 mg/lit. (*c*) 100 mg/lit. (*d*) 1000 mg/lit.

114. Bacterias which can survive with or without free oxygen are called
(*a*) aerobic bacteria (*b*) anaerobic bacteria
(*c*) faculative bacteria (*d*) none of the above

115. If the coliform bacteria is present in water, then the test carried out is
(*a*) presumptive coliform test (*b*) confirmed coliform test
(*c*) completed coliform test (*d*) all of the above

116. Bacterias which are 60μ in length and 5μ in diameter, are known as
(*a*) B. Butschlii (*b*) Dialister pneumonsintes
(*c*) Rotifers (*d*) None of the above

117. Organisms present in water causing infectious diseases are
(*a*) viruses (*b*) protozoa
(*c*) fungi (*d*) all of the above

118. The most harmful constituent of water is
(*a*) manganese (*b*) bacteria
(*c*) alkali (*d*) all of the above

119. Bacteria coli or Escherichia coli are harmless organisms but their presence in water indicates the
(*a*) presence of pathogenic bacteria
(*b*) absence of pathogenic bacteria
(*c*) presence of non pathogenic bacteria
(*d*) absence of non pathogenic bacteria

120. Disease which is not a water borne disease is
(*a*) typhoid (*b*) cholera
(*c*) dysentery (*d*) malaria

121. Disease caused by protozoal infection is
(*a*) poliomyelitis (*b*) amoebic dysentery
(*c*) bacillary dysentery (*d*) maleria

122. Pathogenic bacteria
(*a*) are harmless
(*b*) cause diseases
(*c*) are called coliform organisms
(*d*) are visible to the naked eye

123. The total count of bacteria in potable water should not exceed
(*a*) 10000 per cc (*b*) 1000 per cc
(*c*) 100 per cc (*d*) 10 per cc

124. In one litre of drink water, coliform organism should not excced
(*a*) 10,000 (*b*) 1000 (*c*) 100 (*d*) 10

125. The presence of algea in water indicates that water is
(*a*) hard (*b*) soft (*c*) acidic (*d*) turbid

126. A chemical used to inhabit the growth of algae is
(*a*) chlorine (*b*) iodine (*c*) sunlight (*d*) algicides

127. The system which collect the water from the source and then discharge the collected water by means of pumps or directly to the treatment system of water is known as
(*a*) intake (*b*) conduit (*c*) reservoir (*d*) pumping

128. The device installed for drawing water from the sources are called
(*a*) filters (*b*) intakes
(*c*) aquifers (*d*) none of the above

129. Factor considered in selection of site for intake point is that it should be
(*a*) nearer to the treatment plant
(*b*) in pure zone of the water source
(*c*) upstream of the point of disposal of waste water
(*d*) all the above

130. When there are no wide fluctuations and constant head is required at the intake and where river is also not to wide, type of intake used is called

(*a*) river intake (*b*) cross weir intake
(*c*) side weir intake (*d*) lake intake

131. The intake which is not suitable for big water supply schemes or on rivers or reservoirs and draw only small wuantity of water is called
(*a*) pipe intake in lakes (*b*) canal intake
(*c*) river intake (*d*) reservoir intake

132. The structure constructed for transportation of water from the source of supply situated at a distance apart from the town or city is known as
(*a*) intake (*b*) conduit (*c*) reservoir (*d*) suction pipe

133. The allowable average velocity in canal gravity canal suction is
(*a*) 0.5 to 0.6 m/sec (*b*) 0.5 to 0.9 m/sec
(*c*) 1.8 to 2.4 m/sec (*d*) 2 to 3 m/sec

134. Trapezoidal section is best suited for
(*a*) gravity canal (*b*) flume gravity conduit
(*c*) aqueduct gravity conduit (*d*) all the above are correct

135. Gravity conduits are generally in the form of
(*a*) open channel (*b*) tunnels
(*c*) aqueducts (*d*) all the above

136. Gravity conduits
(*a*) carry water under gravity
(*b*) follow the hydraulic gradient line
(*c*) are carried through tunnels in deep cuttings
(*d*) all the above

137. The open channels supported above the ground over trestles, is called
(*a*) tunnel (*b*) aqueduct
(*c*) flume (*d*) all the above

138. Most commonly used section in the grade aqueduct is
(*a*) circular (*b*) rectangular
(*c*) parabolic (*d*) all the above

139. Grade aqueducts are not allowed to run
(*a*) one fourth full (*b*) half
(*c*) three fourth full (*d*) full

140. A circular gravity aqueduct is preferred because of
(*a*) it has maximum hydraulic mean depth
(*b*) maximum area per unit of wetted perimeter.
(*c*) minimum cost of construction
(*d*) all the above

141. Advantage of pressure conduit, is
(*a*) flow is independent of grade of hydraulic grade line
(*b*) economical, since it follows shorter routes

(*c*) less chances of water pollution, as it is closed
(*d*) all the above

142. Selection of material for the pipe depends on
(*a*) carrying capacity of the pipe
(*b*) durability and life of the pipe
(*c*) effect of water to be conveyed on the pipe material
(*d*) all the above

143. The permissible velocity in pipe pressure conduit is
(*a*) 0.6 to 0.9 m/sec (*b*) 0.9 to 1.8 m/sec
(*c*) 0.6 to 1.8 m/sec (*d*) 6 to 9 m/sec

144. The screens are arranged at an angle
(*a*) 45° to 60° (*b*) 70° to 85°
(*c*) vertically (*d*) 30° to 40°

145. The area of openings in screens is so kept that the velocity of flow through them does not exceed
(*a*) 0.75 to 1 m/sec (*b*) 1.5 to 3 m/sec
(*c*) 0.3 to 0.9 m/sec (*d*) less than 0.5 m/sec

146. The first stage in water treatment is
(*a*) sedimentation (*b*) filtration
(*c*) disinfection (*d*) coagulation and mixing

147. The process of causing heavier solid particles in suspension to settle by holding the water in a basin is called
(*a*) sedimentation (*b*) co-agulation
(*c*) filtration (*d*) disinfection

148. The rate of settling of a particle depends upon the
(*a*) viscosity of water (*b*) specific gravity of particle
(*c*) shape and size of particle (*d*) all the above

149. The theoretical time for which water is detained in settling basins (or sedimentation tank) is called
(*a*) basis period (*b*) detention period
(*c*) settling period (*d*) all the above are correct

150. The zone which is designed in such a fashion so that the incoming water should be uniformly distributed on the ful width of settling tank is known as
(*a*) inlet zone (*b*) outlet zone
(*c*) settling zone (*d*) approach

151. The path of a particle in sedimentation tank is
(*a*) parabolic (*b*) straight (*c*) elliptical (*d*) circular

152. In sedimentation tank, the velocity of flow generally does not exceed
(*a*) 30 cm/min (*b*) 3 m/min
(*c*) 30 mm/min (*d*) 30 cm/sec

153. In rectangular Sedimentation tank of ratio in between width and length should not be less than

(*a*) 1 : 4 (*b*) 1 : 1 (*c*) 1 : 2 (*d*) 1 : 3

154. The tank which is generally used for plain Sedimentation is of

(*a*) circular shape (*b*) rectangular shape
(*c*) hopper bottom type (*d*) hectagonal shape

155. In plain sedimentation process the suspended particles settle down as a result of the

(*a*) action of sun rays
(*b*) action of gravity
(*c*) action of velocity of the particles
(*d*) none of the above

156. The overflow rate of plain sedimentation is

(*a*) 500-750 litres/hour/sq.m
(*b*) 1000-1500 litres/hour/sq.m
(*c*) 1500-2000 litres/hour/sq.m
(*d*) 2500-3000 litres/hour/sq.m

157. The detention time for an ordinary plain sedimentation tank varies from

(*a*) 1 to 2 hours (*b*) 2 to 4 hours
(*c*) 6 to 8 hours (*d*) 20 to 24 hours.

158. In the design of a sedimentation tank, the essential factor to be considered is

(*a*) period of detention (*b*) velocity of flow
(*c*) depth of tank (*d*) all the above

159. Coagulants are used when turbidity of water is more than

(*a*) 50 ppm (*b*) 60 ppm (*c*) 70 ppm (*d*) 80 ppm

160. The best coagulant for removing the colour of water is

(*a*) alum (*b*) lime
(*c*) iron sulphate (*d*) copper sulphate

161. Alum increases

(*a*) hardness of water (*b*) acidity of water
(*c*) carbonates in water (*d*) sulphates in water

162. The dosage of coagulant depends upon

(*a*) turbidity (*b*) pH value
(*c*) colour (*d*) all the above are correct

163. For Sedimentation with coagulation, the detention period varies from

(*a*) l-2hours (*b*) 2-4 hours
(*c*) 4-8 hours (*d*) 8-12 hours
(e) 12-24 hours

164. Fine suspended particles are removed by
(*a*) filtration (*b*) agitation
(*c*) aeration (*d*) bleaching

165. The organic impurities are converted into simple and harmless compounds due to
(*a*) mechanical straining (*b*) sedimentation
(*c*) biological action (*d*) electrolytic action

166. The process of purifying water by passing it through a bed of fine granulär material is called
(*a*) filtration (*b*) coagulation (*c*) sedimentation (*d*) screening

167. The void spaces in the filtering material act like.
(*a*) drain (*b*) inlet
(*c*) tiny settling basins (*d*) outlet

168. The organic impurities in water form a layer on the top of a filtering media. Such a layer is called
(*a*) permeable layer (*b*) filtering layer
(*c*) dirty skin (*d*) none of the above

169. The effective size of sand particles for slow sand filters varies from
(*a*) 0.30 to 0.35 mm (*b*) 0.35 to 0.50 mm
(*c*) 0.50 to 0.65 mm (*d*) 0.65 to 0.75 mm

170. The effective size of sand particles for rapid sand filters varies from
(*a*) 0.30 to 0.35 mm (*b*) 0.35 to 0.50 mm
(*c*) 0.50 to 0.65 mm (*d*) 0.65 to 0.75 mm

171. The required depth of sand bed in slow sand filter should be
(*a*) 0.6 to 0.9 m (*b*) 0.6 to 0.75 m
(*c*) 0.3 to 0.45 m (*d*) 1 to 2 m

172. The required depth of sand bed in rapid sand filter should be
(*a*) 0.6 to 0.9 m (*b*) 0.6 to 0.75 m
(*c*) 0.3 to 0.45 m (*d*) 1 to 2 m

173. In rapid sand filters, the depth of the tank varies in between
(*a*) 1 to 2 m (*b*) 2.5 to 3.5 m
(*c*) 3 to 3.5 m (*d*) 0.6 to 0.9 m

174. Slow sand filter can be used for filtering water if its turbidity is
(*a*) not more than 50 ppm (*b*) not less than 50 ppm
(*c*) not more than 80 ppm (*d*) of any value

175. The size of one unit for slow sand filter is
(*a*) 30 m × 60 m (*b*) 8 m × 10 m
(*c*) 6 m × 8 m (*d*) 15 m × 20 m

176. The period of cleaning for slow sand filter is
(*a*) 2 to 3 months (*b*) 24 to 48 months
(*c*) 24 to 48 hours (*d*) 24 to 48 weeks

177. The period of cleaning of rapid sand filter is
(*a*) 2–3 months (*b*) 24–48 months
(*c*) 24–48 hours (*d*) 24–48 weeks.

178. The required percentage of total water available for washing the filter of slow and sand filter is
(*a*) 0.2 to 0.5% (*b*) 2 to 4%
(*c*) 0.2 to 4% (*d*) zero%

179. The turbidity of water that can be removed in a rapid gravity filter is generally in the rängе
(*a*) 35 - 45 gm/litre (*b*) 100 - 125 gm/litre
(*c*) 200 - 300 gm/litre (*d*) 500 - 600 gm/litre

180. The advantage of rapid sand filter is
(*a*) slow but efficient filteration
(*b*) small space required
(*c*) highly efficient in bacteria removal
(*d*) useful for low turbidity waters

181. Air binding can be prevented by
(*a*) avoiding increase in water temperature
(*b*) avoiding excessive negative head
(*c*) the control of algae growth
(*d*) all of the above are correct

182. The coefficient of uniformity for slows and filters is
(*a*) 1.35 (*b*) 1.55 (*c*) 1.75 (*d*) 2.05

183. The slow sand filter should be cleaned if the loss of head becomes more than
(*a*) 0.75 m (*b*) 1.2 m (*c*) 2.2 m (*d*) 3.5 m

184. After cleaning the slow sand filter, the filtered water should not be used for a period of
(*a*) upto 6 hours (*b*) 6 hours to 12 hours.
(*c*) 12 hours to 24 hours (*d*) 24 hours to 48 hours.

185. Rate of filteration of a slow sand filter ranges from
(*a*) 10 to 100 litres/hr/m^2 (*b*) 100 to 200 litres/hr/m^2
(*c*) 200 to 400 litres/hr/m^2 (*d*) 400 to 1000 litres/hr/m^2

186. Slow sand filter is efficient to remove the bacteria from the raw water to an extent of
(*a*) 50% (*b*) 70% (*c*) 85 % (*d*) 99%

187. The process of killing pathogenic bacteria of water borne diseases to make it safe for use in hospital is called
(*a*) sterilization (*b*) aeration
(*c*) chlorination (*d*) disinfection

188. Pre-chlorination
(*a*) improves coagulation (*b*) reduces odour and taste
(*c*) reduces organisms (*d*) all the above

189. Swimming pool water requires
(*a*) prechlorination (*b*) super-chlorination
(*c*) dual-chlorination (*d*) de-chlorination

190. The maximum permissible chlorine contents for public supplies should be between
(*a*) 0.1 to 0.2 ppm (*b*) 0.3 to 0.4 ppm
(*c*) 1.2 to 4 ppm (*d*) 6.5 to 8 ppm

191. Disinfection of water is done by
(*a*) filteration (*b*) passing chlorine
(*c*) ozone (*d*) alum

192. The bacteria that can survive without oxygen is called
(*a*) faculative bacteria (*b*) anerobic bacteria
(*c*) aerobic bacteria (*d*) furobic bacteria

193. Aeration is done for the removal of
(*a*) colour (*b*) turbidity
(*c*) hardness (*d*) bad odour

194. Alum increases
(*a*) hardness of water (*b*) carbonates in water
(*c*) sulphates in water (*d*) acidity of water

195. Aeration is done by
(*a*) cascading (*b*) spray nozzles
(*c*) tickling bed method (*d*) tank method

196. Chlorination of water is done for
(*a*) removal of sediments (*b*) removal of suspended solids
(*c*) removal of hardness (*d*) removal of bacteria

197. When bleaching powder is added to water its pH value
(*a*) increase (*b*) decreases
(*c*) remains unaffected
(*d*) increase or decrease will depend on characteristics of water

198. Bleaching powder is
(*a*) hypo-chlorite of lime (*b*) hypo-chloride of lime
(*c*) chloride of lime (*d*) slaked lime

199. The most common cause of acidity of water is
(*a*) nitrogen (*b*) carbon dioxide
(*c*) oxygen (*d*) alkalies

200. The process used for the removal of dissolved carbon dioxide from the supply water is
(*a*) coagulation (*b*) agitation
(*c*) aeration (*d*) sedimentation

201. During epidemics the best and economic method of disinfection of water treatment is
(*a*) by ozone (*b*) potassium permanganate
(*c*) by boiling of water (*d*) by chlorination

202. The presence of ozone in water is indicated by
(*a*) red colour (*b*) light yellow colour
(*c*) blue colour (*d*) black colour

203. The chlorine added beyond the back point for the safety of public against some disease is knows as
(*a*) super-chlorination
(*b*) plain-chlorination
(*c*) post chlorination

204. The water sample not having iron, manganese, nitrites as impurities should be tested for residual chlorine by
(*a*) orthotolidine test (*b*) orthotolidine-arsenite test
(*c*) starch iodine test (*d*) ozone test

205. Disinfection of swimming pool water is usually done by
(*a*) chlorine (*b*) alum
(*c*) ultraviolet rays (*d*) ozone

206. A chemical generally used for controlling algae is
(*a*) alum (*b*) lime
(*c*) bleaching powder (*d*) copper sulphate

207. The radio-activity of water due to strontium can be removed by
(*a*) aeration (*b*) lime-soda solution
(*c*) phosphate coagulation (*d*) alum coagulation

208. Potassium permanganate is used for
(*a*) improving colour (*b*) removing organic matter
(*c*) reducing acidity (*d*) reducing hardness

209. If by the addition of the minimum quantity of potassium permanganate the pink colour disappears soon, then it shows that
(*a*) water is hard
(*b*) water is alkaline
(*c*) water is acidic
(*d*) organic matter is present in water.

210. For preventing tooth decay, the substance added to water is
(*a*) chlorine (*b*) fluorine
(*c*) bromine (*d*) iodine

211. For drinking water, iron content is usually restricted to
(*a*) 0.01 mg per litre (*b*) 0.1 mg per litre
(*c*) 0.3 mg per litre (*d*) 1 mg per litre

212. The suitable layout of a distribution system for haphazardly developing city is
(*a*) dead and system (*b*) grid iron system
(*c*) ring system (*d*) radial system

213. For a well planned city, the suitable layout of distribution system is
(*a*) dead end system (*b*) grid iron system
(*c*) ring system (*d*) radial system

214. For towns and cities having well planned roads, the suitable layout of distribution system is
(*a*) dead end system (*b*) grid iron system
(*c*) ring system (*d*) radial system

215. The layout of a distribution system, in which water flows towards the outer periphery, is
(*a*) dead end system (*b*) grid iron system
(*c*) ring system (*d*) radial system

216. The type of layout system of pipes to be adopted depends upon
(*a*) topography of the town
(*b*) location and elevation of source of supply
(*c*) quantity to be supplied
(*d*) all of the above

217. When the level of the source is higher than the level of the consumer place, water is generally supplied by
(*a*) gravitational system
(*b*) pumping system
(*c*) combined gravity and pumping system
(*d*) any of the above

218. The advantage of the "pumping without storage system" is
(*a*) during fire, large volume of water under high pressure can be forced in the required direction.
(*b*) high lift pumps are required
(*c*) no storage is required
(*d*) all of the above

219. The system which requires uniform rate of pumping, is
(*a*) pumping without storage system
(*b*) gravitational system
(*c*) pumping and storage system
(*d*) all of the above

220. The safe and most reliable system in distribution of water is
(*a*) gravity system
(*b*) pumping system
(*c*) pumping and storage system
(*d*) all the above are correct

221. The system, which is cheap in initial cost, easy in determination of discharge and pressure at any point in the system is
(*a*) grid iron system (*b*) dead end system
(*c*) ring system (*d*) radial system

222. The system which is more advantageous with the "direct-indirect system" is
(*a*) dead end system (*b*) grid iron system
(*c*) ring system (*d*) radial system

223. For centrifugal pumps
(*a*) limited space is required
(*b*) initial cost is low
(*c*) steady discharge is obtained
(*d*) all the above

224. The discharge of a centrifugal pump
(*a*) increases with the increase of head
(*b*) decreases with the increase of head
(*c*) independent of head
(*d*) none of the above

225. The main draw-back of centrifugal pump, is
(*a*) low efficiency upto 50%, for high heads
(*b*) necessity of priming
(*c*) suction lift is limited to 6 m
(*d*) all the above

226. Reciprocating pumps
(*a*) are not suitable for variable heads
(*b*) are costlier than centrifugal pump
(*c*) produce pulsating flow
(*d*) all the above

227. For a centrifugal pump at zero discharge the head developed is
(*a*) maximum (*b*) minimum
(*c*) zero (*d*) none of the above

228. Pump commonly used for lifting water in water supply mains is
(*a*) axial flow pump (*b*) centrifugal pump
(*c*) reciprocating pump (*d*) rotary type pump

229. The source of power for pumping is mostly
(*a*) steam engine (*b*) diesel engine
(*c*) electric motor (*d*) all the above

230. Pump efficiency is 85%, motor efficiency is 94%, the combined efficiency of the two is
(*a*) 90% (*b*) 80% (*c*) 70% (*d*) 47%

231. If W is the density of water in kg/m^3, Q is discharge in m^3/sec and H is total head, then the Break horse power of the pump is given by

(*a*) $\frac{WQH}{75\eta}$ (*b*) $\frac{WQ}{75H}$ (*c*) $\frac{WQH}{75}\eta$ (*d*) $\frac{WQ\eta}{75H}$

where η is the efficiency.

232. According to Lee economic diameter of the pipe is given by

(*a*) $D = 0.67$ to $0.87\sqrt{Q}$ (*b*) $D = 0.67$ to $0.97\sqrt{Q}$

(*c*) $D = 0.97$ to $1.22\sqrt{Q}$ (*d*) $D = 1.22$ to $1.32\sqrt{Q}$

233. If the diameter of the main pipe is taken less than the economical diameter then

(*a*) head loss will be high
(*b*) cost of pipe will be less
(*c*) cost of pumping will be more
(*d*) all the above

234. Suction lift of the pump depends upon

(*a*) atmospheric pressure
(*b*) water temperature
(*c*) frictional losses
(*d*) velocity of water in suction pipe
(e) all the above

235. The efficiency of a pump set is generally taken as

(*a*) 90% (*b*) 80%
(*c*) 65 to 70% (*d*) 50 to 60%

236. The approximate diameter of a water main for supplying 7.15 mld with 1.21 m/sec velocity flow will be

(*a*) 10 cm (*b*) 20 cm (*c*) 30 cm (*d*) 50 cm

237. The depth of water table at a place is 50 m below the general ground level. The pump recommended for the tube well will be

(*a*) centrifugal pump (*b*) reciprocating pump
(*c*) deep well turbine pump (*d*) screw pump

238. A pump is delivering at less than the expected rate of discharge, it means

(*a*) pump is not primed (*b*) speed of motor is too low
(*c*) Impeller is clogged (*d*) all the above

239. A valve which operates automatically when the pressure in the pipe exceeds the set pressure, is called

(*a*) safety valve (*b*) relief valve
(*c*) sluice valve (*d*) reflux valve

240. Air relief valves in water supply pipe lines are provided at
(*a*) pipe bends (*b*) pipe joints
(*c*) summits (*d*) low points

241. The valve provided after the pump in the rising mains to prevent the back flow of water, when the pump is stopped, is
(*a*) air relief valve (*b*) reflux or check valve
(*c*) sluice valve (*d*) all the above

242. Pressure relief valve is provided to prevent
(*a*) the increase of pressure after certain safe limit
(*b*) back flow, when the pump is stopped
(*c*) the water flowing out of the suction pipe
(*d*) none of the above

243. Sluice valves in main water supplies are
(*a*) used to regulate the flow
(*b*) spaced about 5 km apart
(*c*) generally provided at the summits
(*d*) all the above

244. Pressure relief valve is provided in water main
(*a*) to reduce the pressure (*b*) upstream of sluice
(*c*) af low points (*d*) all the above

245. Air inlet valve is provided in water mains at
(*a*) summit of the pipe
(*b*) downstream of the sluice valve
(*c*) both (*a*) and (*b*) above
(*d*) upstream of the sluice valve

246. The valve provided at low points of pipes to drain off water quickly under gravity is called
(*a*) drain valve (*b*) blow off valve
(*c*) sewer valve (*d*) all the above

247. The commonly used material for water supply pipes having property of not easily getting corroded, long life but heavy and brittle is
(*a*) steel (*b*) cast iron
(*c*) wrought iron (*d*) reinforced cement concrete

248. As compared to cast iron pipes, steel pipes are
(*a*) strenger (*b*) heavier
(*c*) costlier (*d*) all the above

249. The least thickness of B class cast iron (spun) pipe is
(*a*) 7.2 mm (*b*) 7.9 mm (*c*) 8.3 mm (*d*) 8.6 mm

250. The maximum pressure which the pipe can withstand without any leakage during hydrostatic pressure test, is called
(*a*) working pressure (*b*) test pressure
(*c*) design pressure (*d*) none of the above

251. The type of joint used in joining the plain ends of cast iron pipes, is
(*a*) flanged joint (*b*) flexible joint
(*c*) dresser coupling joint (*d*) socket and spigot joint

252. The type of joint generally used in cast iron pipes, is
(*a*) flanged joint (*b*) socket and spigot joint
(*c*) dresser coupling joint (*d*) all the above

253. Steel pipes are
(*a*) generally laid Underground and no expansion joint is required
(*b*) connected by riveted or welded joints
(*c*) suitable for withstanding high internal pressure
(*d*) all the above

254. Asbestos pipes are not only smooth but also
(*a*) highly resistant to corrosion
(*b*) light in weight and easy to transport
(*c*) highly flexible to resist deflection up to 12°
(*d*) all the above

255. Asbestos pipes are joined by means of
(*a*) flanged joint (*b*) flexible joint
(*c*) simplex joint (*d*) dresser coupling joint

256. The asbestos cement pipes are generally laid
(*a*) horizontal (*b*) vertical
(*c*) in any position (*d*) inclined at 60°

257. PVC pipes can withstand pressure upto
(*a*) 20 m head of water (*b*) 50 m head of water
(*c*) 100 m head of water (*d*) 150 m head of water

258. Service connections to consumer houses, are generally provided with
(*a*) copper pipes (*b*) PVC pipes
(*c*) galvanized iron pipes (*d*) cast iron pipes

259. In order to consider temperature changes, metal pipes are usually provided with
(*a*) expansion joints (*b*) contraction joints
(*c*) anchorages (*d*) pressure relief valves

260. A GI pipe has a coating of
(*a*) zinc (*b*) lead (*c*) silver (*d*) aluminium

261. Pipes for water supply are tested for
(*a*) pressure (*b*) leakage (*c*) dimensions (*d*) all the above

262. Thrust blocks are installed whenever there is a
(*a*) change in the direction or size of the pipe line
(*b*) change in the pressure line diagram

(*c*) pipe line ends at a dead end
(*d*) all the above

263. Stand pipes are provided with
(*a*) single inlet (*b*) single outlet
(*c*) single inlet and outlet pipes
(*d*) double inlet and single outlet pipes

264. Generally the height of stand pipes should be
(a 15 m (*b*) 20 m
(*c*) more than 15 m (*d*) less than 15 m

265. The meter commonly used for domestic service and small trade supplies involving small quantities of water to be measured is
(*a*) inferential meter (*b*) positive meter.
(*c*) semi-positive meter (*d*) none of the above are correct

266. The name of the valve vvhich is provided at all dead ends in distribution system as well as at all depressions in the mains, is called as
(*a*) drain valve (*b*) air relief valve
(*c*) check valve (*d*) hydrant

267. The treatment that should be given to be water from a deep tube well is
(*a*) pre-settling only
(*b*) coagulation and flocculation only
(*c*) filtration only
(*d*) disinfection only

268. Methemoglobinemia, the 'blue baby' syndrome is caused by consuming water containing excess of
(*a*) fluoride (*b*) phosphate
(*c*) nitrate (*d*) nitrite

269. Hardness of water is caused by the presence of the following in water
(*a*) chlorides and sulphates (*b*) calcium and magnesium
(*c*) nitrates and nitrates (*d*) sodium and potassium

270. A small filter of 0.05 m depth removes 90% of particles present in water. If the particle removal required is 99% what should be the depth of filter ?
(*a*) 0.10 m (*b*) 0.50 m (*c*) 0.75 m (*d*) 1.00 m

271. Bacteriological examination of drinking water for Escherischia Coliforms (E. Coli) is performed because
(*a*) they are pathogenic causing intestinal diseases.
(*b*) their presence indicates viral contamination of water.
(*c*) they are used as indicator organisms for probable presence of pathogens.
(*d*) they represent unique indicator organism of sewage pollution.

272. The most important water quality parameter for domestic use of water is
(*a*) carbonate hardness
(*b*) non-carbonate hardness
(*c*) coliform group of organisms
(*d*) chlorides

273. Presence of fluoride in water greater than permissible level of 1.5 mg/L causes
(*a*) Jaundice (*b*) Diarrhoea
(*c*) Hepatitis (*d*) dental fluorosis

274. The removal of dissolved organic matter occurs in
(*a*) slow and filters (*b*) trickling filters
(*c*) rapid and filters (*d*) dual media filters

275. The temporary hardness of water is caused by
(*a*) dissolved carbon dioxide
(*b*) bicarbonates and carbonates of calcium and magnesium
(*c*) bicarbonates of sodium and potassium
(*d*) carbonates of calcium and magnesium

276. Breakpoint chlorination of water involves addition of chlorine in an amount sufficient to
(*a*) react with any ammonia and readily oxidisable organic matter
(*b*) kill Giardia cysts
(*c*) react with inorganic matter
(*d*) reduce bacterial growth in filters

277. If present in water, chlorination of water does not reduce the
(*a*) ammonia content (*b*) organic matter content
(*c*) BOD (*d*) dissolved oxygen content

278. Dechlorination of water is achieved by adding
(*a*) Sodium thiosulphate
(*b*) Sodium sulphate
(*c*) Sodium hexameta phosphate
(*d*) Sodiurn bisulphate

279. The water supply to house begins with the connection of the service pipe with the municipal water mains. The connection comprises
1. Stopcock 2. goodseneck 3. ferrule 4. water meter
The correct sequence of these connections is
(*a*) 1,2,3,4 (*b*) 3,2,1,4 (*c*) 3,1,2,4 (*d*) 1,2,4,3

280. Which of the following treatments reduce salinity of water
1. Flocculation and sedimentation
2. Filtration
3. Reverse osmosis
4. Electrodyalysis

Select the correct answer using the codes given below:

Codes:

(*a*) 1 and 2 (*b*) 3 and 4 (*c*) 2 and 3 (*d*) 1 and 4

281. A city supply of 15000 cubic metres of water per day treated with a chlorine dosage of 0.5 ppm. For this purpose, the requirement of 25% bleaching power per day would be

(*a*) 300 kg (*b*) 75 kg (*c*) 30 kg (*d*) 7.5 kg

282. The distribution system configuration shown in the above figure is called

(*a*) Deadend system (*b*) Grid iron system
(*c*) Radial system (*d*) Ring system

283. The yield of a well depends upon

(*a*) permeability of soil
(*b*) area of aquifer opening in to the wells
(*c*) actual flow velocity
(*d*) all of the above

284. The cleaning of slow sand filter is done by

(*a*) reversing the direction of flow of water
(*b*) passing air through the filter
(*c*) passing a solution of alum and time through the filter
(*d*) scrapping off the top layers of sand and admitting water

285. Which one of the following pairs is not correctly matched ?

(*a*) Check—To check water flow in all directions.
(*b*) Sluice valve—To control flow of water through pipelines.
(*c*) Air valve—To release the accumulated air.
(*d*) Scour valve—To remove silt in a pipeline.

286. Which one of the following would contain water with maximum amount of turbidity ?

(*a*) Lakes (*b*) Oceans (*c*) Rivers (*d*) Wells

Answers

1. (*b*)	2. (*a*)	3. (*d*)	4. (*d*)	5. (*b*)	6. (*d*)
7. (*b*)	8. (*d*)	9. (*a*)	10. (*b*)	11. (*c*)	12. (*a*)
13. (*a*)	14. (*a*)	15. (*a*)	16. (*a*)	17. (*a*)	18 (*a*)
19. (*c*)	20. (*a*)	21. (*b*)	22. (*a*)	23. (*d*)	24. (*b*)
25. (*c*)	26. (*c*)	27. (*c*)	28. (*a*)	29. (*b*)	30. (*b*)
31. (*d*)	32. (*b*)	33. (*c*)	34. (*a*)	35. (*c*)	36. (*d*)
37. (*a*)	38. (*a*)	39. (*c*)	40. (*d*)	41. (6)	42. (*d*)
43. (*d*)	44. (*b*)	45. (*b*)	46. (*b*)	47. (*d*)	48. (*a*)
49. (*b*)	50. (*a*)	51. (*b*)	52. (*a*)	53. (*c*)	54. (*b*)
55. (*b*)	56. (*d*)	57. (*c*)	58. (*a*)	59. (*b*)	60. (*a*)

61. (*b*) 62. (*a*) 63. (*d*) 64. (*d*) 65. (*b*) 66. (*b*)
67. (*c*) 68. (*c*) 69. (*d*) 70. (*d*) 71. (*b*) 72. (*b*)
73. (*a*) 74. (*c*) 75. (*d*) 76. (*a*) 77. (*a*) 78. (*c*)
79. (*c*) 80. (*d*) 81. (*d*) 82. (*c*) 83. (*b*) 84. (*c*)
85. (*b*) 86. (*a*) 87. (*b*) 88. (*c*) 89. (*d*) 90. (*d*)
91. (*a*) 92. (*c*) 93. (*b*) 94. (*b*) 95. (*a*) 96. (*d*)
97. (*a*) 98. (*d*) 99. (*a*) 100. (*a*) 101. (*b*) 102. (*c*)
103. (*b*) 104. (*c*) 105. (*d*) 106. (*c*) 107. (*d*) 108. (*c*)
109. (*d*) 110. (*c*) 111. (*a*) 112. (*d*) 113. (*b*) 114. (*c*)
115. (*d*) 116. (*a*) 117. (*d*) 118. (*b*) 119. (*a*) 120. (*d*)
121. (*b*) 122. (*b*) 123. (*d*) 124. (*d*) 125. (*c*) 126. (*d*)
127. (*a*) 128. (*b*) 129. (*d*) 130. (*b*) 131. (*a*) 132. (*b*)
133. (*b*) 134. (*a*) 135. (*d*) 136. (*d*) 137. (*c*) 138. (*b*)
139. (*d*) 140. (*d*) 141. (*d*) 142. (*d*) 143. (*a*) 144. (*a*)
145. (*a*) 146. (*a*) 147. (*a*) 148. (*d*) 149. (*b*) 150. (*b*)
151. (*a*) 152. (*a*) 153. (*a*) 154. (*b*) 155. (*b*) 156. (*a*)
157. (*c*) 158. (*d*) 159. (*a*) 160. (*a*) 161. (*b*) 162. (*d*)
163. (*b*) 164. (*a*) 165. (*c*) 166. (*a*) 167. (*c*) 168. (*c*)
169. (*a*) 170. (*a*) 171. (*a*) 172. (*b*) 173. (*c*) 174. (*a*)
175. (*a*) 176. (*a*) 177. (*c*) 178. (*a*) 179. (*a*) 180. (*d*)
181. (*d*) 182. (*c*) 183. (*b*) 184. (*d*) 185. (*b*) 186. (*b*)
187. (*a*) 188. (*d*) 189. (*b*) 190. (*a*) 191. (*b*) 192. (*a*)
193. (*d*) 194. (*d*) 195. (*b*) 196. (*d*) 197. (*a*) 198. (*a*)
199. (*b*) 200. (*c*) 201. (*c*) 202. (*b*) 203. (*a*) 204. (*a*)
205. (*c*) 206. (*d*) 207. (*c*) 208. (*a*) 209. (*d*) 210. (*b*)
211. (*c*) 212. (*a*) 213. (*b*) 214. (*c*) 215. (*d*) 216. (*d*)
217. (*a*) 218. (*a*) 219. (*c*) 220. (*a*) 221. (*b*) 222. (*d*)
223. (*d*) 224. (*b*) 225. (*d*) 226. (*a*) 227. (*a*) 228. (*a*)
229. (*c*) 230. (*b*) 231. (*a*) 232. (*b*) 233. (*d*) 234. (*d*)
235. (*c*) 236. (*c*) 237. (*c*) 238. (*d*) 239. (*b*) 240. (*c*)
241. (*b*) 242. (*a*) 243. (*d*) 244. (*d*) 245. (*c*) 246. (*d*)
247. (*b*) 248. (*a*) 249. (*b*) 250. (*b*) 251. (*c*) 252. (*d*)
253. (*d*) 254. (*d*) 255. (*c*) 256. (*d*) 257. (*c*) 258. (*c*)
259. (*a*) 260. (*a*) 261. (*d*) 262. (*d*) 263. (*c*) 264. (*c*)
265. (*c*) 266. (*a*) 267. (*d*) 268. (*c*) 269. (*b*) 270. (*a*)
271 (*c*) 272. (*b*) 273. (*d*) 274. (*a*) 275. (*b*) 276. (*a*)
277. (*a*) 278. (*c*) 279. (*b*) 280. (*b*) 281. (*c*) 282. (*b*)
283. (*d*) 284. (*d*) 285. (*a*) 286. (*c*)

TRUE OR FALSE TYPE QUESTIONS

1. Arithmetical increase method for population forecast should be used for those large cities, which luu- icached iheii salutation population.
2. In Geometrical Increase method, it is assumed that the percentage increase in population from decade-to-decase is constant.
3. Incremental increase method combines both the arithmetic average method and the geometrical average method
4. The fire demand for a city having population 140000 using Buston's formula is 0.627 cumecs.
5. Maximum monthly consumption is 180% of annual average daily rate of demand.
6. Maximum hourly consumption is 150% of average for the day.
7. The temperature of water to be supplied should be between 10°C to 20°C.
8. Turbidity is a measure of the resistance of water to the passage of light through it.
9. Turbidity rod is used for measuring turbidity of water in a laboratory.
10. The MPN is determined by applying laws of probabilities to the available test data.
11. The quantity of water passing per hour per unit horizontal area is known as the surface loading.
12. The main disadvantage of alum is that it is difficult to dewater the sludge formed.
13. Anionic polyelectrolytes can be used independently as effective coagulants.
14. Filtration is one of the most important operations in the water purification process.
15. In bitflow filters, filteration is in two opposite directions at the same time.
16. In the dual media bitflow filter, coal sand is provided for the downflow bed.
17. The aim of disinfection is to kill all the microorganisms in water.
18. The pH of the water to be treated should be maintained at less than 7 to prevent great ionization of HOCI.
19 The effective sterilising compound HOCI is formed in greater quantities of low pH than at high pH.
20. The permanent hardness is also known as non-carbonates hardness.
21. The amount of pure CaO required to precipitate 1 ppm of magnesium is 2.30 kg per million litres of water.
22. An accelerator, combines the miking, flocculation and clarification in the same structure.
23. Zeolite process is suitable for iron and manganese removal.

24. Temperature stresses are caused in metal pipes when they are laid above ground.
25. Air relief valves are provided at the summits along the water pipe, to provide an exit for such accumulated air.
26. The spacing of sluice valves may be between 50 to 150 metres.
27. Altitude valves are mainly used on those lines which supply water to elevated tanks.
28. Combined gravity and pumping system is the most common system adopted in most of the cases.
29. Surface reservoirs should be located at high points in the distribution system, so that gravity supply can be done directly.
30. Elevated reservoirs are commonly known as overhead tanks.
31. Circular or Ring system is most suitable for the town or area having well planned streets and roads.
32. Hardy cross method is essentially a relaxation technique involving controlled trial and error.
33. The physical tests carried on water include temperature test, colour test, turbidity test and taste and odour test.
34. Taste and colour of water is due to calcium, manganese, iron oxide and chlorides of sodium.
35. Presence of calcium and magnesium chloride in water causes hardness.
36. Fungi grows in absence of sunlight.
37. Algae grous in presence of sunlight.
38. Sump well is a lower portion of the intake in which water enters from the conduits.
39. Screens are used for removing, bushed, branches, debries etc., from water.
40. When the pH value of water is less than optimum, the floc formed by alum disappears.
41. The coefficient of uniformity for rapid sand filters is less as compared to slow and filters.
42. The yeild of a rapid sand filter is thirty times to that of slow and filter.
43. Disinfection of drinking water involves removal of faculative bacteria.
44. The permissible amount of residual chlorine is 0.2 ppm
45. At the socket and spigot joint—enlarge end of the pipe is called socket; normal end of the pipe is called spigot which is fitted into the socket; and molten lead is used for water proofing.

Answers

1. True	2. True	3. True	4. False	5. False
6. True	7. True	8. True	9. False	10. True
11. True	12. True	13. False	14. True	15. True
16. True	17. False	18. True	19. True	20. True

21. True	22. True	23. False	24. True	25. True
26. False	27. True	28. True	29. True	30. True
31. True	32. True	33. True	34. True	35. True
36. True	37. True	38. True	39. True	40. True
41. True	42. True	43. True	44. True	45. True

FILL IN THE BLANK TYPE QUESTIONS

1. The enumeration of the total bacteria is known as ________ count.
2. Amount of water for drinking purposes in litres per head per day is ________.
3. The difference between birth rate and death rate gives the ________ increase in the population.
4. For Indian conditions, a provision of ________ litre per head per day will be sufficient for fire fighting.
5. Maximum seasonal consumption is ________ % annual average daily rate of demand.
6. Maximum daily consumption is ________ % annual average daily consumption.
7. The intensity of colour in water is measured in terms of arbitrary unit of colour on the ________ scale.
8. Odour is measured in terms of ________ odour number.
9. Odour test may be carried by an instrument called ________.
10. Turbidity is imparted by the ________ matter present in water.
11. MPN is related with bacterial ________
12. The pH of water admitted into a treatment plant was 6.0 in the morning. Consequent to inflow of raw water from a different source, it changed to 8.0 in the next 24 hours. Assuming linear variation in time of the hydrogen ion concentration, the time mean pH value of the water over this 24 hour period is ________
13. The population of a city at previous consecutive census year was 4,00,000, 5,58,500, 7,76,000 and 10,98, 500. The anticipated population at the next census to the nearest 5, 000 would be ________ by incremental increase method.
14. During filtration, the turbidity and ________ matter of non-settlement type are removed.
15. Formation of mud balls can be eliminated by providing an adequate ________ flow rate.
16. ________ is adopted to remove all the floating matter from surface waters.

17. The purpose of sedimentation is to remove the ________ impurities.
18. In water, organic suspended solids having specific gravity of about ________ .
19. A particle that does not alter its shape, size and weight while setting or rising in water is known as ________ particle.
20. In the horizontal flow type, the tank is generally ________ in plan.
21. ________ time is the theoretical time taken by a particle of water to pass between entry and exit of a settling tank.
22. In many of rectangular tank, ________ are provided with the object of preventing short circuiting.
23. The chemical assisted sedimentation comprises several separate processes of treatment which go to make up the complete system known as ________ .
24. Generally the break point lies between 3 to ________ ppm of chlorine.
25. The ________ method is the common test employed to measure chlorine.
26. ________ treatment is specially for swimming pools or troops or army etc.
27. A wastage of ________ ppm of soap takes place per one ppm hardness of water removed by it.
28. By addition of line and ________ both the temporary and permanent hardness can be removed.
29. ________ is a mechanical process of mixing water and air intimately.
30. To kill all the micro organisms in water so as to make it streile, the process is known as
31. The oxidizing power of a hypochlorite is represented in terms of ________ chlorine.
32. At break point, nearly all, the residual chlorine is ________ chlorine.
33. In towns with single storeyed buildings, a minimum of ________ metres of residual pressure should be available.
34. ________ are normally employed where the construction of a surface reservoir would not provide sufficient head.
35. A mass curve of demand is the cumulative demand curve, and is obtained by continuously adding the ________ demands and plotting these against time of the maximum day.
36. Galvanizing is the process by which a ________ coating is applied to the metal.
37. ________ value is an automatic device which allows water to flow in one direction only.

38. Pressure relief valves are located at those points where pressure is likely to be ________

39. The method of distribution depends upon the ________ of the area.

Answers

1. colony	2. five	3. natural	4. one
5. 130	6. 180	7. cobalt	8. threshold
9. osmoscope	10. colloidal	11. density	
12. 6.3 (6 + Log 2)	13. 141,5000	14. colloidal	15. backwash
16. screening	17. Suspended	18. 1.04	19. discrete
20. rectangular	21. Detention	22. baffles	23. clarification
24. 7	25. orthotolidine	26. Bromice	27. 25
28. soda ash	29. Aeration	30. sterilization	31. available
32. free	33. 6	34. Stand pipes	35. hourly
36. zinc	37. Reflux	38. maximum	39. topography

❐❐❐

PART-B

WASTE WATER ENGINEERING

1. Introduction
2. Quantity of Sewage
3. Sewage System
4. Laying and Construction of Sewers
5. Quality of Sewage
6. Natural Methods of Sewage Disposal
7. Sewage Treatment
8. Activated Sludge Process
9. Building Drainage
10. Rural Santation
11. Maintenance

Objective Type Questions with Answers

CHAPTER 1

INTRODUCTION

Learning Objectives

- General
- Waste
- Type of Sewage
- Common Terms used in Sanitary Engineering
- Principles of Sanitation
- Necessity of Systematic Collection and Disposal of Waste
- Methods of Collection
- Patterns of Collection Systems
- Sewage Disposal System

1.1. GENERAL

The puropose of sanitation is to maintain such environments which do not effect the public health. The correct siting and proper orientation of buildings provide better living for the inhabitants.

There is always the problem of generation of some undesirable matter, especially that which is of excremental nature.

The development of Sanitary Engineering has contributed to the growth of cities. Now the need of sanitation is very well realized in our country and every city is now trying to have closed sewers and efficient disposal system of sewage, industrial wastes and solid refuse. The efficient scheme can be partly financed from the cost obtained from its products.

The 'Public Health Engineer' plays a major role in the field of control. The methods have been developed by him for the treatment of water and sewage. He should see that the effluent from any community does not pollute the natural water supplies.

1.1.1. Sanitary Engineering

The branch of engineering which deals with the removal and disposal of the sewage (liquid waste) without causing any nuisance to the community is called **Sanitary Engineering.**

1.1.2. Importance of Sanitary Engineering

Sanitary Engineering deals with the removal and disposal of waste of the entire city. Its importance is because it helps in:

(*i*) General developments of the city.

(*ii*) Protecting water supplies from pollution.

(*iii*) Collecting and disposing off the waste of the city.

(*iv*) Removing rainwater from a town.

(*v*) Maintaining good environments for public.

(*vi*) Preventing the pollution of natural streams etc.

(*vii*) Preventing the occurrence of disease e.g., malaria, typhoid etc.

Table 1.1. Layout of Sanitary Engineering

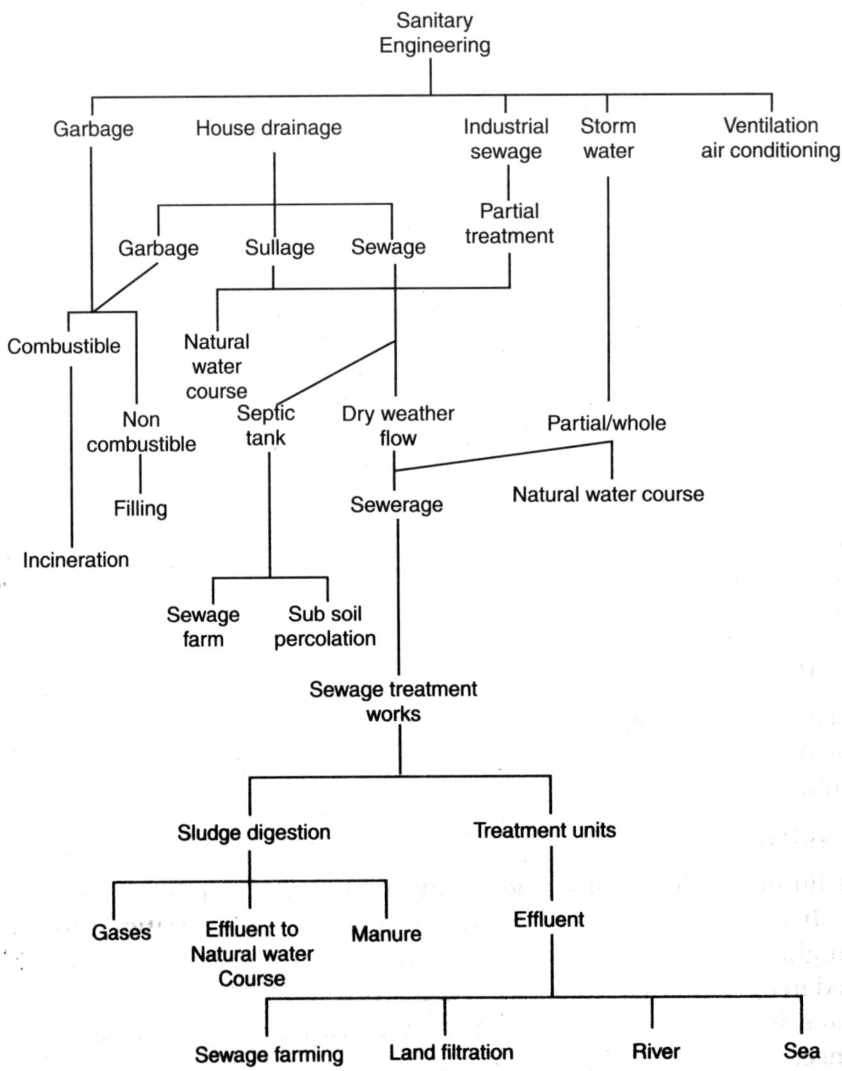

1.2. WASTE

Any thing which is not completely utilized and finally wasted in one or the other form is called waste. The waste may be in any of the following states : (*i*) dry, (*ii*) semi liquid, and (*iii*) liquid.

1.2.1. Dry Waste

The waste which does not contain moisture is called dry-waste. It is comprised of the following items:

(*i*) **Rubbish:** It means all sundry solid wastes as paper, leaves, grass, broken furniture, pottery breakage, waste building materials etc., comparatively inert in chemical action.

(*ii*) **Ashes:** These are the residues which remain after the combustion of coal, coke, timber in the hearths and furnaces of industries and houses.

(*iii*) **Garbage:** It includes all types of semi-solid and solid waste food products as vegetables, peelings of fruits, waste meat etc.

The dry-waste is never mixed with sewage. It is carried away from road-sides by means of carts, trucks or tractor-trollys to safe places outside the city where it is sorted into inflammable matter. The inflammable matter consists of sundried waste such as paper, leaves, grass and broken furniture pieces. This is burnt in incinerators. The non-inflammable matter consisting of ashes, building materials, vegetables, peelings of fruits and waste meat are buried underground.

1.2.2. Semi-liquid

The semi liquid waste is that waste which contains organic matter. The semi-liquid waste i.e., human excreta or night soil is collected in a separate pan in the same latrine from where it is removed through human agency. It is then taken outside the town in closed carts, trucks or tanks mounted on the tractor-trailors. The night-soil is buried in trenches and after 2-3 years it is converted into good manure which can be used for growing crops.

1.2.3. Liquid Waste

It mainly consists of water and very less organic matter. It is a discharge from kitchens, bathrooms, washbasins and water-closets (flush type latrines) and urinals.

1.2.3.1. Sullage

The liquid wastes from kitchens, bathrooms and wash-basins is called sullage. It does not include discharge from hospitals, operation theaters and slaughter-houses. It is only waste water and not very foul-smelling. It is carried in open drains and is given no treatment before disposal. Amount of sullage is 100 kg to 250 kg/capita depending on the water supply allowance.

1.2.3.2 Sewage

It is liquid waste consisting of sullage, discharge from water-closets, urinal and hospitals including man and animal night-soil. It is easily decomposed and creates foul-smelling gases and so it is conveyed in covered sewers.

Table 1.2. Different states of waste

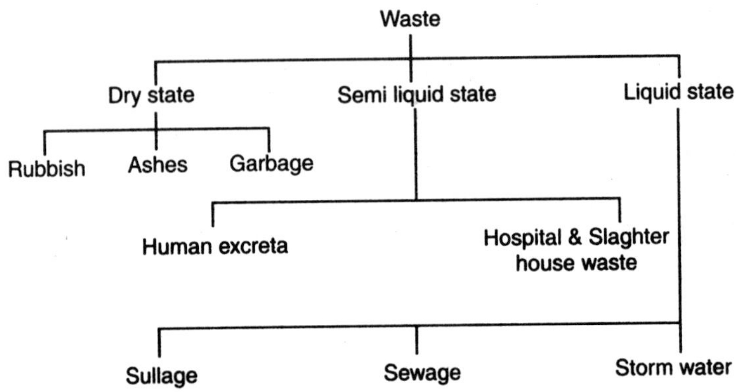

1.3. TYPES OF SEWAGE

(*i*) Sanitary sewage

(*ii*) Storm sewage

1. **Sanitary sewage** is further divided into two classes:

   (*a*) **Domestic Sewage:** It is liquid waste from kitchens and bathrooms and is commonly known as *sullage* and excremental matter such as urine and night-soil. This is from residential areas, offices and institutions.

   (*b*) **Industrial Waste:** It is waste from industries and manufacturing processes. This is also called trade waste in general.

The sewage consisting of the above two classes (*a*) and (*b*) is called Dry Weather Flow (DWF). It does not contain any storm water i.e., rainwater. The quantity of DWF is assumed same as water supply allowance per capita. In India an average value is 150 litre per capita per day.

2. **Storm Sewage:** It is any surface water including rainwater of the city, which may be admitted into underground conduits.

1.4. COMMON TERMS USED IN SANITARY ENGINEERING

Sewage: The liquid waste containing suspended, colloidal and dissolved solid is called sewage.

Sanitary Sewage: Sanitary sewage, also known as *domestic* sewage is that which originates in the sanitary conveniences of a dwelling, business building, factory or institution.

Industrial Waste: It is a liquid waste from an industrial process, as dyeing, brewing, paper making, textile mills, tanneries etc.

Storm Sewage: It is liquid flowing in sewers during or following a period of rainfall and resulting therefrom.

Sewer: A pipe or conduit which carries sewage

Common Sewer: A common sewer is one in which all abutting properties have equal rights of use.

Sanitary Sewer: It is one that carries sanitary sewage and excludes so far as possible storm sewage, surface water and ground water.

Storm Sewer: It is that which carries storm sewage including surface run off and street wash.

Sewerage: Network of sewers or pipes used for the purpose of transporting sewage.

Sludge: A water formed sedimentary deposit usually in a very wet condition.

House Sewer: It is a pipe conveying sewage from the plumbing system of a single building to a common sewer or a point of immediate disposal.

Lateral Sewer: It gets discharge directly from building through house drain. It has no other common sewer discharging into it.

Sub-main Sewer or Branch Sewer: It is one that receives the discharge of a number of lateral sewers.

Main Sewer: A main sewer also known as a trunk sewer, receives the discharge of one or more submaid sewers.

Sewer Outfall: A sewer outfall receives the discharge from the collecting system and conducts it to a treatment plant or point of final disposal.

Intercepting Sewer: Is one that cuts transversely a number of other sewers to intercept dry weather-flow, with or without a determined quantity of storm water from a combined system.

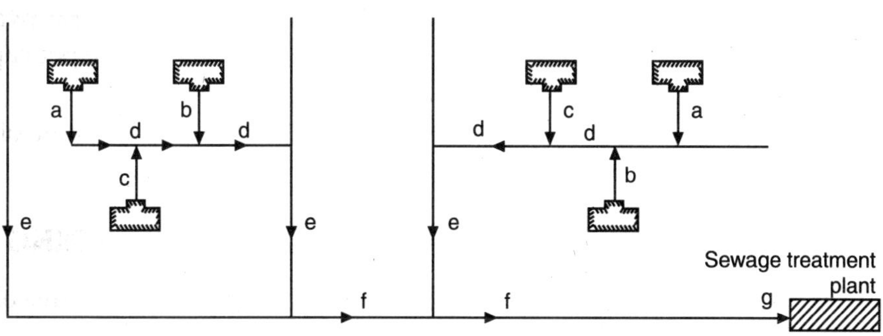

Fig. 1.1. Nomenclature of sewers

1. Lateral sewers: *a*, *b* & *c*.
2. Sub-main sewer or Branch sewer: *d*
3. Main sewer: *e*

4. Intercepting sewer: *f*
5. Trunk sewer or outfall sewer: *g*

Relief Sewer: A relief sewer is one that has been built to relieve an existing sewer of inadequate capacity.

Drain: It is pipe carrying sewage plus rainwater. Open drains, however, carrying only rainwater and sullage.

Building Connection or House Drain: It is a pipe carrying sewage from inside a building to a sewer.

Sewage Treatment: This covers any artificial process to which sewage is subjected in order to remove or change its objectionable constitutents so as to render it less dangerous/offensive.

Sewage Disposal: This term applies to the act of disposing of sewage by any method. It may be done with or without previous treatment of the sewage.

Invert: It is the lowest point of the interior of a drain or sewer at any cross-section.

Anti-siphonage: It is the device which preserves the water seal in traps by providing ventilation.

Barrel: The portion of a pipe having diameter and wall thickness uniform throughout is called barrel.

Benching: The sloped floor of a manhole or an inspection chamber on both sides and above the top of the channel.

Cover: The vertical distance from the ground level to the top of the barrel of a buried sewer is termed as cover.

Drainage: Any system constructed for the removal of any liquid is called drainage.

Manhole: This is an opening by which a man may enter or leave a drain, sewer or any other closed structure for inspection, cleaning and other maintenance operations, fitted with a sutable cover.

Cleaning Eye: This is an access opening which has a removable cover to enable obstructions to be cleared by means of a drain-rod.

Soakaway: A pit which is prepared to receive soil waste water for seepage into the surrounding ground.

Soil Waste: The discharge from water closets, urinals, sinks, stable or cowshed gullies etc., is called soil waste.

Soil Pipe. Any pipe receiving soil waste is called soil pipe.

Vent Pipe: A pipe line erected to provide flow of air to or from a drainage system or to provide circulation of air within such system to protect trap seals from syphonage and back flow.

Ventilating Pipe or Shaft: This is a pipe through which the fresh air comes inside the sewers and it provides a safe outlet into the atmosphere for the foul gases.

Self Cleaning Velocity: It is that velocity at which the solid matter in sewage does not settle down at the bottom or the sides of the sewer. In India, for small sewers, 0.9 m/sec., is suitable and for large sewers, a self-cleaning velocity 0.75 m/sec. may be sufficient.

1.5. PRINCIPLES OF SANITATION

The Principles which provide better living conditions for the inhabitants are called principles of sanitation. These may include:

(*i*) **Collection:** The waste must be collected as soon as it is formed as explained above.

(*ii*) **Conveyance:** The collected waste is removed by providing underground conduits up to the disposal works so that it may not create nuisance to the public.

(*iii*) **Building Orientation:** It should provide fresh air and sufficient light in all the rooms of building.

(*iv*) It must be prevented.

(*v*) **Water Requirements:** It must be sufficient otherwise it will create unhealthy conditions in the area.

(*vi*) **Disposal:** The sewage must be disposed off in a satisfactory manner after making it harmless.

1.6. NECESSITY OF SYSTEMATIC COLLECTION AND DISPOSAL OF WASTE

The collection works are principally meant for collecting the all types of waste products of the town. The waste products—refuse and sewage both are collected separately. If the house refuse is not collected and removed promptly away from the town, it stored in large heaps thus inviting rats, flies and vermin which give out germs of disease. Proper maintenance and use of the dust-bin in the house is the key to the satisfactory solution of the problem of sanitary storage and collection of refuse without causing nuisance.

Both the quality and quantity of house refuse vary from place to place and season to season depending upon the habits of the people. The problem of dealing with house refuse resolves itself into four parts: storage, collection, transport and disposal.

The ideal requirements for storage and collection are:

(*i*) There should be a bin-preferably two bins, maintained by every house, one for garbage and organic matter and the other for mineral matter.

(*ii*) Each house-holder should be compelled under a law to use a bin of approved type and size.

(*iii*) All waste papers should be bundled up and kept separately to be collected by the dust-man.

(*iv*) The receptacle must be kept in a good condition and sanitarily cleaned by frequent washing and disinfecting.

(*v*) The bin should have a close-fitting cover otherwise stray dogs, pigs and cattle might scatter the refuse.

(*vi*) The place where the receptacle is located should be fixed and be easily accessible to the collector. As far as possible, Muncipality should insist on the owner to show it on the building plan while submitting the same for sanction.

In addition to the bin to be provided by each house owner the muncipality should provide a number of them on foot-paths of streets for facility of pedstrains to throw whatever rubbish such as paper wrapping, tram tickets, cigarette ends, etc., into them instead of scattering them on the road surface. These bins should better be fixed to posts 30 or 60 cm above the ground so as not to be within reach of dogs, etc.

The emptying of refuse from dust-bins should be done by employed dustmen either at night or early in the morning between 5 and 6 A.M.

1.7. METHODS OF COLLECTION

Two methods of sewage-collection are in use:

1. Conservancy system or dry system
2. Water carriage system

1.7.1. Conservancy System

This system is also called dry system. In this system different typs of wastes, wet or dry, foul or non-foul are collected manually and trasported in vehicles to the points of ultimate disposal. The waste water from kitchens, baths etc., called *sullage* and *storm water* are carried separately in open or closed drain upto the point of disposal.

1.7.1.1. Advantages and Disadvantages of Conservancy System

Advantages

(*a*) It is cheaper in initial cost.

(*b*) The quantity of sewage reaching at the treatment plant is less.

Disadvantages

(*a*) There is chance of health hazards to the personnel handling the sewage.

(*b*) It is inhuman practice and social injustice.

(*c*) Nuisance due to bad smell resulting from decomposition of the accumulated night soil in cesspools and privies.

(*d*) There is chance that underground water gets polluted.

1.7.2. Water Carriage System

In this system, the wastes from water closets get mixed with sufficient quantity of water by flushing and this water is used to transport the wastes for subsequent treatment and disposal in harmless manner without any nuisance. This system is very advantageous and is universally employed in practice.

1.7.2.1. Advantages and Disadvantages of Water Carriage System

Advantages

(*a*) This is a hygienic method of collection.

(*b*) The risk of epidemics is reduced.

(*c*) Due to reduced strength of waste due to dilution, the waste becomes more amenable for treatment.

(*d*) Self cleaning velocity can be obtained even at less gradients due to more quantity of sewage.

Disadvantages

(*a*) Initial cost is vert high.

(*b*) The maintenance cost is more.

1.7.3. Comparison of Water Carriage and Conservancy System

Water carriage system	Conservancy system
(*i*) It is hygienic method because it avoids the use of human agency as far as possible for handling the wastes.	(*i*) Unhygienic method because manual labour involved for removal of human excreta.
(*ii*) No risk of spreading of disease germs.	(*ii*) Here the carts used for conveyance of night soil are accessible to flies and insects who may transmit and spread out any diease germs in the locality.
(*iii*) The system is very costly in initial cost	(*iii*) It is cheaper in initial cost because storm water passes in open drains and conservancy latrines are much economical.
(*iv*) There is no smell or sight nuisance.	(*iv*) Nuisance due to bad smell resulting from decomposition of the accumulated night soil.
(*v*) Sewers being underground do not effect the beauty of the place and cause any inconvenience to the traffic.	(*v*) Here the open surface drains will be responsible for all such troubles.
(*vi*) It is very useful for compact design and multistoreyed houses, hence universally employed in practice.	(*vi*) Not possible here.

(Contd...)

(*vii*) The maintenance cost is high and require high technical knowledge or skill.	(*vii*) No technical knowledge required and maintenance cost almost negligible.
(*viii*) Ample quantity of water is required for the adoption of this system.	(*viii*) On the other hand this does not need much of water and is quite useful at places where water supply is properly arranged.
(*ix*) Less land is required for disposal works.	(*ix*) More land is required for burying human excreta.
(*x*) During floods, pumping cost of sewage for disposal will be much more.	(*x*) Here pumping cost is very nominal.

1.8. PATTERNS OF COLLECTION SYSTEMS

The factors which affect the pattern of collection system are as follows:

(*a*) Topography and hydrological features of the area.

(*b*) Location of treatment and disposal works.

(*c*) Types of sewerage system (combined or separate)

(*d*) Area to be served.

Following are the most commonly used patterns of collection system:

(*i*) Perpendicular pattern

(*ii*) Radial pattern

(*iii*) Interceptor pattern

(*iv*) Zonal pattern

(*v*) Fan pattern

(*i*) Perpendicular Pattern: In this patten the storm water sewers are laid through shortest route to the natural water course. The shortest route is achieved if the sewers are laid perpendicular to each other. This pattern is suitable for separate and partially separate system. Fig. 1.2 shows perpendicular pattern.

(*ii*) Radial Pattern: Radial pattern is employed if the sewage is to be disposed off from heart of the city to outside periphery. In this pattern sewers are laid radially outward from the centre of the city and are provided with large number of outlets. This system requires two or more treatment units.

(*iii*) Interceptor Pattern: Interceptor pattern is an improvement over perpendicular pattern. In this pattern interceptor sewers are laid to collect sewage from all the main sewers and transport it either to a treatment plant or natural water couse at a downstream point, as shown in Fig. 1.4. During flood season there is a chance of backflow of sewers due to overflowing of natural water course and hence overflows should be provided to allow excess sewage to spill over through outlets.

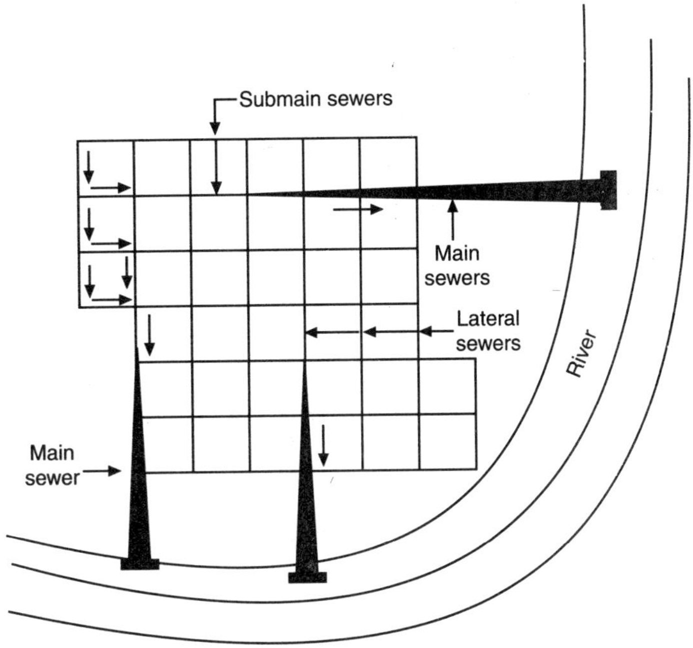

Fig. 1.2. Perpendicular pattern

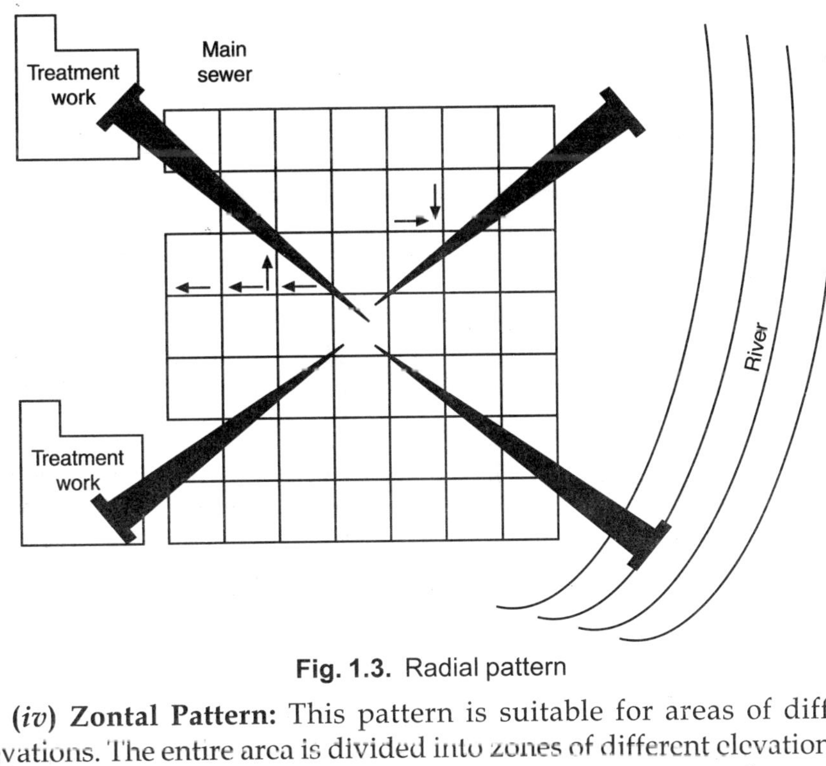

Fig. 1.3. Radial pattern

(*iv*) Zontal Pattern: This pattern is suitable for areas of different elevations. The entire area is divided into zones of different elevations and separate interceptors are provided to collect sewage from each zone. This is called zonal pattern.

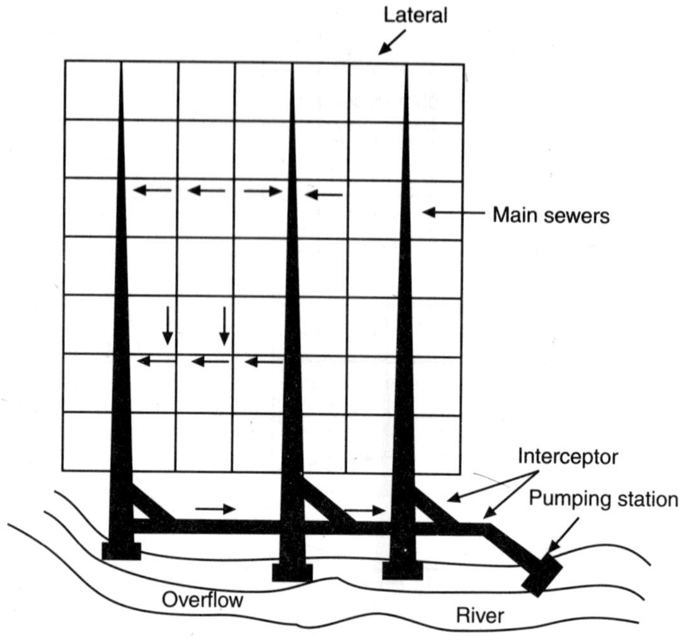

Fig. 1.4. Interceptor pattern

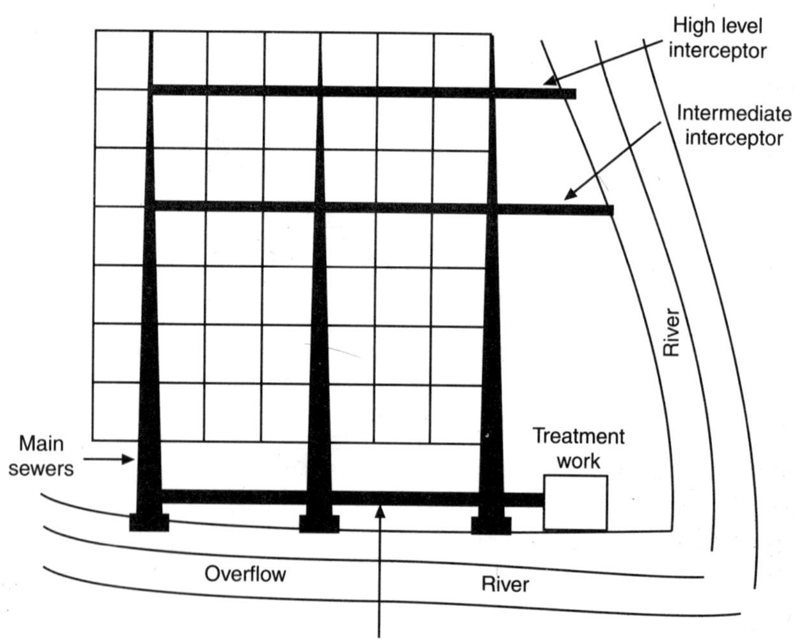

Fig. 1.5. Zonal pattern

(*v*) **Fan Pattern:** This pattern is suitable for a city which is situated near a river and the river is on side of it. In this pattern the flow is collected from outward to inward to a central main sewer. This pattern equires only a single treatment work.

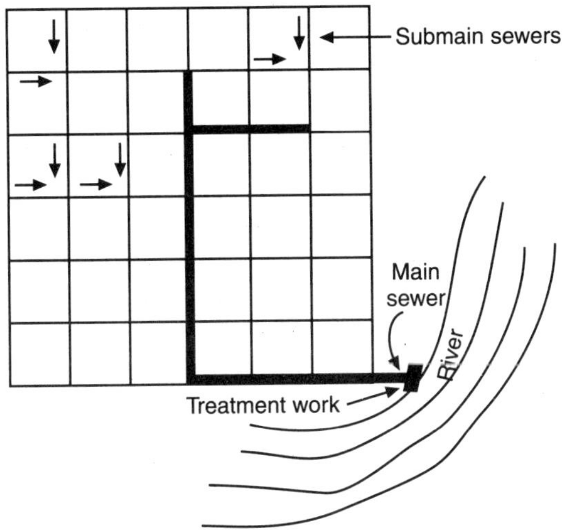

Fig. 1.6. Fan pattern

1.9. SEWAGE DISPOSAL SYSTEM

The sewage disposal system is effected in natural and artificial ways.

Further, the natural methods comprise: (*i*) Land treatment and (*ii*) Dilution into a nearby river or a drain. In land treatment, the sewage is spread uniformly on the surface of land for filtration through the pores of soil, or for irrigation purpose. The theory underlying this method is that the water in the sewage is soaked (absorbed) by soil and the organic residues (solids) left on the surface are oxidised partly by air, light and heat of the sun and partly by biological action of bacteria present in the upper most layer of the soil.

In dilution method, the sewage before or after primary treatment is allowed to fall down into a mass of water e.g., into sea, river or lake etc. The sewage has dissolved oxygen for bacteriological action.

The treatment of sewage by artificial methods is done in two ways: (*i*) Primary, (*ii*) Secondary. The primary treatment comprises, in succession, of separating the solid matter from the liquid by screnning, by grit chambers or detritus tanks, by sedimentation tanks for settling the suspended solids and by partial digestion in Imhoff tanks.

The secondary treatment is given to the effluent (liquid coming out from) passing out of the primary treatment stage, which still contains organic matter and this should be removed by land treatment, or by intermittent sand filters, contact beds, trickling filters or by activated sludge process.

The Sewage Disposal System is represented by the Table 1.3.

Table 1.3.

- Sewage disposal system
  - Natural methods
    - Dilution into water with or without pre-treatment
    - Land treatment
      - Filtration through soil medium
      - Irrigation (sewage farming)
  - Artificial methods
    - Primary Treatment
      - (i) Removal of coarse, solids such as sand, gravel etc., by grit chambers.
      - (ii) Removal of coarse, floating and heavy solids such as tins, coal, paper, etc., by screens.
      - (iii) Settlement of fine organic solids by
        - (a) Plain sedimentation with coagulation
        - (b) Sedimentation
      - (iv) Separation and partial digestion of fine organic solids by
        - (a) Septic tanks
        - (b) Imhoff's tanks
    - Secondary treatment i.e., oxidation of organic matter
      - (i) Filteration by
        - (a) Intermittent sand filters
        - (b) Contact bed filters
        - (c) Trickling filters
      - (ii) Aeration with the help of activated sludge process

MODEL QUESTIONS

1. Put the correct words in the blank:
   (*a*) Kitchen waste is ______. (Sullage/Sewage)
   (*b*) Discharge from W.C. ______. (Sullage/Sewage)
   (*c*) A pipe carrying a liquid waste is termed as ______.
   (Sewer, sewage, sewerage)
   [**Ans.** (*a*) sullage, (*b*) sewage, (*c*) sewerage]
2. (*a*) Differentiate between the following terms:
   Sewage, Sullage, Garbage
   (*b*) What is the difference between a Sewer and a Drain ?
3. Why is it necessary to dispose the waste ?
4. Describe briefly the sanitary works under the following classifications:
   (*a*) Collection works
   (b) Treatment works
   (c) Disposal works
5. Explain the term 'industrial Sewage' ?
6. Distinguish between
   (*i*) Sewerage and sewage
   (*ii*) Sludge and sullage
   (*iii*) Storm water and dry weather flow.
7. Explain the terms
   (*i*) Sewer
   (*ii*) Sewerage
   (*iii*) Sewage

(*iv*) Sullage

(*v*) Anti-syphonage pipe

8. Write short note on :

Sewage, sewerage and sullage.

9. Differentiate between soil pipe, waste pipe and vent pipe.

10. Define the following:

Garbage, DFW, Sewage, Sewerage, Sullage.

11. What is sanitary Engineering? Explain necessity of systematic to collection and disposal of waste.

12. Explain briefly the following:

(*i*) Sewage disposal system

(*ii*) Sewage treatment works

(*iii*) Waste and its types

(*iv*) Sewer and its types

(*v*) Venti pipe, soil pipe and shaft

13. (*a*) What are the principles of sanitation.

(b) What are ideal requirement for storage and collection of house refuse.

14. What are different patterns of collection of sewage? Explain briefly with figures.

15. Write the factors affecting the selection of a particular pattern of sewage collection.

CHAPTER 2

QUANTITY OF SEWAGE

Learning Objectives

- Necessity of Quantity of Sewage
- Classification of Sewage
- Sources of Sanitary Sewage
- Dry-weather Flow (D.W.F.)
- Factors Affecting Sanitary Sewage (or DWF)
- Determination of Quantity of Dry Weather Flow
- Variation in Dry Weather Flow
- Quantity of Storm Water
- Factors Affecting Storm Water Drainage
- Estimation of Storm Water Flow
- Hydraulic Formulae
- Minimum and Maximum Velocity
- Minimum Size of Sewer
- Gradients
- Shape of Sewers

2.1. NECESSITY OF QUANTITY OF SEWAGE

The necessity of determining the quantity of sewage is most important and vital for the following purposes:

(*a*) The design of sewerage scheme and sewage disposal works are based on it.

(*b*) The shape, size gradient and depth of sewers and drainage lines are fixed on this quantity.

(*c*) The size of pumping installation units depend on this quantity.

(*d*) The sizes of the various units in the sewage disposal work are directly dependent upon the quantity of sewage.

2.2. CLASSIFICATION OF SEWAGE

Based on its source, sewage is of two types:

1. **Sanitary sewage:** sanitary sewage includes :

   (*a*) Domestic sewage from residential areas

   (*b*) Industrial sewage from industries

2. **Storm sewage or storm water** includes run-off from roofs, pavements, streets, courtyard and open spaces.

2.3. SOURCES OF SANITARY SEWAGE

(*a*) The water supplied to the people in litres per head per day for their domestic use by the local authority. This amount of water after certain losses becomes the sources of sewage and reaches the sewers.

(*b*) The water supplied by the local authority to different public places like schools, colleges, bus-stands, cinema halls, hotels, fire-hydrants etc.

(*c*) The water supply made available by the local authority to various trades and industrial concerns.

(*d*) Water drawn from well and tube-wells by some individual houses for their domestic use.

(*e*) Water supply derived by some industries from local ponds, wells, canals lakes etc., by pumping for their trade purposes.

(*f*) Infiltration of ground water into the sewers through leaky joints.

2.4. DRY-WEATHER FLOW (DWF)

It is the flow of only sanitary sewage in the absence of storms in dry season. It is generally expressed in litres per capita per day.

DWF = Population × per capita rate of sewage contributed per day

or, DWF = (Density of population × Area served by the sewer) × per capita rate of flow

2.5. FACTORS AFFECTING SANITARY SEWAGE (OR DWF)

The sanitary sewage or dry weather flow depends upon the following four factors:

1. Population
2. Rate of water supply
3. Type of area served
4. Ground water infiltration

2.5.1. Population

The quantity of sanitary sewage or DWF depends upon the population served at the end of the design period. The quality of sewage to be removed from any locality will depend upon the population and contribution of sewage per capita, both are variable factors. The DWF increases with the increase in population served. Sewerage system is usually designed for 25-30 years, hence the estimation of population for this design period is very essential.

In the absence of population data, the following population densities (given in Table 2.1.) can be used for design purpose for different sizes of town/city.

Table 2.1. Population density

S.No.	Population of town/city	Density (Population/hectare)
1.	Upto 5,000	75–150
2.	5,000–20,000	150–250
3.	20,000–50,000	250–300
4.	50,000–1,00,000	300–350
5.	Above 1,00,000	350–1000

2.5.2. Rate of Water Supply

The quantity of used water discharged into a sewer system is less than the amount of water to the community.

This is because of losses due to leakage from pipes, lawn sprinkling and manufacturing process etc. Hence the rate of consumption both public as well as private supplies must be taken into account. in estimation of quantity of sewage the anticipated rate of water consumption at the end of the design period must also be studied.

2.5.3. Type of Area Served

The quantity of sewage depends on the type of area such as residential, industrial or commercial. In residential areas the quantity of sewage produced depends directly on the quantity of water supply to that area. This type of sewage is generally expressed as litres per capita per day.

In industrial area the quantity of sewage produced depends on their various industrial processes and it is different for different industry. The quantity of sewage produced from commercial places can be determined by studying the development of such places. The rate of flow of sewage in such areas is expressed in litres per day per sq. metre of area.

2.5.4. Ground Water Infiltration

In case of sewers which are laid below ground water table and in water logged areas, an allowance should be made for the water entering the sewers through leaky joints. The quantity of ground water infiltration depends upon the following factors:

(*a*) the nature of soil
(*b*) the materials of sewers
(*c*) number and condition of sewer joints
(*d*) the depth of the sewer below ground water table

The recommended quantity of ground water infiltration in case of sewers below ground water table are as follows:

(*i*) 250–500 litres/day/man-hole
(*ii*) 500–50,000 litres/day/hectare of drainage ara
(*iii*) 500–5000 litres/day/km length of sewer/cm diameter of sewer

2.6. DETERMINATION OF QUANTITY OF DRY WEATHER FLOW

The quantity of sewage produced should be equal to the quantity of water supply by water works. But for design purpose quantity of sewage is calculated by some addition and subtraction to the quantity of water supplied. The addition is mainly due to private water supply i.e., from tube well, ponds, lakes etc. Subtraction is done for loss of water due to leakage and water consumed for drinking, cooking and sprinkling. These subtractions generally vary from 20% to 30% of water supplied.

∴ Quantity of sanitary Sewage = (Total quantity of water supplied)
+ (additions due private supply)
– subtractions

Generally quantity of sanitary sewage is 75 to 80% of total quantity of water supplied.

2.7. VARIATION IN DRY WEATHER FLOW

Dry weather flow depends on the quantity of water used and as the rate of water consumption is not constant in practice, there will be variation in dry weather flow also. The variation may be seasonal or monthly, daily and hourly.

The seasonal variations are due to climatic effect. The consumption of water is more in summer than in winter and this change in consumption of water directly affect the quantity of sewage.

The daily variations are due to the local conditions, involving habits and customs of people. In India, the consumption of water is more in Sunday or other holidays as compared to working days due to cloth washing, house cleaning etc., and hence the quantity of sewage produced is more on Sundays or holidays.

Hourly variation is due to varying rate of water consumption in different hours of the day. Fig. 2.1. shows a graph showing the hourly variations in the sewage flow.

The variation in the rate of dry weather flow affects the design of sewers. The sewer must be designed for the maximum discharge. The increased quantity of flow is determined by multiplying the average daily quantity by a constant known as multiplying factor. Table 2.2 shows the values of multiplying factor.

Table. 2.2. Multiplying factor for servers under normal conditions

Type of sewer	Multiplying factor
Lateral sewers	4
Branch sewers	3
Trunk and outfall sewers	1.5–2
Main sewers	1.5

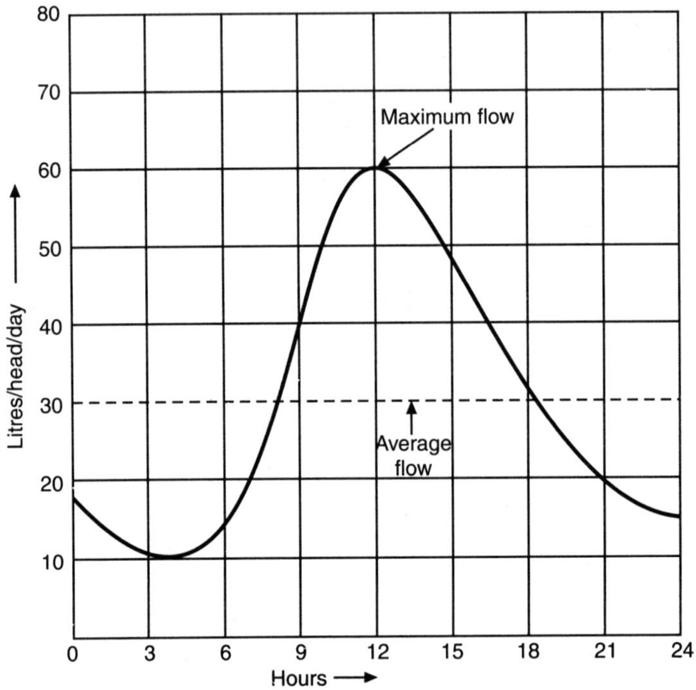

Fig. 2.1. Hourly variation in sewage flow.

2.8. QUANTITY OF STORM WATER

When rain falls over the ground surface, a portion of it is lost as evapouration and percolation. The remaining portion flows over the ground surface as storm water. The quantity of storm water reaching the combined sewers is many times more than the dry weather flow. It is necessary to collect and remove the storm water from the area the following reasons:

(*a*) For aesthetic reason

(*b*) Health of inhabitants

(*c*) For safety of structures

(*d*) For safety of traffic

2.9. FACTORS AFFECTING STORM WATER DRAINAGE

The quantity of storm water that goes to the sewers depends on :

(*i*) Intensity and duration of rainfall
(*ii*) Extent of the area drained
(*iii*) Shape and slope of catchement area
(*iv*) Nature of the soil
(*v*) Atmospheric temperature, humidity and wind
(*vi*) Number of available ditches in the area

2.10. ESTIMATION OF STORM WATER FLOW

The quantity of storm water is estimated from the following two methods:

(*i*) Rational method
(*ii*) Empirical formulae method

2.10.1 Rational Method

This method is mostly used in estimating the quantity of storm water.

$$Q = \frac{\text{AIR}}{360}$$

where Q = rate of run-off in m^3/s
A = area of watershed being drained in hectares
I = Impervious factor
R = intensity of rainfall in mm/hr

This formula gives better results for small areas due to the definite information obtainable of the factors involved.

The value of R can be calculated by using British Ministy of Health Formula

$$R = \frac{760}{t+20} \text{ mm/hr} \qquad \text{for 5–20 minutes storm}$$

$$R = \frac{1020}{t+20} \text{ mm/hr} \qquad \text{for 20–100 minutes storm}$$

2.10.2. Empirical Formula Method

This method is useful for large areas. All these empirical formulae are applicable only under certain specific conditions peculiar to the region, i.e., area to be drained, slope of land, imperviousness of surface, rate of rainfall and so these should be used cautiously

A. Burkli-Zeiglar formula

$$Q = \frac{\text{CRA}}{455}\sqrt[4]{\frac{S}{A}}$$

where Q = run off in cu-m/sec

C = run off coefficient

R = intensity of rain fall in mm/hr

A = drainage area in hectares

S = slope of the area in metre per thousand

B. Talbot's formula

$$Q = 87\ M^{\frac{1}{4}}$$

C. Fanning's formula

$$Q = 3.12\ M^{\frac{5}{8}}$$

where Q = run off in cu-m/sec

M = drainage area in sq. km.

The talbot's and Fanning's formula are used for areas larger than about 650 hectares.

2.11. HYDRAULIC FORMULAE

The flow of sewage in sewers occurs under two conditions:

(*a*) Open channel flow, and

(*b*) Closed channel flow.

When the hydraulic grade line lies in the surface of the flowing liquid, the open channel flow occurs. When sewage flows in a conduit at a pressure above or below atmospheric pressure, the flow is called closed channel flow.

The hydraulic formulae are based upon these two conditions of flow. These formulae are generally adopted for the determination of the slopes and designed velocity of flow during the design of sewers. The formulae generally used are:

A. The Chezy Formulae

$$V = C\sqrt{mi}$$

where V = velocity of flow in m/sec.

i = hydraulic gradient $\dfrac{\text{fall of sewer}}{\text{length of sewer}}$

m = hydraulic mean depth in metre

C = a constant

The value of C depends upon the hydraulic mean depth and roughness of channel. Kutter formula is used to evaluate the value of C.

B. The Kutter Formulae

$$C = \frac{23 + \frac{1}{n} + \frac{0.00155}{i}}{\left(1 + 23 + \frac{0.00155}{i}\right)\frac{n}{\sqrt{m}}}$$

where n = co-efficient of roughness of the sewer surface.

Table 2.3 gives the values of n.

Table 2.3. Values of n in Kutter's formula

S.No.	Type of material of the sewer	Value of n
1.	Cast iron pipes	0.013
2.	Brick sewers	
	(*i*) Smooth plastered	0.013
	(*ii*) Flush pointed	0.015
3.	Vitrified clay or stoneware pipe	0.014
4.	Smooth earthen channel	0.020
5.	Natural rock channels	0.030

C. The Crimp and Burge's Formula

$$V = 83.5\, m^{2/3} i^{1/2}$$

where V = velocity of flow in m/s

m = hydraulic mean depth in m

i = slope of the sewer

This formula is very popular in England and is useful for the purpose of design because of simple calculations due to availability of accurate tables and charts. This has been obtained by putting n = 0.012 in Manning's formula.

D. The Manning Formula

$$V = \frac{1}{n} m^{2/3} i^{1/2}$$

where, V, n, m and i have the same meaning as in Crimp and Burge's formula.

This formula is used for open channel flow.

E. The Hazen William's Formula

$$V = 0.85\, C m^{0.63} i^{0.54}$$

where C – a constant whose value depends upon the roughness of channel.

Table 2.4 gives values of C.

Table 2.4. Value of *C* in the Hazen-William's Formula

S.No.	Type of material of the sewer	Value of *C*
1.	Cast iron (old)	130
2.	Cast iron (new)	170
3.	Brick	130
4.	Vitrified clay	145
5.	Steel, riveted joint	110
6.	Concrete	140
7.	Plastic	150

All these hydraulic formulae are used for the design of sewer.

Ready-made charts shown in Fig. 2.2 and Fig. 2.3 are also available by which design of sewers flowing full can be done very easily. These readymade charts are known as *Nomogram*.

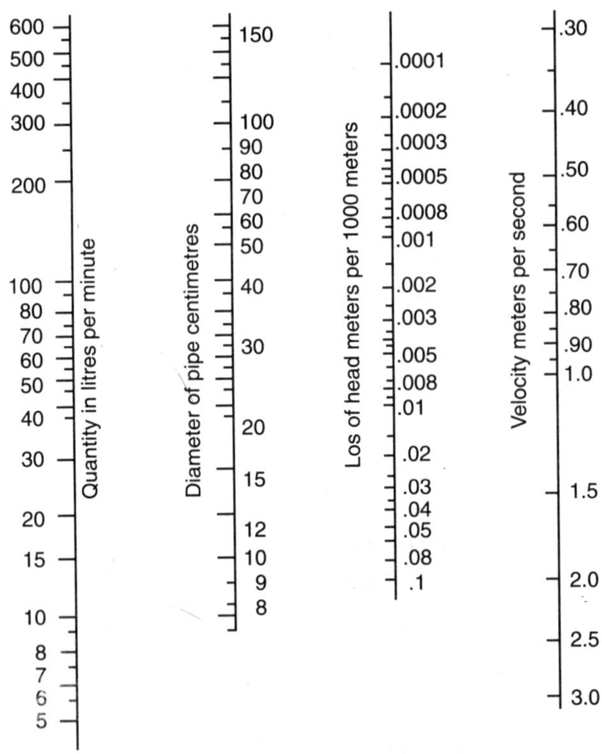

Fig. 2.2. Nomogram based on Hazen-William's formula

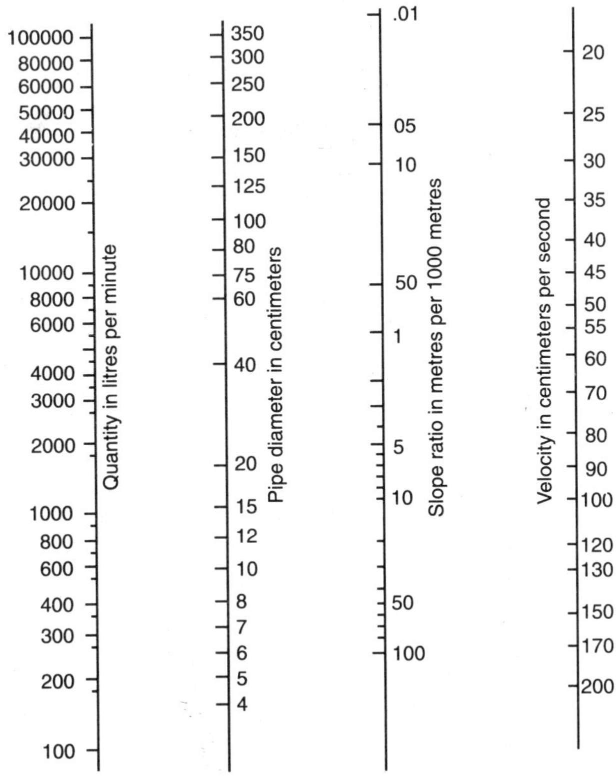

Fig. 2.3. Nomogram based on Manning's formula

2.12. MINIMUM AND MAXIMUM VELOCITY

Minimum Velocity: The minimum velocity at which no solids get deposited in the invert of the sewer is called *self-cleaning velocity*. While designing the sewer, the velocity should be maintained such that no solids get deposited in the sewer. In case of separate sewers a minimum velocity of 0.6 m/s should be maintained. A velocity of 0.9 m/s would be very acceptable. The velocity of flow should not be less than 0.75 m/s in case of combined sewer. Table 2.5 gives minimum velocity for various diamters of sewer pipes.

Table 2.5. Minimum velocity in pipe sewers

S.No.	Pipe size in (mm)	Velocity of flow (in m/s)
1.	150 to 250	1.00
2.	300 to 600	0.75
3.	Above 600	0.60

Maximum Velocity: The minimum velocity in the sewer should be equal to the self cleaning velocity so that suspended and floating solids should flow with the sewage. The velocity in the sewer cannot be increased beyond a certain limit, because if the velocity is increased beyond the limit the solids carried in suspension will move in contact with the sewer material and will cause wear of the contact surface. Due to wear the inner surface of sewar becomes rough resulting in reduction in carrying capacity of sewer as well as reduction in life of sewer. The maximum velocity is thus limited to about 3 m/s though in practice it should preferably not to exceed 2.25 m/s. Table 2.6 gives maximum permissible velocities for different material of the sewer.

Table 2.6. Maximum permissible velocity in sewers

S.No.	Type of material of sewer	Maximum permissible velocity (in m/s)
1.	Brick lined	1.5–2.4
2.	Concrete	2.4–3.0
3.	Stoneware	3.0–4.5
4.	Cast iron	3.5–4.5
5.	Earthen channel	0.6–1.2

2.13. MINIMUM SIZE OF SEWER

The minimum size of public sewer should not be less then 150 mm. In practice 200 mm size sewer is provided in plain area and 100 mm in hilly area.

EXAMPLE 2.1. Find the diameter of a circular sanitary sewer for the following data:

(*i*) Population = 12500 persons

(*ii*) Rate of supply = 135/phd.

(*iii*) Velocity of flow in sewer = 1 m/sec.

Solution:

$$\text{Average flow} = 135 \times 12500 = 16{,}90{,}000\ l/\text{day}$$

or $$= 29.6\ l/\text{sec.}$$

Considering variation, max. intensity of flow

$$= 29.6 \times 3 = 88.8\ l/\text{sec.}$$

$$= 0.09\ \text{cu. m/sec.}$$

For designing a section take 1.5 times because this quantity represents only 2/3 rd full.

$\therefore$ full capacity of sanitary sewer = 0.09 × 1.5 = 0.315 cu.m/sec.

$\because$ velocity = 1 m/sec.

$$\therefore \quad \text{Area} = \frac{\pi d^2}{4} = \frac{0.135}{1.0}$$

or $$d^2 = 0.173$$

$$d = \sqrt{0.173} = 0.421 \text{ m}$$

Hence sewer of 450 mm may be used.

2.14. GRADIENTS

Sewage flows in sewers under the action of gravity only, which is obtained by laying the sewers on slope. The slope of sewer generally follows the natural slope of the ground. The gradient depends on the following factors:

(*a*) Natural ground slope

(*b*) Diameter of the sewer

(*c*) Allowable maximum and minimum velocities

(*d*) Quantity of sewage to be discharged.

No pipe sewer should be laid at a gradient steeper then 1 in 20. A slope of 1 in 40 to 1 in 80 is provided to the house-sewer which is connected to the public sewer. The depth of sewer below ground level is generally 1 to 5 metres under traffic roads. The minimum cover at the start of any lateral sewer should not be less than 500 mm. Sewers are generally laid parallel to the ground surface where slope of ground is sufficient. Sewers may be laid at flatter slopes in steep grounds provided they are not near the surface. If ground slope suddenly drops, then a drop man-hole is to be provided.

Table 2.7 gives allowable minimum gradients for different sizes of sewers necessary to ensure a self cleaning velocity of atleast 0.6 m/s.

Table 2.7. Values of gradients for different sizes of sewers

S.No.	Diameter of sewer (in mm)	Gradient
1.	200	0.003
2.	300	0.002
3.	450	0.0015
4.	600	0.0008
5.	750	0.0006
6.	1000	0.005

2.15. SHAPES OF SEWERS

The cross-section of sewers depend upon:

(*a*) Structural stability

(*b*) Efficiency of flow

(*c*) Ease of maintenance and operation

(*d*) Resistance to internal and external pressures

(*e*) Cost.

Based on these factors, sewer-cross sections may be classified as:

(*i*) Circular section

(*ii*) Non-circular section

(*i*) Circular Section: Circular sewers are economical upto a diameter of 2500 mm and are commonly used for separate sewers. They are less effective for combined sewers because of low velocity conditions during dry weather flow. Circular sections gives the least perimeter for a given area and therefore has the maximum hydraulic mean depth for running full and half-full conditions. These also prevents possibility of deposit anywhere being of circular curvature.

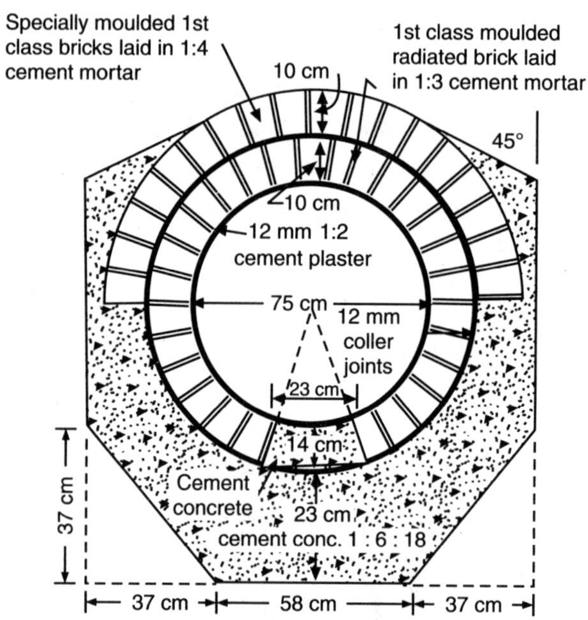

Fig. 2.4. Circular section of sewer

(*ii*) Non-circular Section: Non-cicular section is suitable for use for diameter above 2500 mm and because they accommodate large variation in discharge by maintaining the required self cleaning velocity at the same time. Non-circular sections commonly used are:

(*a*) Egg shaped (Fig. 2.5 and 2.6)

(*b*) Semi-elliptical

(*c*) Rectangular

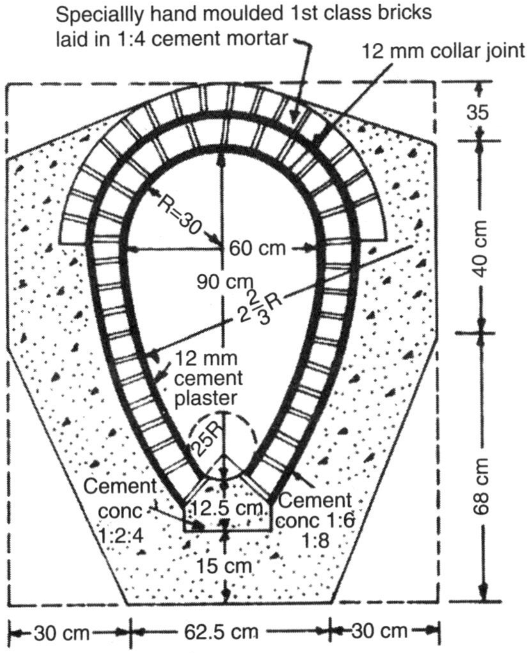

Fig. 2.5. Egg shaped sewer

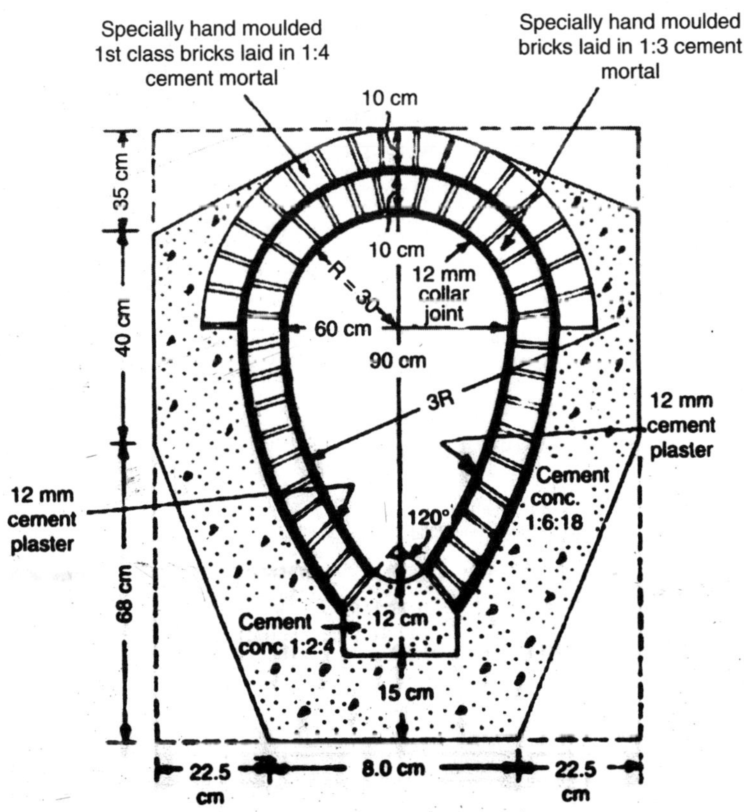

Fig. 2.6. Egg shaped sewer

MODEL QUESTIONS

1. Fill in the blanks:
   (*i*) DWF depends on ______. (Sewage, storm water, water consumption)
   (*ii*) The co-efficient of runoff means the percentage of ______ that enters into the sewer. (Rainfall, sullage, DWF.
   (*iii*) Sewage is the ______ waste of community.
   (*iv*) Storm water means ______.
   (*v*) Dry weather flow is also called.
   [Ans. (*i*) Water consumption, (*ii*) Rainfall, (*iii*) liquid, (*iv*) Rainwater (*v*) Sanitary sewage]
2. Differentiate between:
   (*i*) Sullage and Garbage
   (*ii*) Dry waste and liquid waste
   (*iii*) Dry weather flow and storm water.
3. What is the necessity of classifying the sewage as Domestic, Industrial and Storm Water?
4. What do you understand by self-cleaning velocity of sewers? Give the usual gradients for 10 cm and 15 cm diameter sewers? Why is self-cleaning velocity necessary to maintain it in the sewers?
5. Discuss the rational method of estimating the flow of storm water runn-off.
6. State how a rational estimate of storm run-off can be obtained for the design of storm sewers.
7. Explain briefly the difference between :
   (*i*) Sewage, sullage and storm water and
   (*ii*) Domestic and industrial sewage.
8. Explain the factors which affects the quantity of sewage flow in a sewer.
9. What do you mean by variation in flow of sewage? Explain average flow, dry weather flow and maximum flow?
10. What is self cleaning velocity? Why is it necessary to maintain it in sewers? What is the max. velocity permissible in sewers?
11. Explain the meaning of dry flow and how is it determined from a district.
12. What is Dry Weather flow? Explain variability of dry weather flow.

❐❐❐

CHAPTER

3

SEWERAGE SYSTEM

Learning Objectives

- Sewerage System
- Types of Sewerage System
- Comparison Between Combined and Separate System
- Requirements of Materials for Sewers
- Material for Sewers
- Joint for Sewers
- Appurtenances
- Storm Water Regulator (Storm Relief Work)
- Sewer Ventilation

3.1. SEWERAGE SYSTEM

The entire system of conduits and appurtenances for collecting sewage and delivering it to a disposal point is called Sewerage System.

3.2. TYPES OF SEWERAGE SYSTEM

Sewerage systems are of three types:

1. Separate system
2. Combined system
3. Partially separate system

3.2.1. Seperate System

In this system two separate sets of sewers are installed, one for carrying sanitary sewage and other for storm water. This system is suitable for use under the following conditions:

(*a*) When the topography of the area is flat and deep excavation is necessary for combined sewers.

(*b*) Where rainfall is uneven.

(*c*) When only sanitary sewage is to be treated and storm water is not treated.

(*d*) When sewers are to be constructed in rocky strata and cost of cutting is more for large combined sewer.

(*e*) When there is a chance of combined sewers backing up the flow into the house sewers.

(*f*) When funds are limited but sanitary drainage is imperative.

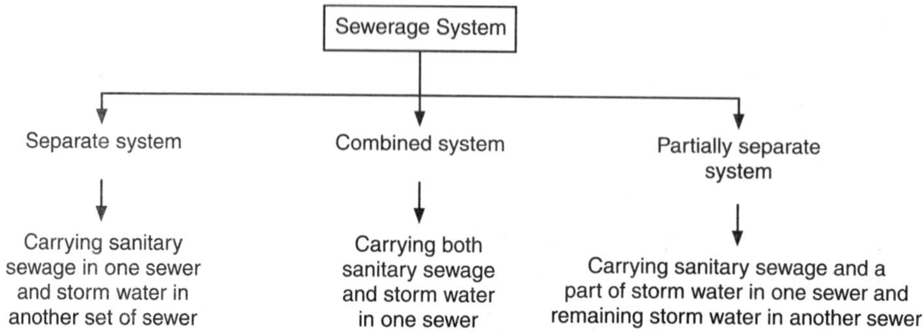

Fig. 3.1. Types of sewerage system

3.2.1.1. *Advantages and Disadvantages of Separate System*

Advantages

(*a*) Smaller size of sewer is required.

(*b*) Storm water, which is less harmless, can be disposed off by dilution.

(*c*) There is no chance of stream pollution as no overflow weirs are used.

(*d*) The quantity of sewage to be treated being small, loading on treatment plant is less.

(*e*) Capacity of pump required for pumping of sewage and their running cost would be much less.

Disadvantages

(*a*) Self-cleaning velocity cannot be attained unless laid at a steep slope and hence flushing is required at various points.

(*b*) Risk of entering of storm water into sanitary sewer is always there and it may cause overflowing of sewer.

(*c*) As two separate set of sewers are used, initial cost is high.

(*d*) Maintenance cost of sewer is high because of two separate sets.

3.2.2. Combined System

In this system both sanitary sewage and storm water are carried in single sewer. Combined system is used under follwoing circumstances:

(*a*) When rainfall is even throughout the year.

(*b*) Both sanitary sewage and storm water require pimping.

(*c*) When the quantity of sanitary sewage is small and can be admitted to already existing storm water sewer.

3.2.2.1 *Advantages and Disadvantages of Combined System*

Advantages

(*a*) House plumbing is economical due to lesser degree of complication and expense.

(*b*) Rainwater dilutes the sewage making it easier and more economical for treatment purposes.

(*c*) Flushing of sewage is not required.

Disadvantages

(*a*) Due to large size of sewer initial cost of laying is more.

(*b*) Cost of pumping is more and size of treatment plant becomes larger.

(*c*) During heavy rains, the overflowing of sewer may endanger public health.

3.2.3. Partially Separate System

In partially separate system, in addition to sanitary sewage, a portion of storm water is allowed to enter in the sewers carrying sanitary sewage. The remaining portion of storm water is diverted into a separate set of sewers.

3.2.3.1. *Advantages and Disadvantages of Partially Separate System*

Advantages

(*a*) Resonable sizes of sewers are required.

(*b*) House plumbing connections are simplified.

(*c*) No flushing is required.

Disadvantages

(*a*) The cost of pumping is more.

(*b*) There is always chance of overflow, hence requires storm water overflows.

(*c*) Self cleaning velocity may not develop during dry weather.

3.3. COMPARISON BETWEEN COMBINED AND SEPARATE SYSTEM

Combined system	Separate system
(*i*) This system is simpler in arrangement and operation as only one lateral sewer is required in each street.	(*i*) In this system two laterals are laid in each street, therefore its maintenence cost is more.
(*ii*) House plumbing is easy as only one house connection for each building is required.	(*ii*) House plumbing is difficult as two house connections for each building are required.
(*iii*) Automatic flushing is provided by water.	(*iii*) Not possible here.
(*iv*) The bigger size of sewer would involve larger excavation.	(*iv*) Here main sewers are smaller, since the rainwater is taken by open drains.
(*v*) Costs of pumping and treatment is more due to large quantity of sewerage to be handled.	(*v*) Cost is much less
(*vi*) In this system, initial cost is high.	(*vi*) Initial cost is less.
(*vii*) Suitable in congested areas because it is easy to lay one large sewer.	(*vii*) It is difficult to lay two sewers in busy lanes which also causes great inconvenience to the traffic during repairs.
(*viii*) During heavy rains, the overflowing of sewers will endanger the public health.	(*viii*) There is no fear of stream pollution.
(*ix*) Cleaning is easier due to bigger size of sewer.	(*ix*) Cleaning is difficult due to smaller size of sewers.

3.4. REQUIREMENTS OF MATERIALS FOR SEWERS

The materials used for pipe sewers are required to satisfy the following requirements:

(*i*) As sewers are normally buried below ground, highways etc., therefore they should be strong enough to withstand the stresses to which they may be subjected to.

(*ii*) It should be resistant to corrosion and abrasion caused by the sewage.

(*iii*) Sewer should be durable to resist sudden changes of temperature caused by alternate freezing and thawing.

(*iv*) It should have minimum possible weight so that it can be handled, transported and laid in position easily.

(*v*) Uniformly in size and shape, in order to avoid projection etc., at the joints.

(*vi*) Proper jointing of pipes, so as to ensure absolute water-tightness.

(*vii*) Economy in cost and construction.

(*viii*) The material should be impervious.

3.5 MATERIAL FOR SEWERS

Following materials are generally used in the construction of sewer:

(*i*) Bricks

(*ii*) Vitrified clay or stoneware pipes

(*iii*) Cement concrete (Plain or reinforced)

(*iv*) Cast iron

(*v*) Asbestos cement

(*vi*) Plastic materials

(*i*) Brick Sewers: Brick as a sewer material is quite cheap and readily available at almost all places. These sewers are made durable by plastering from outside, which also prevents the entry of tree roots inside. These are lines inside with stoneware or ceramic blocks. The disadvantage of brick sewers is that the progress of construction is very slow. Brick sewers are generally found suitable for large size storm water drains.

(*ii*) Vitrified Clay or Stoneware Pipes: These pipes are suitably used for house dramage work and small branch sewers. These pipes are made from vitrified clay or stoneware. Special clay is used for its manufacturing. Clay is first processed and mould to the shape of the pipes and then heated slowly upto 1200°C in the kiln. During burning the glazing is done by throwing sodium chloride on the surface of the pipe.

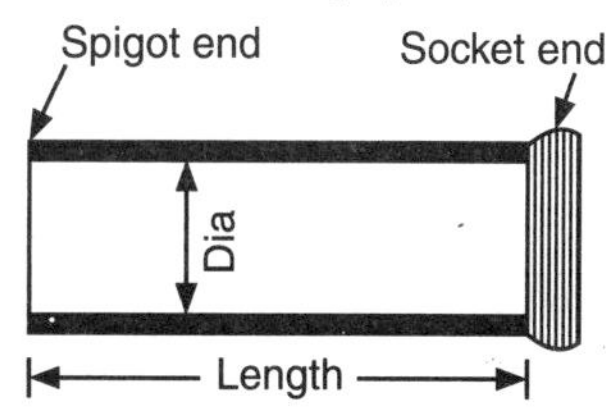

Fig. 3.2. Stoneware pipe sewer

These pipes are cheap and provide smooth and durable surface. These are resistant to corrosion and abrasion. But these are difficult to handle because of its heavy weight and brittleness.

Vitrified clay pipes are manufactured in size upto 600 mm in diameter and 600 to 900 mm in length and are able to withstand 1.5 kg/cm^2 internal pressure.

(*iii*) Cement Concrete Pipes: Cement concrete pipes are used extensively in the construction of sewer lines. These pipes may be plain or reinforced, may be precast or cast *in situ*. Precast pipes are manufactured by centrifugal process using reinforcement and are known as 'Hume pipes'. Hume pipes are manufactured in 100 to 250 mm in diameter and 1000 to 3000 mm in length. These pipes are light in weight, strong and smooth. Collar joints are generally used for joining these pipes.

Plain cement concrete pipes may be used upto 600 mm diameter.

Durability of cement concrete pipe is not more because they are corroded by acid effluents.

***(iv)* Cast Iron Pipes (CI Pipes):** Cast iron pipes are used under special cases where these have to withstand high internal and external loads. These pipes are manufactured in sizes ranging from 150 mm to 750 mm diameter. Cast iron pipes provide greater water-tight joints. These are suitable as sewer under the following conditions:

(*a*) When sewer is to be laid under the buildings.

(*b*) When sewer is to be laid on unstable ground.

(*c*) When sewer is required to be pumped.

(*d*) When sewer is to be at insufficient depth.

(*e*) Where sewer line is to be carried in exposed position over trestles.

Cast iron pipes are not suitable for use in soils containing peaty waters containing sulphates, bacause they are liable to corrosion if laid in such areas.

***(v)* Asbestos Cement Pipes:** Asbestos cement pipes are manufactured from a mixture of cement and asbestos fibre. These pipes are manufactured in sizes ranging from 100 mm to 600 mm diameter and can withstand pressure from 0.035 N/mm^2 to 1.4 N/mm^2. Asbestos cement pipes are very light in weight and can be easily transported and handled. These can be easily cut, fitted, drilled, tapped and joined. These are most suitable for domestic sanitary fittings.

***(vi)* Plastic Pipes:** Nowadays plastic, polythylene or unplasticised PVC pipes are used for carrying sewage. These are resistance to corrosion and have excellent flow characteristics. Plastic pipes are quite durable, strong against internal pressure and easy to handle.

3.6. JOINTS FOR SEWER

3.6.1. Requirements of a Good Sewer Joint

The type of joint depends on the material of pipe, internal pressure, external loads, type of support etc. The requirements of a good sewer joints are as follows:

(*i*) It should be water-tight.

(*ii*) It should be easy in construction and economical.

(*iii*) It should be flexible and offer resistance to the penetration of the tree roots etc.

(*iv*) It should be durable.

(*v*) It should be resistant to sewage gases and acids etc. which flow in sewers.

(*vi*) It should be non-absorbent.

3.6.2. Types of Sewer Joints

Commonly used sewer joints are explained below:

1. Socket and Spigot Joint: Joints for small sewers are of socket and spigot type made either in 1 : 1 cement mortar or by providing gasket and filling the joint with bituminous material.

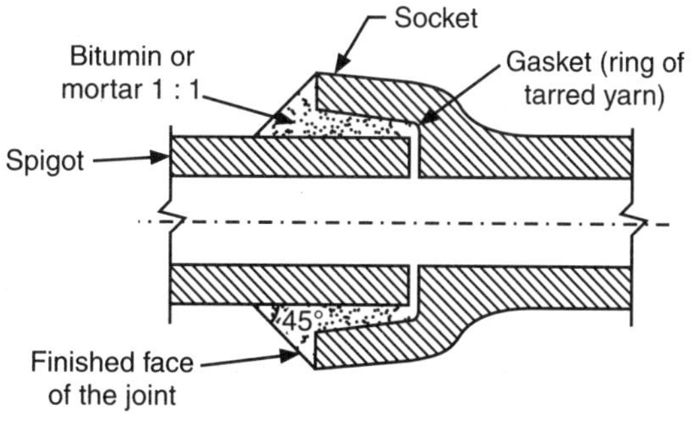

Fig. 3.3. Socket and spigot joint for SW pipes

The joint is used mainly for cast iron pipes of all sizes and concrete pipes below 600 mm diameter. Method of construction of this joint for cast iron pipes has been given in the part, "Water Supply Engineering" of the book.

2. Collar Joint: Larger sewers have collar joints i.e., having diameter over 600 mm. The pipe lengths have plain ends and are joined together by placing a collar of slightly larger diameter over the ends of the two pipes. Collars are 150–200 mm wide. The annular space between the collar and pipes filled by cement mortar 1 : 1.

Collar joint may be rigid and semi-flexible type. Rigid type collar joints are used for light hydraulic pressure pipes whereas semi-flexible type joints are used for large diameter hydraulic pipes.

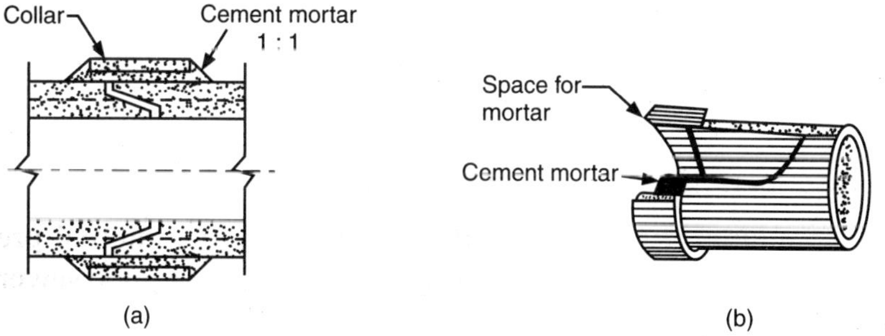

Fig. 3.4. Collar joint in cement mortar

3. Other Joints for Sewers: Where subsidence of ground is expected, flexible joints are used. The bitumen joints are flexible. Here hot bitumen is poured in the joint after placing the gasket properly.

Flush joints may be internal and external flush joints. Internal flush joints are usually used for sewers passing below culverts. Cement mortar 1 : 1 is used for filling the jointing space and the finished joint flush with both inside and outside with the pipe wall. The external flush joints suitable for small diameter pipes which cannot be jointed from inside.

3.7. APPURTENANCES

The different devices required for the efficient working of sewerage system are called appurtenances.

3.7.1. Classifications of Appurtenances

These callsified as under:

(*i*) Manholes
(*ii*) Drop manholes
(*iii*) Lamp hole
(*iv*) Clean outs
(*v*) Street inlets
  (*a*) Curb inlets
  (*b*) Gutter inlet
(*vi*) Catch basins
(*vii*) Flushing tanks
(*viii*) Grease and oil traps
(*ix*) Inverted syphon or depressed sewers
(*x*) Ventilating shafts.

3.7.1.1. Manholes

The opening constructed on the alignment of sewer line for the purposes of permitting in a man to enter for inspection and cleaning of sewers is called a Manhole.

3.7.1.1.1. Objects of Providing Manholes

(*i*) To provide an access for a man for inspection and cleaning of sewer.
(*ii*) To allow joining of sewers in case of change in direction or level or both.
(*iii*) To lay sewer lines in convient lengths.
(*iv*) To achieve the ventilation of sewers.

3.7.1.1.2. Location of Manholes

(*i*) At distances varying from 75 m to 300 m if the length is straight.
(*ii*) At the junctions of main and branch sewers.
(*iii*) At all the changes of gradients and sizes.
(*iv*) At all changes of directions.

3.7.1.1.3. Construction of Manhols

A manhole may be considered to be made of following parts:

(*i*) Cover and frame
(*ii*) Access shaft
(*iii*) Working chamber
(*iv*) Bottom invert
(*v*) Steps of ladder

(*a*) **Cover and Frame:** The manhole cover and frame are made of cast iron or RCC. These are classified as light, medium and heavy types and are used depending on type of traffic to be borne. The size of opening should be 500–550 mm diameter, so that a man can enter easily into the manhole. The frame and cover should conform of IS : 1726–1960.

(*b*) **Access Shaft:** The upper portion of the manhole is known as access shaft, which provides an access to the working chamber. It is built in brick masonary.

(*c*) **Working Chamber:** The lower portion of the manhole is called as working chamber, which provides a working space for cleaning and inspection. The size of working chamber should be less than 0.9 m × 1.2 m for rectangular shape and 1.2 m diameter for circular shape, the large dimension being in the direction of flow.

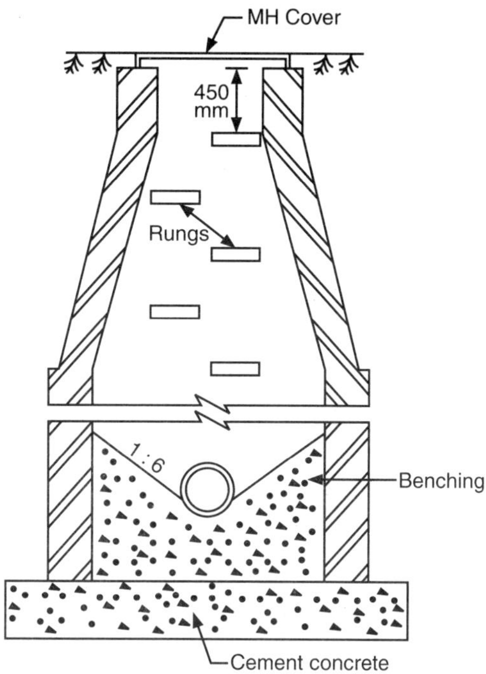

Fig. 3.5. Manhole

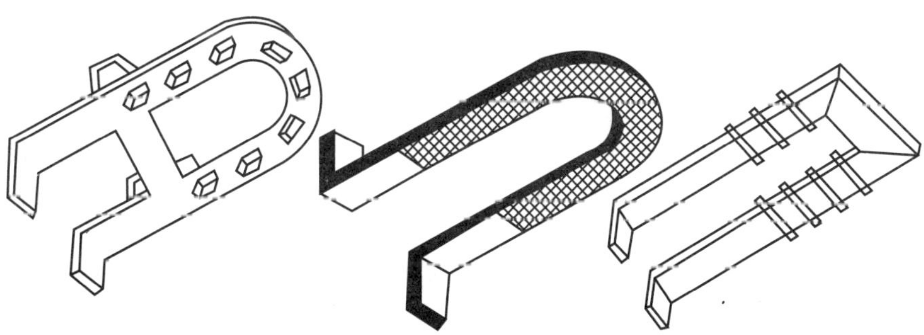

Fig. 3.6. Various kinds of manhole steps

(*d*) **Bottom Invert:** The bottom of the manhole is made of concrete bed 150–300 mm thick. The top surface of concrete floor is given a slope in 1 in 6 towards the centre to form benching. A semicircular or U-shaped channel is formed on the floor along the centre joining both incoming and outgoing sewers.

(*e*) **Steps of Ladder:** Cast iron steps or rungs are fixed in the masonary wall at vertical intervals of 300 mm in a zig-zag fashion.

3.7.1.2. *Drop Manhole*

Drop maholes are provided for joining sewers running at different levels and where it is desired to drop the levels of the invert of the incoming sewer by 600 mm or above that of the manhole. The connection is made by constructing a vertical pipe outside the manhole and then joining it to the manhole chamber. The drop manhole is shown in Fig. 3.7. The main purpose is to avoid the splashing of sewage on the man working and on the masonry work.

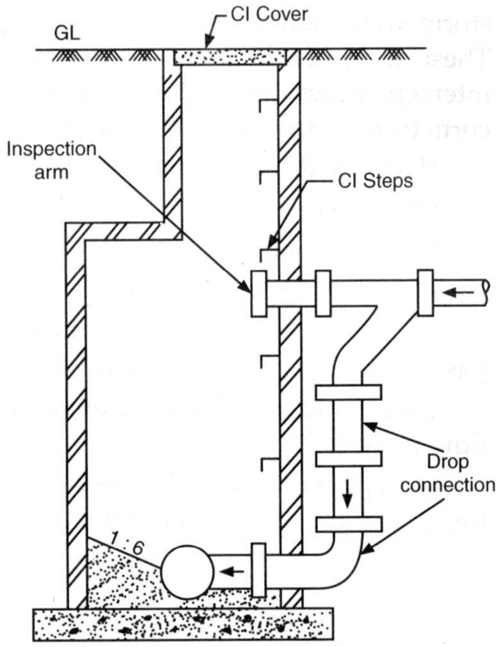

Fig. 3.7. Drop manhole

3.7.1.3. Lampholes

These are openings, ususlly of small size, constructed on straight sewer lines between two manholes which are far apart to permit the insertion of lamp into the sewer. The lamp light is seen from the adjcent manholes upstream and downstream to detect any obstruction inside the sewer. Lamphole consists of a stoneware or cast iron pipe of 225-300 mm in diameter. This pipe is connected to sewer line through a T-junction as shown in Fig. 3.8.

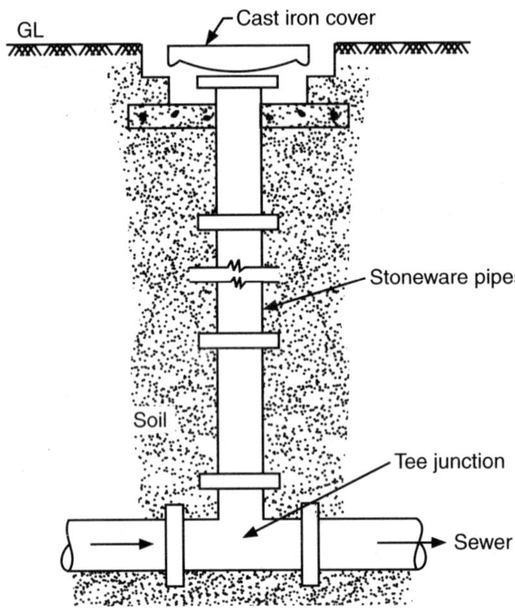

Fig. 3.8. Lamphole

3.7.1.4. Street Inlets

Street inlets are openings in the street curb or gutter through which storm water is admitted and conveyed to the storm or combined sewer. These are generally located at street intersections in such a way that they intercept water before it reaches the crossing used by pedestrains. Street corners are avoided for this purposes.

There are three types of street inlets:

(*a*) Curb inlets

(*b*) Gutter inlets

(*c*) Combined inlets

Curb inlets are vertical openings in the road curb through which flow passes. The openings are fitted with gratings.

Gutter inlets are the horizontal openings in the gutter through which flow passes.

Combination inlet includes a curb opening and a gutter opening, with the gutter directly infront of the curb opening.

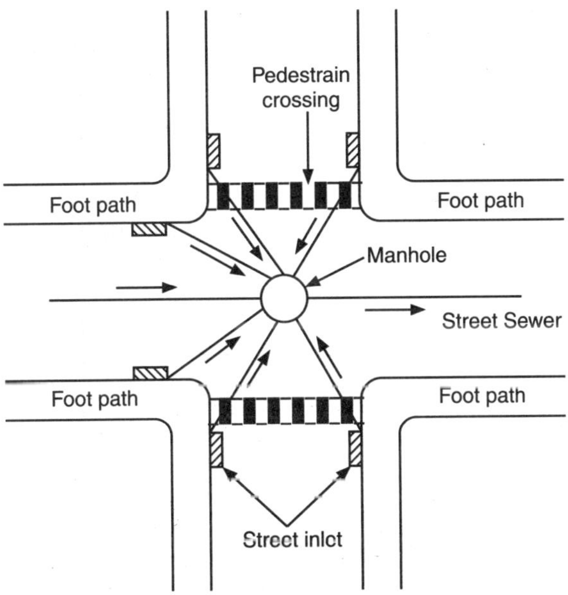

Fig. 3.9. Street inlet

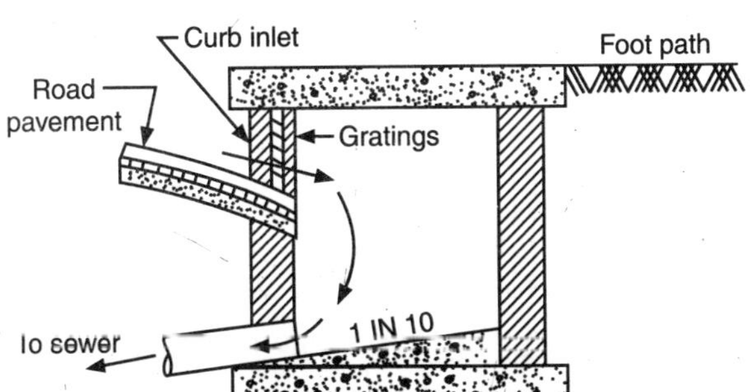

Fig. 3.10. Curb inlet

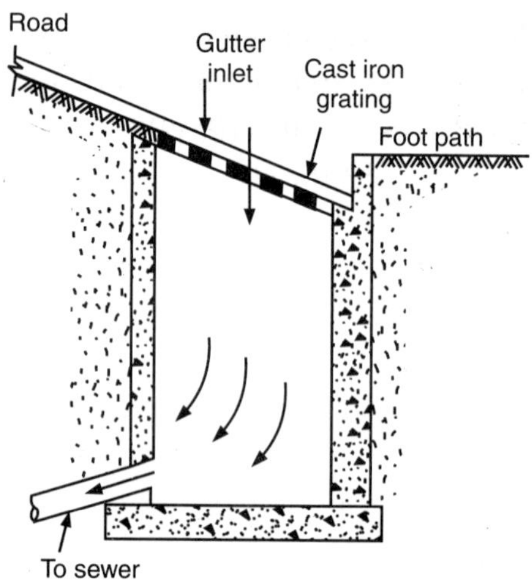

Fig. 3.11. Gutter inlet

3.7.1.5. Catch Basin

Catch basins are small settling chambers constructed below the street inlets to allow grit, sand or debris to settle down and thus preventing them from entering into the sewer. The catch basins are generally 600–900 mm in diameter and 600–750 mm in depth. It is shown in Fig. 3.12. The outlet pipe of catch basin is fixed about 600 mm above the bottom of the basin and is usually trapped to prevent escape of odour from the sewers and to retain floating matter.

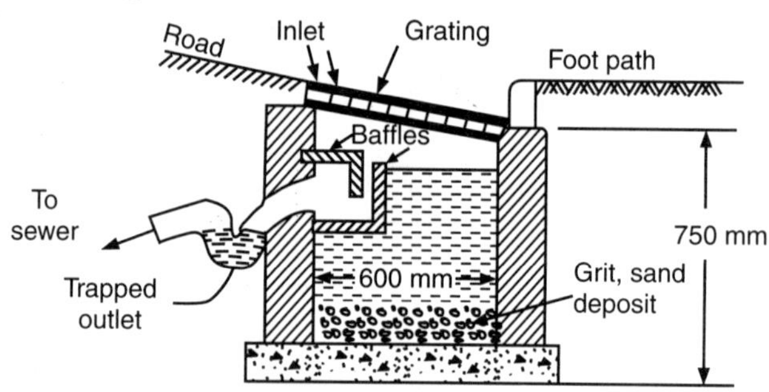

Fig. 3.12. Catch basin

3.7.1.6. Inverted Siphon or Depressed Sewer

In areas having more depressions in the ground, the sewer line can be laid above the ground by supported on piers. But it is not possible when a Railway line, canal or road crosses the sewer line. In these cases inverted siphons are provided.

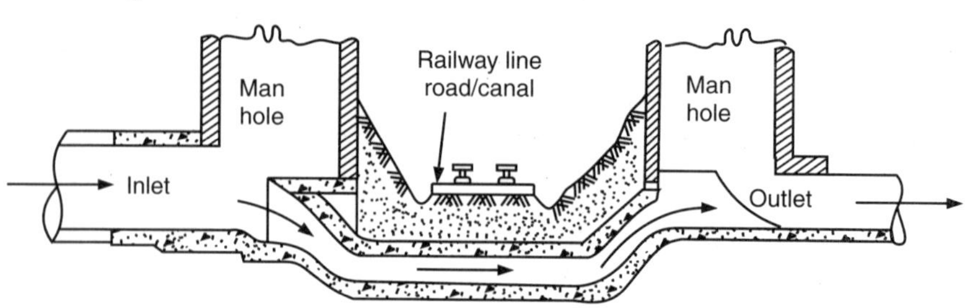

Fig. 3.13. Inverted siphon or depressed sewer

It is also called depressed sewer. Here the hydraulic gradient line is above the flow line whereas it is below the flow line in true siphon.

The inverted siphons are normally constructed of RCC or Cast Iron. At the ends manholes are provided for cleaning and inspection. The pipe of inverted siphon should be able to withstand the internal pressure. Moreover pipe diameter must be such that sewage may flow with great velocity to avoid silting.

3.7.1.7. Flushing Tanks

These are necessary to flush sewers laid at flat slopes in whcih deposits are taking place. These are necessary for combined and separate sewers only. The tanks are usually provided at the dead end or at head of sewers. Flushing tanks may be operated manually or automatically. Automatic flushing tanks are most commonly used.

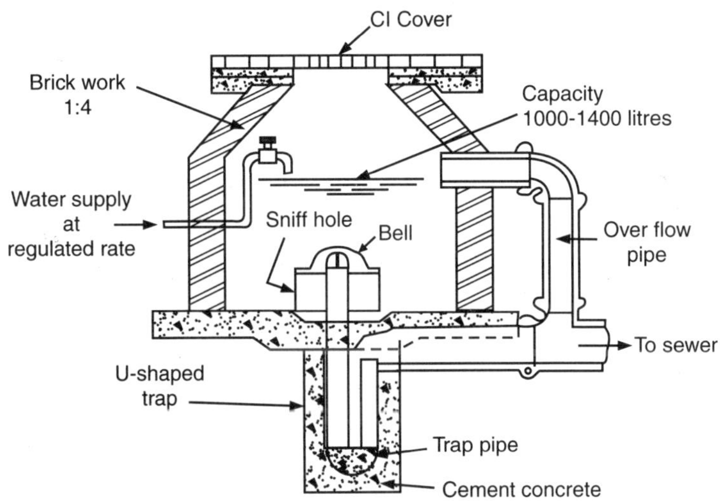

Fig. 3.14. Automatic flushing tank

In manually operated flushing arrangement sewage or water is held up in the sewer by plugging the inlet and outlet ends of the sewer. When the flushing tank is filled up with water by tap, the outlet plug is removed, hence the whole water rushes towards outlet and flushes the sewer.

Working: An automatic flushing tank consists of a concrete or masonry chamber fitted with a tap for filling the tank with water [See Fig. 3.14.]. A U-tube with a bell cap at its one end connected the chamber with sewer. When the water level increases in the chamber, it also increases in the bell cap. When it reaches a certain level, siphonic action takes place and all the water of the chamber rushes to the sewer-pipe and flushes it.

Table 3.1. Water requirement for flushing sewer lines

Slope	Quantity of water required for flushing in litres		
	200 mm/sewer	250 mm/sewer	300 mm/sewer
1 in 200	2000	2500	3000
1 in 133	1500	1800	2300
1 in 100	1300	1500	1800
1 in 50	500	800	1000
1 in 32	400	500	750

3.7.1.8. Grease and Oil Traps

The sewage from industries, hotels, restaurants, kitchen etc. contain oils and grease. These must be removed before they enter the sewer. Otherwise these will stick to the interior surface of the conduit and will become hard causing obstruction in the movement of the sewage. To check them, grease traps are provided in the pipe connec-ting the kitchen with sewer line. A simple grease-trap is shown in Fig. 3.15. Oil and grease being lighter float at the top and are carried in the effluent, the outlet drawing water from about the bottom.

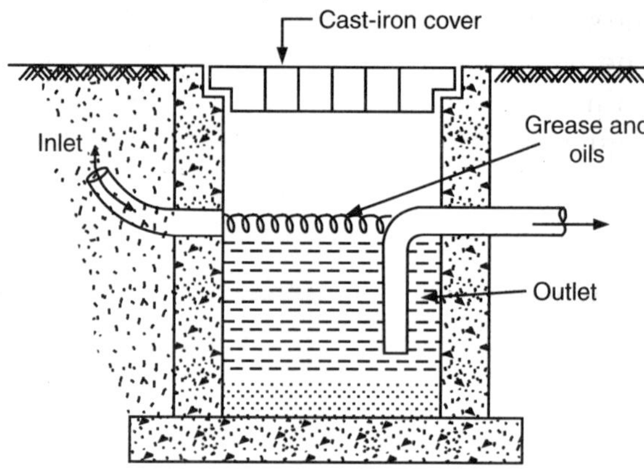

Fig. 3.15. Simple grease trap

Fig. 3.16 shows a combined sand, oil and grease trap. Here entry is through the grating and it has two chambers. Sand settle down at the bottom and grease floats at the top. Sand is not carried to the next chamber due to baffle walls dividing the chambers. The outlet is provided in the second chamber where sand and grease separate out similarly. These traps must be regularly cleaned for their proper working.

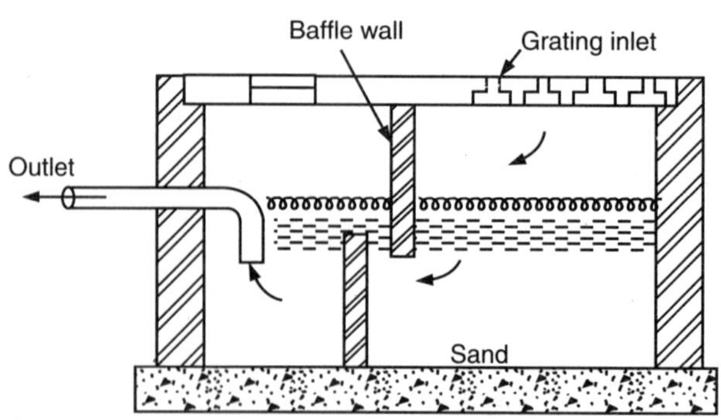

Fig. 3.16. Combined sand, oil and grease trap

3.8. STORM WATER REGULATOR (STORM RELIEF WORK)

The storm relief works are constructed in case of combined sewers to permit the excess storm water to divert and pass off into a nearby water course. This diverted sewage flow is sufficiently diluted by the water course to avoid any potential danger of stream pollution. The storm relief works are of three types:

(*a*) Overflow weir

(*b*) Leaping weir

(*c*) Shiphon spillway

(*a*) Overflow Weir: This is also called side flow weir. Overflow weir consists of a masonary chamber with a weir built in the side of the sewer channel so that the excess strom water spills over the weir to pass off into a relief channel leaving behind the normal dry-weather flow flowing into the sewer outlet. This is the simplest method of regulating flow in sewers. The weir should be so designed that it should be able to discharge the estimated overflow. The length of the weir should be sufficient for effective regulation. The length of weir is calculated by using Babbit's formula.

$$L = 7.6 \times 10^{-3} VD \log \frac{h_1}{h_2}$$

where L = length of the weir

V = velocity of approach in m/s

D = diameter of sewer in mm

h_1 = head above the crest of weir u/s in m

h_2 = head above the crest of weir d/s in m

The limitation of the formula is that it was devised for pipes between 450–600 mm in diameter and in that the depth of flow above the weir should not exceed 3D/4.

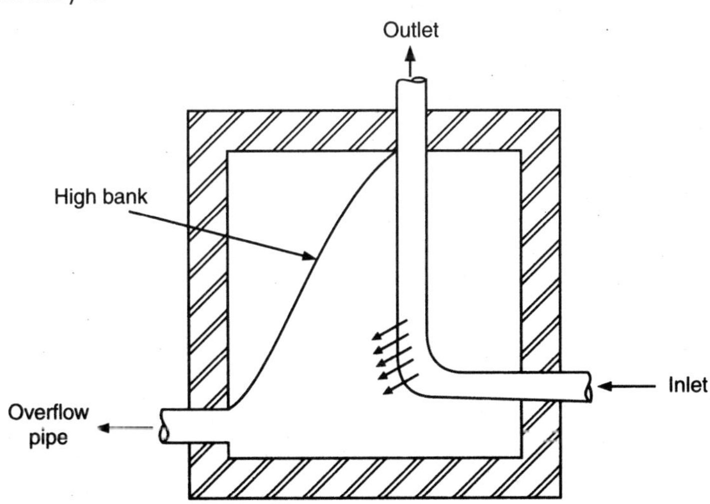

Fig. 3.17. Overflow weir

(*b*) **Leaping Weir:** This is also known as jumping weir. Leaping weir consists of an opening in the invert of the storm water drain through which the dry weather flow falls in intercepting sewer and the excess flow will leap across the opening to the water course. These are good regulator but the disadvantage is that sometime during heavy storn, most of the flow may leap over the combined sewer and only a small quantity is left in the sewer. Leaping weir is shown in Fig. 3.18.

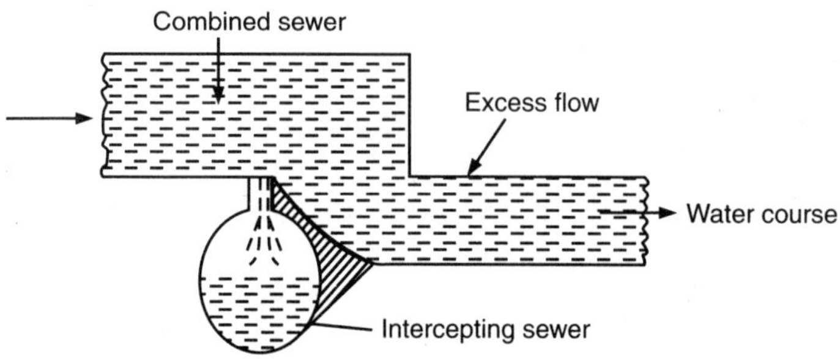

Fig. 3.18. Leaping weir

(*c*) **Siphon Spillway:** Siphon spillway is very sensitive and automatic in its action and it removes more volume of extra sewage in less time. This is provided in the side walls of a sewer. When the sewage in the sewer rises above the crest level, entrance of the priming pipe is sealed and stops the entrance of air in the pipe. Due to this the level of sewage rises simultaneoulsy both in the sewer and above the crest. As the level rises further, the air in the priming pipe is also driven out and siphonic action beings. It continues till the level falls below the entrance of priming pipe. As soon as level falls below the entrance of priming pipe the air starts entering the crest and stops siphonic action.

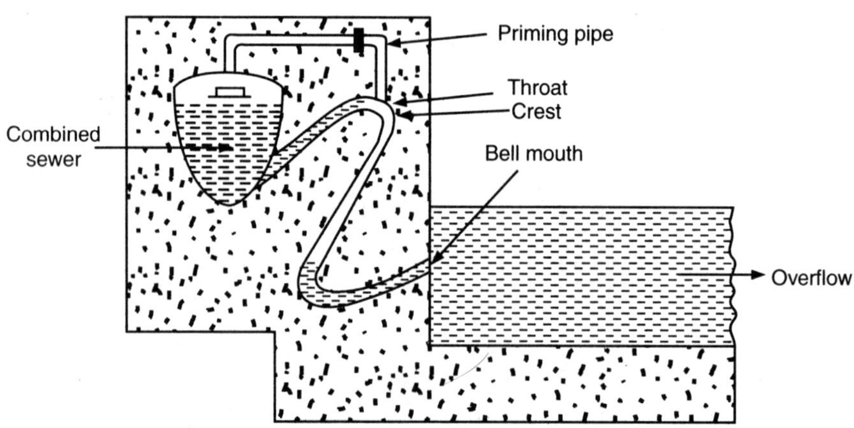

Fig. 3.19. Siphon spillway

3.9. SEWER VENTILATION

3.9.1. Purpose

Sewers are ventilated to fulfil the following purposes:

(*i*) To prevent the accumulation of explosive, corrosive or poisonous gases and vapours e.g., methane, petrol vapour etc.

(*ii*) To relieve pressure above or below atmospheric. It may be due to sudden rise or fall of sewerage.

(*iii*) To help free flow of sewerage.

(*iv*) To prevent concentration of unpleasant odours causing nuisance.

3.9.2. Methods of Ventilation

Ventilation is done by the following means:

1. Ventilating shaft: Ventilation is provided to the sewer lines at every 300 m which will provide fresh air to the workers working in the manholes.

Fig. 3.20 shows a ventilating shaft or column commonly used in practice. These are made up of cast iron or steel pipe jointed together to form a vertical shaft with a cowl of copper wire domes at the top. The height of the ventilating shaft must be more than the roof of the tallest building in its neighbourhood. The base of the shaft is surrounded in a concrete block. A rust box is also provided to collect the rust falling from the inside of the shaft. The rust can be periodically emptied through an air-tight cover.

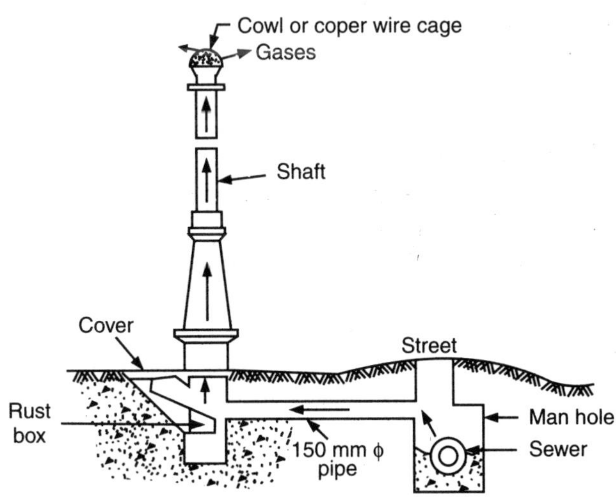

Fig. 3.20. Ventilating shaft

The shafts are made in different sizes = 100 mm, 150 mm, 225 mm and 300 mm depending upon the respective sizes of the circular sewers. The ventilating column is generally connected to the manholes by 150 mm dia. pipe as shown in the figure above.

2. By providing unobstructed outlets in the case of storm-water drains or sewers.

3. House soil and vent pipes will ventilate house drains and public sewers.

4. Sometimes forced draught is provided with the help of exhaust fans to expel out the foul gases from the sewer. This method is costly and requires mechanical attention and supervision.

5. By providing perforations in the manhole covers. The practical utility of ventilating manhole covers is less. On the other hand they cause more of nuisance and admit surface water in large quantities into the sewers, which is another disadvantage.

MODEL QUESTIONS

1. (*i*) Maximum distance between successive manholes is ______.

(*ii*) Automatic flushing tank is used for sewers with ______ gradient. (Less/More)

(*iii*) What is minimum size of a manhole?

(*iv*) What is a drop manhole?

(*v*) Why is it necessary to provide a ventilating shaft in the sewerage system?

(*vi*) Ventilating shaft is provided to the sewer line at every ______ meters. (100/150/300)

2. By drawing a neat sketch explain the working and function of a ventilating shaft on a sewer line.

3. Describe the location, functions and construction of any two of the following:

(*i*) Manholes (*ii*) Grease trap

(*iii*) Screening chamber

4. Write short notes on any two of the following:

(*i*) Grease trap and gully trap. (*ii*) Automatic flushing tank.

(*iii*) Catch basins (*iv*) Inlets

(*v*) Manholes.

5. What do you understand by the term "Flushing of Sewers"? Explain with the help of a sketch the location construction and working of an automatic flushing tank in this connection.

6. Describe, with a suitable diagram, a manhole.

7. (*a*) Write a brief specification for the construction of a manhole.

(*b*) Sketch neatly a drop manhole and explain its function.

8. Describe briefly and explain the function of the following

(*i*) Street inlet, (*ii*) Flushing manhole, (*iii*) Drop manhole

9. Write notes on:

(*i*) Manhole (*ii*) Catch Basin

10. Explain with sketches the function of the following:

(*i*) Ventilating shaft (*ii*) Storm water overflows.

11. Write a note on the flushing of town sewers. Sketch and describe the working of an automatic flushing tank for a two sewer.
12. Write short notes on any five of the following.
    (*a*) Sewer ventilation (*b*) Anti siphonage pipe
    (*c*) Soil pipe (*d*) Intercepting trap
    (*e*) Grease trap (*f*) Drop manhole
    (*g*)Self cleaning velocity (*h*) Lamp hole
13. (*a*) Write a note on sewer ventilation.
    (*b*) Discuss briefly flushing manholes.
14. Sketch the following:
    (*i*) Drop manhole
    (*ii*) Automatic flushing tank
15. Explain the necessity of ventilating the sewers. Also describe how the sewers are ventilated. Give necessary sketches and other details.
16. Write short note on Inverted Siphon.
17. (*i*) What sewerage system will you adopt for an Indian town.
    (*ii*) ______ system is a hygienic method. (Conservancy/Water carriage)
    (*iii*) Water carriage system is ______. (Costly/Economical)
    (*iv*) When sanitary sewage and storm water flows in one sewer, it is called ______. (Seperate/Combined system)
    (v) House plumbing is difficult in case of ______. (Separate/Combined System)
18. Write the merits and demerits of water carriages and conservancy system of refuse collection.
19. What is meant by sewerage and sewer. Descrbe the separate, partially separate and combined systems of sewerage. Give the various materials used for sewers.
20. Describe the separate and combined systems of sewerage and state the conditions under which one is more suitable to the other.
21. Under what circumstances various sewerage systems are provided? Give your choice of selection of a hill station.
22. Compare the advantages and disadvantages of conservancy system and water carriage system.
23. Discuss the comparative mertis and demerits of the separate system and combined system of sewerage.
24. Write a short note on "Type of Sewerage Systems".
25. Explain clearly the meaning of:
    (*i*) Water carriage system
    (*ii*) Conservancy system
26. Describe the two methods of collection and disposal of sewage from the town and give advantages and disadvantages of each.
27. Compare a separate system and a combined sewerage system.
28. (*a*) Give different types of sewerage systems with suitability of each.
    (*b*) Draw sketches for Man-hole and Inverted System with brief outline of each.

CHAPTER 4

LAYING AND CONSTRUCTION OF SEWERS

Learning Objectives

- Construction of Sewers
- Surface Drains

4.1. CONSTRUCTION OF SEWERS

Before execution of the actual construction the complete design and detailed drawing should be available. The detailed drawing includes : (*a*) sewer alignment plans, (*b*) longitudinal sections of various sewers, (*c*) detailed drawing of all the appurtenances.

4.1.1. Stages of Construction

The various steps to be followed in construction of sewers are:

(*i*) Setting out
(*ii*) Excavation of trenches
(*iii*) Dewatering of trenches
(*iv*) Timbering of trenches
(*v*) Laying of foundation
(*vi*) Checking the gradient
(*vii*) Laying of sewer and jointing
(*viii*) Testing of sewer lines
(*ix*) Backfilling of the trenches.

(*i*) Setting out: Setting out of sewer line is the first step in the construction of sewers. Centre lines of the sewers are marked on the ground by starting from the tail end and proceeding upward. Benchmarks are established at suitable intervals for the entire length of the sewers. The advantage of starting work at the tail end is that the sewers can be put to use in the completed portion without waiting for the completion of the whole sewer line.

(ii) **Excavation of Trenches:** Trench is excavated to correct depth and to proper gradient. This work is usually carried out in open cutting. The excavated material is stacked sufficiently away from the edges of the trench to form spoil bank of ordinary size. The width of trench should be at least 150 mm more than the diameter of sewer for easiness in lowering and adjusting of the sewer pipe. The minimum trench width of 600 mm is necessary for convenient laying and jointing of sewer. At the joints, extra depth and width are cut to accomodate the joints. No trench more than 18 m in advance of the end to the constructed sewer is to be left open at any time.

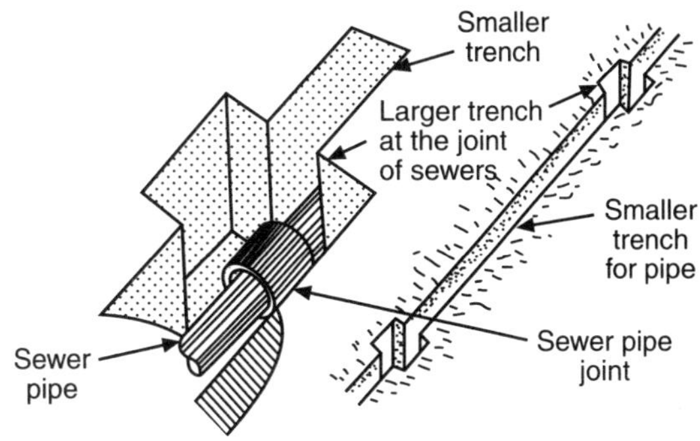

Fig. 4.1. Excavation of trenches

(iii) **Dewatering of Trenches:** Continuous dewatering is needed during trenching operation in water logged areas or in case of deep sewers when the trench goes below groundwater table. Dewatering is done by the following methods:

(a) **Direct Drainage:** Tench water is allowed to collect in a sump in the trench and from where it is pumped out.

(b) **Well Point Drainage:** In case of severe water problem the water table is brought down below the trench level by well point method. In this method pipes with the well strainer are driven vertically at intervals by the side of the trench. Strainer point are taken below the depth of the trench. All vertical pipes are connected to a horizontal pipe which in turn is connected to a pump. So when pump is operated, water from the trench is removed through this well point drainage.

(iv) **Timbering of Trenches:** When the depth of trench exceeds 2 m, the timbering of trench is necessary. The object of timbering is to prevent the sides of trenches from caving in. Type of timbering depends upon the nature of soil. For trenches of limited depth and width, the simple form of timbering with wooden planks and struts are adopted. Steel sheet piling is used in place of timbering in case of badly water logged area and is more water tight, strong and durable. Sheeting or timbering is removed after the sewer has been laid.

Size of Planks and Struts

Poling board – 50 to 75 mm thick timber board.

Whales – 50 mm × 150 mm sizes

Struts – 75 mm × 100 mm sizes at vertical spacing of 1 to 1.25 m

(*v*) Laying of Foundation: Sewers are not directly laid on the soil as unequal settlement of soil may break the pipe. Hence bottom of the trench is prepared to lay the pipe over it. Depending upon the nature of soil and size of sewer pipes, the type of bedding is determined. Concrete bedding is generally provided for poor soil. Fig. 4.3. shows different types of bedding.

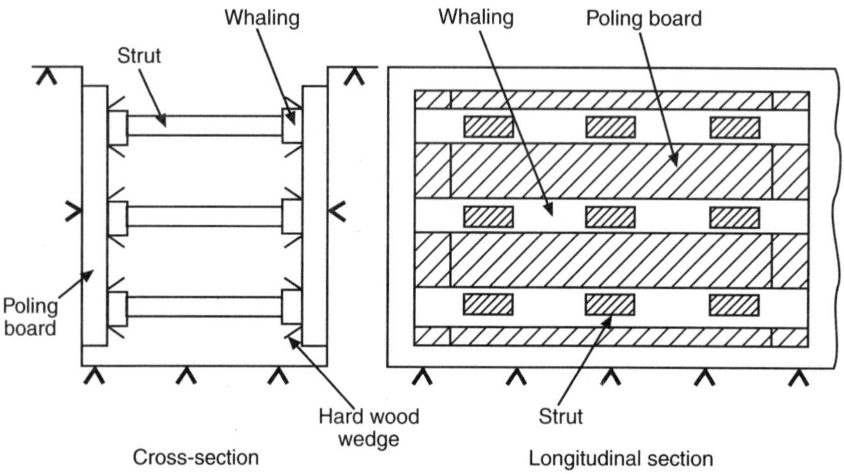

Fig. 4.2. Timbering of trenches

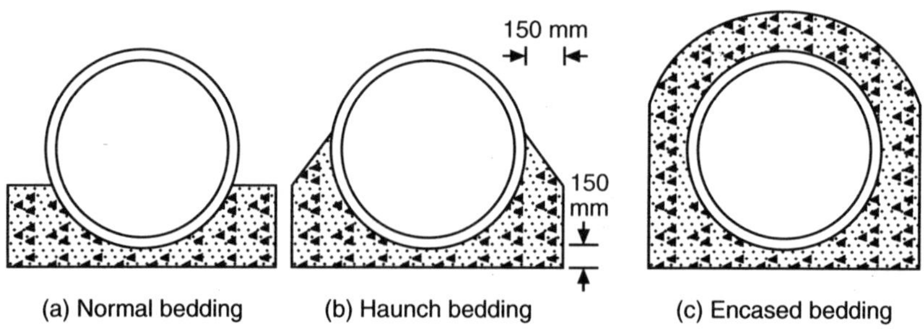

Fig. 4.3. Types of bedding

Normal bedding is used in firm compacted soil. Haunch bedding is used for loose and poor soil. Encased bedding is used under heavy external load.

(*vi*) Checking the gradient of sewer line with boning rod: After preparing bedding for sewers, sewers are laid at correct alignment to ensure gravity flow. Boning rods or travellers and sight rails are used for checking the level of sewer inverts. The boning rod and sight rail are kept a equal height from the invert level of the sewer. At the construction site, at least 3 to 4 sight rails are always maintained in correct level and gradient along the sewer line.

After fixing the sight rails in the trench, pegs at interval of 2.5 m to 3 m are driven in such a way that their top is in level with the invert line of sewer as shown in Fig. 4.4.

In order to maintain the same level, uprights of sight rails are adjusted and a string line 1 – 1 is stretched across the centre marks on sight rails and by moving traveller with its cross head touching the string line. Uprights are adjusted as mentioned below :

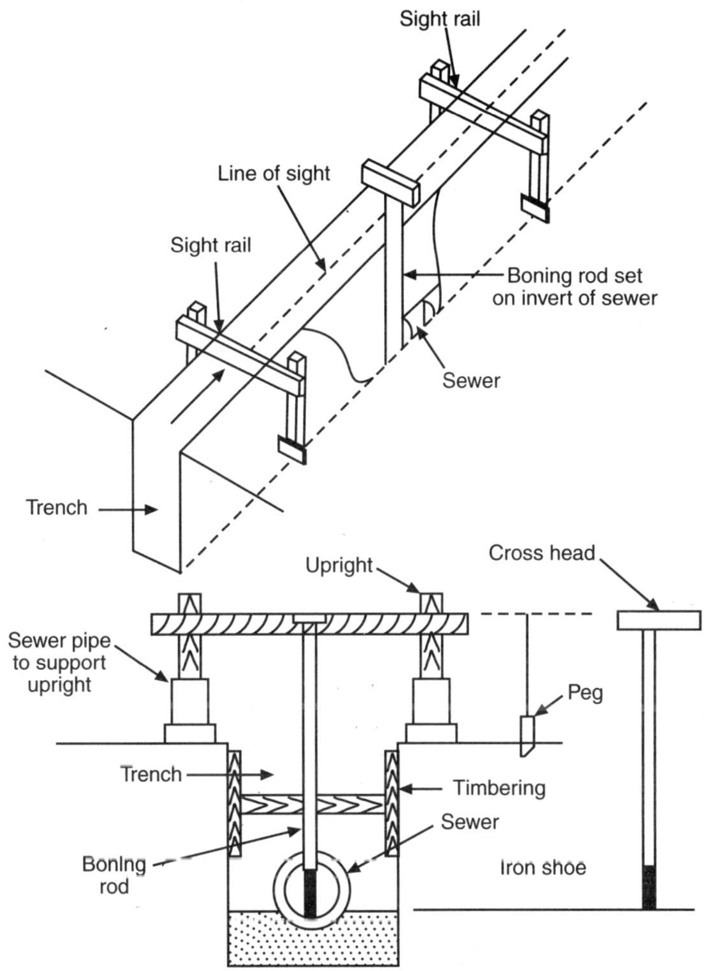

Fig. 4.4. Checking level of sewer invert with boning rod

Let the distance between sight rail (A) and (B) be 20 m and the gradient of the sewer is 1 in 50. The staff reading on sight rail (A) is 0.960 m. Then reading of line of collimation with staff held on sight rail at (B) can be calculated as follows:

$$\text{Fall} = \frac{20}{50} = 0.400 \text{ m}$$

$\therefore$ Reading at sight rail B = 0.960 – 0.400 = 0.560 m

This means that the top level of right rail (B) should be fixed at a height of 0.560 m below the line of collimation. By this method the centre line of sewer is transfered to the bottom of excavated trench.

While laying the pipe, the socket of pipe should face the direction of flow and spigot of another pipe is inserted in the socket of first pipe. The joints of pipes are then sealed.

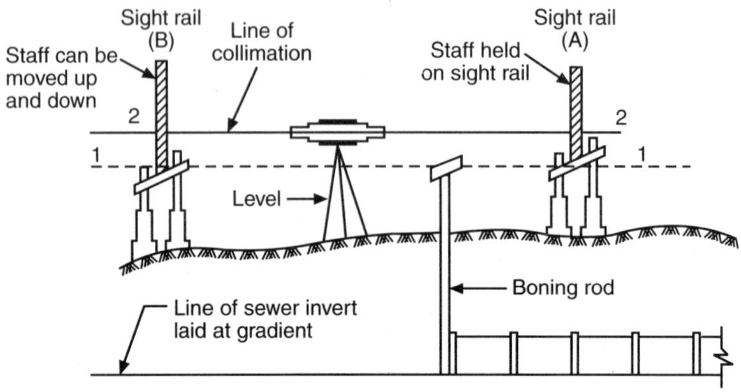

Fig. 4.5. Maintaining same and correct level during laying of sewers.

(*vii*) Laying and Jointing of Sewers

Laying: After the bedding is prepared pipes are laid. Pipes to be laid should be free from any defects and cracks. Pipes are usually laid with their spigot end facing downgrade and are inserted in the sockets. Pipes when laid and pressed in position are tested for level by passing a straight edge through it and seeing that the edge rests squarely on the invert of the pipe throughout its length. Any departure would mean that either the pipe is laid too low or too high.

Jointing: Already discussed in article 3.6.2.

(*viii*) Testing of Sewer Lines: Before commissioning each section of sewer line between two consecutive manholes are normally subjected to the following tests:

1. Water test
2. Air test
3. Tests for straightness and obstruction

1. Water Test: The test is carried out by plugging the sewer opening in the lower manhole. Plugging is done by a rubber bag equipped with a canvas cover. The rubber bag is connected to an air blower. The bag is inflated by blowing air till it tightly fits into the pipe. The other end of the sewer in the upper manhole is plugged with a connection to a nose, ending in a funnel. The sewer is filled with water through the funnel. Water level in the funnel is allowed to rise to 2 metres above the invert of the upper end of sewer. The quantity of water loss is noted after 30 minutes. The water loss should not exceed 30 ml for large size sewers and 15 ml for smaller sewers per cm diameter of pipe per 100 m length. Any section of sewer which does not meet this test attributes to leakage at pipe joints or in the defective length of pipes which should be repaired.

2. Air Test: This test is carried out in large diameter sewers where sufficient quantity of water is not available for testing. It is done by subjecting the sewer section to an air pressure of 100 mm of water by means of hand pump after plugging both the ends. The joints shall be assumed water tight, if the pressure is maintained at 75 mm. If the drop is more than 25 mm of water, the leaking joints are traced out and repaired.

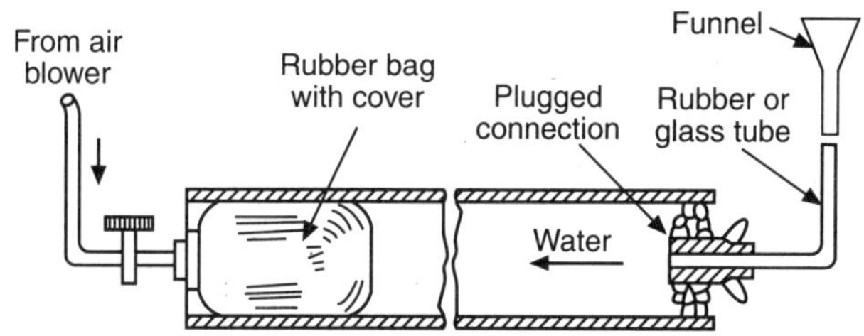

Fig. 4.6. Water test

3. Test for Straightness and Obstruction: A smooth ball of diameter 13 mm less than the internal diameter of sewer is inserted at the upper end of the sewer. If no obstruction is there the ball will come out of the sewer at the lower end.

(*ix*) Backfilling of Trenches: After construction and testing of sewer line, the trenches are required to be refilled. The excavated soil of the trench is used for backfilling after removing the pebbles, stone pieces and lumps. The backfilling is done in layer. Each layer is watered and well rammed. The trench is filled 150 mm above the ground level and is rounded. This is to take care settlement of soil.

4.2. SURFACE DRAINS

These are open drains used for conveying water from kitchens, bathrooms, washing places and rainwater from houses and streets etc. These drains carry away sullage and rainwater upto natural water-courses or discharge it in public sewer.

Surface drains are branch drains in the side streets, while main drainage lines to the city are served by sewers running underground. These drains are mostly laid along the either side of the streets and along the boundary walls of buildings. Surface drains, if properly maintained can give good service.

Surface drains are adopted under following conditions:

(*i*) When the area is not served with sewerage system, i.e., individual houses have their own septic tank.

(*ii*) When the combined sewer is not existing in the area.

(*iii*) When collection and disposal of sullage water is essential although it contains very little organic matter.

(*iv*) When fund available permits only a separate sewer for collecting objectionable waste water.

4.2.1. Requirements of Surface Drains

They should fulfil following requirements:

(*i*) They should develop self-cleaning velocities.
(*ii*) They should be cheap in construction and mainenance.
(*iii*) They should have good free board at the top.
(*iv*) They should be easily cleaned.
(*v*) Their joints should have smooth finish.
(*vi*) Their curves should be easy.
(*vii*) Their inner surfaces should be smooth.

4.2.2. Sections of Drains

In practice the commonly used sections are:

(*a*) Rectangular (*b*) Semi-circular
(*c*) V-shaped or peg-top (*d*) U-shaped

(*a*) Rectangular Section: This type of section is used for large discharges because for small discharge self cleaning velcotity will not develop. This will result in settlement of suspended particles. For small discharges, rectangular drains may have semi-circular inverts (U-section). Refer Fig. 4.7 for construction details. Due to difficulty in cleaning, this section is not usually used.

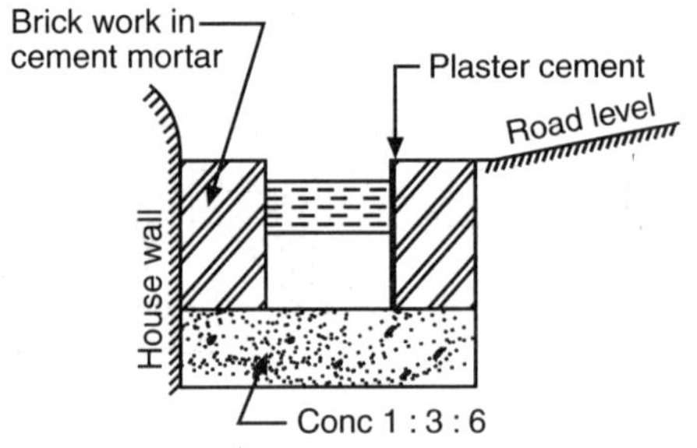

Fig. 4.7. Rectangular section

(*b*) Semi-Circular Section: Drains are running full during monsoons but in dry months the discharge is quite small. Efficient working requires that at least for 3 or 4 hours in a day the drain should run full, so that deposits are scoured off and cleared.

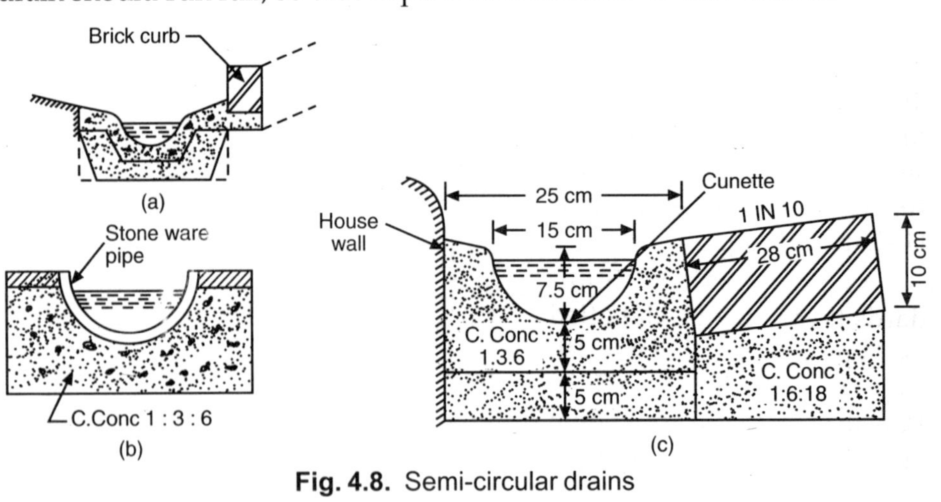

Fig. 4.8. Semi-circular drains

Semi-circular drains are easily made of concrete or bricks. These sections are commonly used for small drains due to its stability, easiness in laying and cheapness. These drains are not suitable for large discharges because in such cases they will occupy more space and if the drains are made deep, there are always chances of some casulaties. The inner surface of the drain is made smooth usually by plastering. Fig. 4.8 shows semi-circular drains.

(c) V-shaped or Peg-Top Section: V-sections with semi-circular inverts are found most suitable. They are also called per-top sections. This section gives self-cleaning velocity even for small discharge because more depth is available in each case. But cleaning is difficult.

(d) U-Section: For small discharge, rectangular drains when provide with semi-circular inverts become U-sections. It may be used at such places where discharge is more and cannot be taken by semi-circular drains. Fig. 4.10 shows U-shaped drain. In sandy areas during high winds, these sections are filled with sand, causing clogging of the drains.

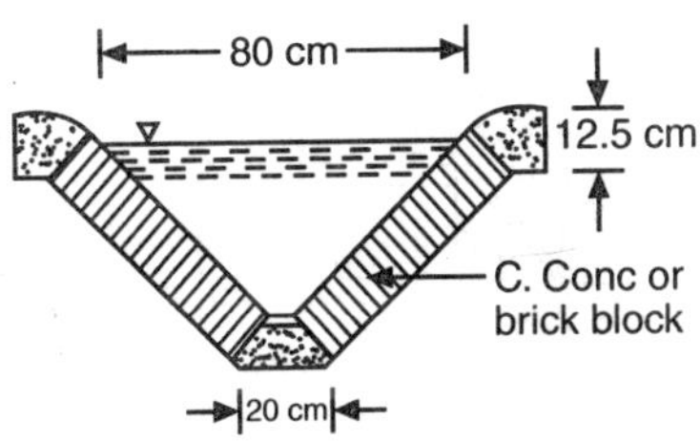

Fig. 4.9. V-section drain

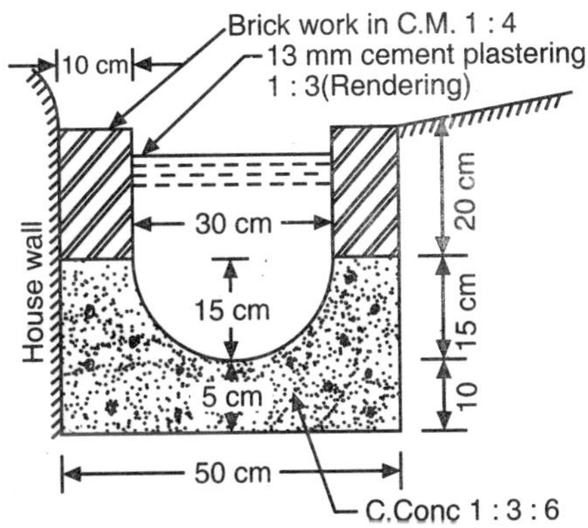

Fig. 4.10. U-section or U-shaped drain

4.2.3. Classification of Surface Drains

Surface drains have been classified into numerous type depending upon that cross-sections and discharge carrying capacities. As discussed earlier they may be classified according to shape of the drains sections *i.e.*, rectangular, semicircular, V-shaped or U-shaped drains.

As per Northern Standards these may be classified as Type I, Type II, Type III, Type IV ... and Type XII. In Punjab state, these are known as *Punjabi Standard Type*

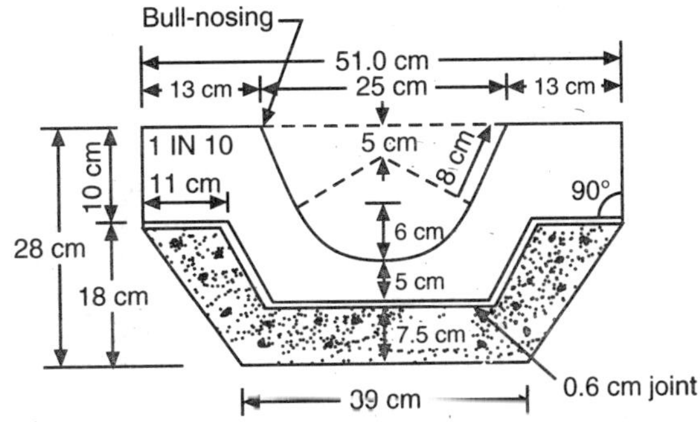

Fig. 4.11.

Drains. Out of these types, Type I to Type VI are commonly used in engineering practice. Other types *i.e.*, type VII to XII are not much favoured due to heavy cost of construction, labour maintenance and difficulties of cleaning. Refer Fig. 4.9, Fig. 4.11 to Fig. 4.15 for typical cross-section of type drains I to VI, and these drains are made of concrete.

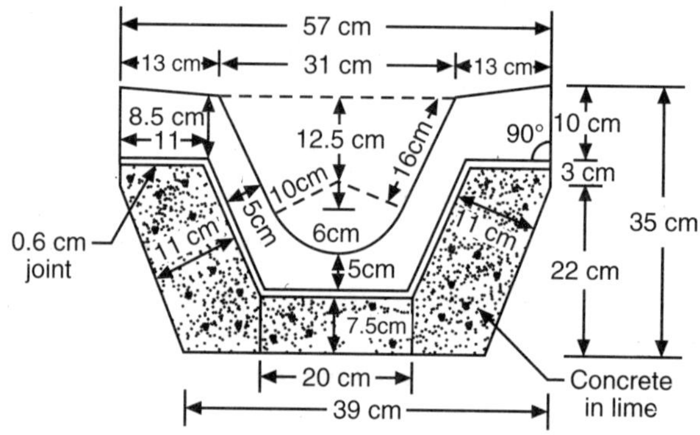

Fig. 4.12.

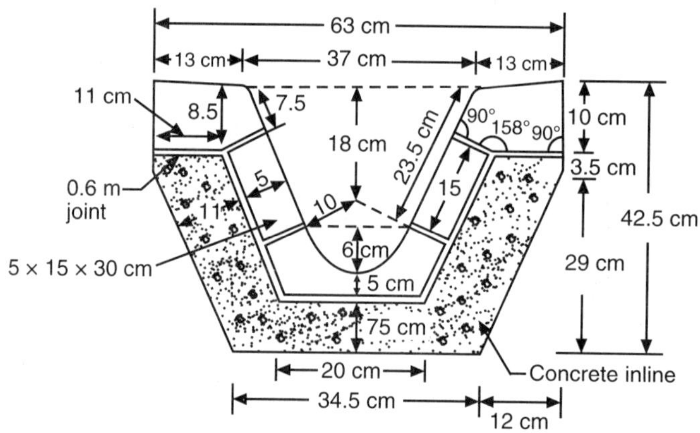

Fig. 4.13.

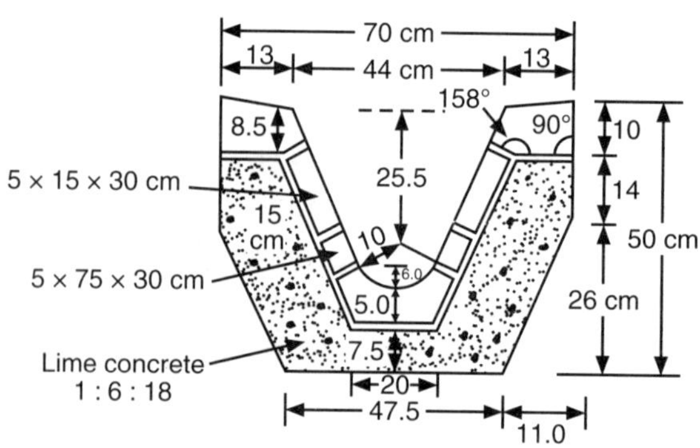

Fig. 4.14. All dimention in cm

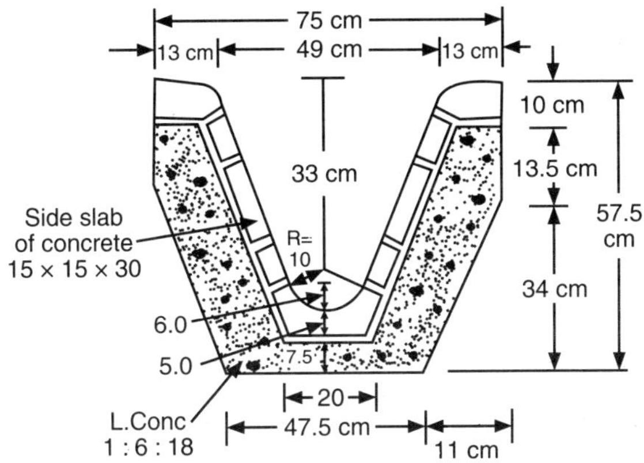

Fig. 4.15. Typical cross-section of drains

4.2.4. Velocity and Gradients of Drains

Discharge carrying capacities of type Drains I to VI are found by using Crimp and Brugs formula. Gradients have been calculated on the basis of self-cleaning velocities i.e., 0.6 m/sec to 1.0 m/sec Table 4.1. gives velocities and gradients of cement concrete type drains I to VI.

Table 4.1. Hydraulic data of common drain sections

Type of drain	Area of flow (m^2)	Hydraulic mean depth in mm	Gradients 1 in		
			For V = 0.6 m/sec	For V = 0.75 m/sec	For V = 1 m/sec
I	0.009	36	250	150	100
II	0.025	62	450	300	200
III	0.047	80	650	400	300
IV	0.070	95	800	550	350
V	0.095	113	–	650	450
VI	0.140	130	–	800	550

4.2.5. Laying and Construction of Surface Drains

In the construction of these drains, alignment of the drains is very important. As far as possible it should be straight and not away. Moreover at every change of direction, the bend must be made as smooth as possible.

Grade line is established on the prepared longitudinal section of the ground. Pegs at 10 to 15 m intervals are fixed. Trenches excavated to the required width, depth, level and grade. To check that the drain is being laid true to line, mason's ball-line is generally used. After laying foundation concrete bed, brick on edge (1 : 5) is constructed on either side.

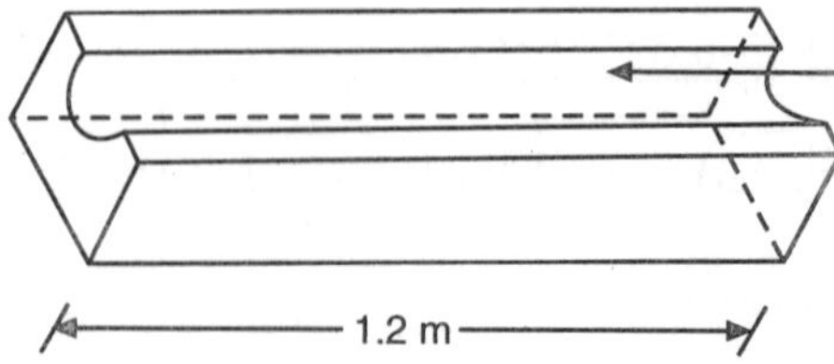

Fig. 4.16. Templates and wooden form work for a drain

The Northern Standard type drains are made up of cement concrete 1:2:4. Here the inverts are laid in situ into 1.2 m lengths. These are separated by 3 mm wide vertical expansion joints. These are formed by accurately shaped metal templates, with the wooden form work (Fig. 4.16) conforming to the cunette of the given drain section.

MODEL QUESTIONS

1. What is the difference between a surface drain and sewer?
2. What is the purpose of providing surface drain?
3. (*a*) Give the general classification of surface drains.
   (*b*) Sketch a class I, II and III type drains and describe the procedure for laying it out.
4. Distinguish between surface drains and underground drains. Why surface drains are neeeded? What are the factors which effect the alignment and section of surface drains?
5. What are the requirements of an efficient system of surface drains?
6. How construction of surface drains is done? Draw neat diagram of template and form work for Type-I shaped surface drains.
7. Draw the X-section of U-shaped surface drain.
8. Write short notes on "Surface Drains."
9. (*a*) (*i*) What is the minimum diameter for a sewer?
   (*ii*) What is the maximum diameter for a stone ware pipe?
   (*iii*) Under what conditions will you use. CI pipe as sewer.
   (*b*) Write short notes on:
   (*i*) Smoke Test
   (*ii*) Water Test
   (*iii*) Considerations for Design of Sewers
   (*iv*) Collar Joint.
10. (*a*) Describe the method of laying and jointing Sewers Lines.
    (*b*) Give the mertis and demerits of differen shapes of sewers in common use.
11. Under what circumstances the following types of sewers as suitable:
    (*i*) Cast Iron Pipe.
    (*ii*) Brick Masonry Egg-shaped Sewers.
    (*iii*) RCC Pipes.
12. During the construction of sewers how the centre line is marked on the ground and excavation of the trenches is done?

13. How the sewer lines are tested? What precautions are taken during the back filling of trenches.
14. Why masonry sewers are necessary. Give the benefit of Egg shape over ciruclar shape sewer.
15. Sketch and describe a large size brick sewer to be laid in deep trenches in a city area. What steps during construction, do you consider necessary to ensure the safety of persons from any untoward accidents.
16. (*a*) What are the various materials used for sewer construction?
   (*b*) Draw a neat sketch of 600 mm. × 900 mm. egg shaped brick sewer.
17. Draw a neat sketch of a standard egg shaped sewer. Under what circumstances is it more advantageous compared to a circular one?
18. Write short note on construction of Egg-shaped sewer and its utility.
19. Give the sketch of any four types of shapes of sewer sections and compare their advantages and disadvantages.
20. (*a*) Describe the method of laying and jointing Sewer Lines.
   (*b*) Give the merits and demerits of different shapes of sewers in common use.
21. Describe the laying of Pipe Sewer by constructing trenches.
22. Write briefly about Smoke Test and Hydraullic Test.
23. During the construction of sewers how the centre line is marked on the ground and excavation of the trenches is done ?
24. (*a*) Sketch the various types of shapes of Sewers. Point out their suitability under different conditions.
   (*b*) List various steps and describe the procedure in detail for construction of a masonry sewer underground.
25. Explain briefly how the invert of a Sewer is set out with sight rails and Boning rods.
26. Compare the advantages and disadvantages of following types of sewers
   (*i*) Stone Ware Pipes
   (*ii*) Asbestos-Cement Pipes,
   (*iii*) Concrete Pipes.
27. Discuss various factors affecting the design of Sanitary Sewers.
28. Describe the various operations involved in laying the sewers at correct gradient and alignment.
29. Explain with sketches setting out alignment of a sewer and checking the gradient with sight rails and boning rod.
30. Describe various tests for testing the sewers.
31. Write short notes on:
   (*i*) Collar Joint
   (*ii*) Setting out Alignment of sewers
   (*iii*) Timbering of Trenches
   (*iv*) Requirements of surface drains.

CHAPTER

5

QUALITY OF SEWAGE

Learning Objectives

- General
- Characteristics of Sewage
- Sampling of Sewage
- Types of Sewage Analysis

5.1. GENERAL

Although sewage is mainly composed of water about 99.9 per cent yet it contains 0.1 per cent of mixture of substances either in solution or in suspension that always go on changing and producing foul smell and creating nuisance to the general health of the public. It is, therefore, imperative to study the various properties and behaviour of sewage.

5.2. CHARACTERISTICS OF SEWAGE

Characteristics of sewage can be classified as:

(*i*) Physical characteristics

(*ii*) Chemical characteristics

(*iii*) Biological characteristics

5.2.1. Physical Characteristics

Sewage has the following physical characteristics:

(*a*) Colour and Odour: Fresh domestic sewage has a soap solution like colour. In 3 to 4 hours decomposition starts and becomes stale with all the oxygen present in the sewage exhausted. When there is no oxygen, the sweage becomes septic and the colour become dark.

Fresh sewage has no odour, but the stale sewage has offensive odour of hydrogen sulphide and other sulphur compounds.

(*b*) **Temperature:** The temperature of sewage is slightly higher than water supply. When sewage flows in closed conduits its temperature further rises, which increases the viscosity and bacterial activity.

(*c*) **Turbidity:** The turbidity of sewage depends directly on the quantity of solid matter present in it in suspension state.

(*d*) **Solids:** Sewage contains more than 99.9% of water and only 0.1% of solids. The solids may be suspended, dissolved or in colloidal state.

5.2.2. Chemical Characteristics of Sewage

Sewage contains inorganic chemical compounds from water supply and many complex organic matters derived from urine, faces and other organic wastes. Fresh sewage is normally alkaline but tends to acidic as it becomes stale. Organic compounds contain both nitrogenous and non-nitrogeneous compounds. Urea, proteins, amino acids are nitrogenous compounds, whereas fats, soaps and carbohydrates are non-nitrogenous or carbonaceous compounds. Inorganic matters present in the sewage are salts and alkalies, which comes from bathrooms, kitchens and industrial plants.

5.2.3. Biological Characteristics

These are due to presence of micro-organisms in the sewage. The following microorganisms are generally present in sewage:

(*a*) Bacteria

(*b*) Algae

(*c*) Fungi

(*d*) Protozoa

Bacteria are the simplest form of plant life. These are single cell organism and are capable of reproduction. In biological treatment the presence of bacteria is very essential. About 22-25 million bacteria may be present in 1 litre of sewage. Most of these bacteria are harmless to men and are engaged in the process of converting the complex organic compounds into simpler and stable organic compound resulting in the purification of sewage. Certain bacteria which are harmful are called 'Pathogenic bacteria'. When the sewage containing these bacteria is discharged in water, they become a source of health hazard.

Algae are unicellular or multicellular autotrophic photosynthetic plants. Algae convert the inorganic materials in the water into organic matter through the process of photosynthesis.

Fungi are multi-cellular, non-photosynthetic plants. The yeasts, bacteria and molds come under the group of fungi. They have the property of decomposing starches, sugars, fats etc., and hence its presence is very important in the biological treatment.

Protozoa are unicellular animals which are reproduced by binary fission and live mainly by eating bacteria. These microscopic animals have complex digestive system which use solid organic matter as an energy and carbon source. These are generally found in oxidation pond, trickling filters and in activated sludge.

5.3. SAMPLING OF SEWAGE

Quality of sewage is never the same throughout the day. It has been noted that the composition of the sewage that reaches the disposal works in the morning is not the same as that of the sewage reaching in the afternoon or at night. Moreover in the same tank, quality of sewage varies at different depths. At the top sewage may contain floating matter whereas at the bottom there may be more sludge.

Samples of sewage are collected over a period of 24 hours at regular intervals say after an hour. The volume of the samples must be 1 to 2 litres. These may be kept in a sterilised bottle. Sample must be taken little below the top surface where due to turbulence, sewage is mixed with impurities.

The sample must bear the date and hour of collection. In any case, the sample must be kept in a very cool place until used for testing. In many cases preservatives like sulphuric acid which do not affect the contents of sewage are used. In the treatment works, number of sampling points at different stages of treatment are provided to study the perfrormance of the unit.

5.4. TYPES OF SEWAGE ANALYSIS

Analysis of sewage are classified into the following three main catagories:

1. Physical analysis
2. Chemical analysis
3. Biological or bacteriological analysis

5.4.1. Physical Analysis

This includes the following tests:

(*a*) **Temperature:** This is an imporant parameter and is measured by using ordinary thermometers. Temperature is recorded upto an accuracy of 0.1°C. If the temperature of sewage is more the biological activities are more, whereas at low temperature these are dormant.

(*b*) **Colour and Odour:** Colour and odour gives the conditions of sewage as fresh, stale or septic. Fresh sewage is light brown and odourless. Whereas septic sewage is dark in colour and gives foul smell. This test is carried out in same way as discussed in Chapter 4 Part-I of this book.

(*c*) **Turbidity:** Turbidity is caused to the sewage due to suspended and colloidal matter. This is determined by using Jackson's turbiditi-meter or turbidity rod as mentioned in Chapter 4 of water supply part of this book.

5.4.2. Chemical Analysis

The following tests are carried out in chemical analysis of sewage.

(*a*) Solids-total, volatile, suspended, settleable and dissolved

(*b*) Dissolved oxygen (D.O.)

(*c*) Bio-chemical oxygen demand (B.O.D.)

(*d*) Chemical oxygen demand (C.O.D.)

(*e*) Relative Stability (R.S.)

(*f*) Nitrogen

(*g*) Chlorides and Sulphides

(*h*) Chlorine demand

(*i*) pH-value

(*j*) Oil and grease

(*a*) Test for Solids: The test for solids helps as a measure of concentration and physical condition of sewage. Also by doing this test some knowledge of organic and inorganic matter present in the sewage is obtained.

Total solids present in the sewage are determined by evaporating a measured volume of sewage at 100°C and weighing the residue.

Volatile solids can be determined by noting the loss in the weight of the total solids on ignition in an electric muffle furnace.

Settleable solids are suspended solids which will settle within one hour to the bottom of the cylinder of specified height. It is determined by using an imhoff cone as shown in Fig. 5.1. Imhoff cone is conical glass of one litre volume and is graduated in militires at the bottom. The cone is filled with the sewage and is allowed is settle for one hour. After one hour the volume of solid settled at the bottom is directly measured. This gives the volume of settleable solids in the sewage.

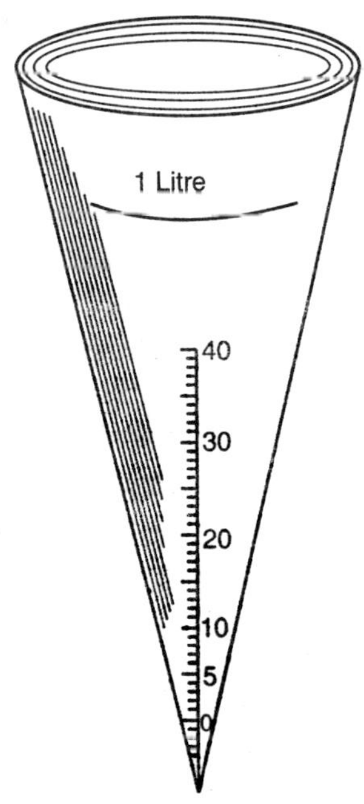

Fig. 5.1. Imhoff cone

(*b*) Test of Dissolved Oxygen (D.O.): Dissolved oxygen is the amount of oxygen dissolved in the sewage. This is very important for the precipitating and dissolution of inorganic matter in water. The presence of dissolved oxygen in a sewage indicates that the sewage is fresh or weak. The solubility of oxygen in sewage depends upon its temperature.

DO is determined by using Winklet's method, an oxidation-reduction process carried out chemically to liberate iodine in amount equivalent to the quantity of DO originally present. In this method, sample is collected in BOD bottle using DO sampler. To this is added 2 ml Manganese suplhate ($MnSO_4$) followed by Alkali-iodide-azide reagent ($NaOH + KI + NaN_3$). After adding the reagent the bottle is closed with stopper. While adding these reagents the tip of the pipet should be below the liquid level. Mix well by inverting the bottle 2-3 times and allow the precipitate to settle down. To this add 2 ml concentrated sulphuric acid (H_2SO_4). Mix well till the precipitate is dissolved. Take 203 ml in a conical flask and titrate against sodium thiosulphate ($Na_3S_2O_3$) using starch as indicator. The volume of sodium thiosulphate required for titration will give the amount of DO in mg/l.

(*c*) Test for Bio-chemical Oxygen Demand (BOD): BOD of a sewage is the quantity of oxygen required for the biochemical oxidation of the decomposible matter at specified temperature within the specified time. In India the standard time and temperature are 5 days and 37°C. This is the most important test in sewage analysis as it indicates the amount of decomposible organic matter in the sewage. The more the organic matter, greater is the BOD and more is the strength of sewage.

The aerobic decomposition of organic matter proceeds in two stages. In the first stage the carbonaceous matters are oxidised and the oxidation of nitrogenous matters takes place in the second stage. The quantity of oxygen taken up for decomposition is rapid in the first stage. The BOD of sewage for the first stage is about 90% of total BOD

For this test, sewage sample is diluted in water in the ratio of 1 : 100 (i.e., 1 part of sewage and 99 parts of water). The water used for dilution should contain excess amount of oxygen which is determined before dilution. The sample is then kept in incubator at temperature of 37°C (for India and hot countries and at 20°C for other countries). After 5 days the quantity of dissolved oxygen in the sample is determined. The difference between the quantity of oxygen present in water in the begining and at the end of incubation is the amount of oxygen consumed by the sewage, i.e., the BOD of sewage.

Table 5.1. Relation between strength of sewage and BOD

5 day-BOD in ppm or mg/*l*	Strength of sewage
(*i*) 5 to 10	Very good filter effluent
(*ii*) 20	Standard filter effluent
(*iii*) 200	Weak sewage
(*iv*) 300	Average sewage
(*v*) 400 to 500	Strong sewage

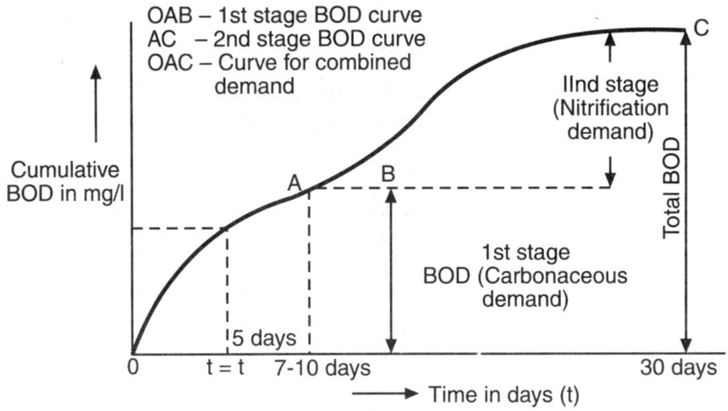

Fig. 5.2. BOD curve

Limitations of BOD Test

(*a*) It measures only the oxygen required for biological oxidizable matter.

(*b*) It takes a long time of 5 days to complete the test.

Mathematical expression of BOD

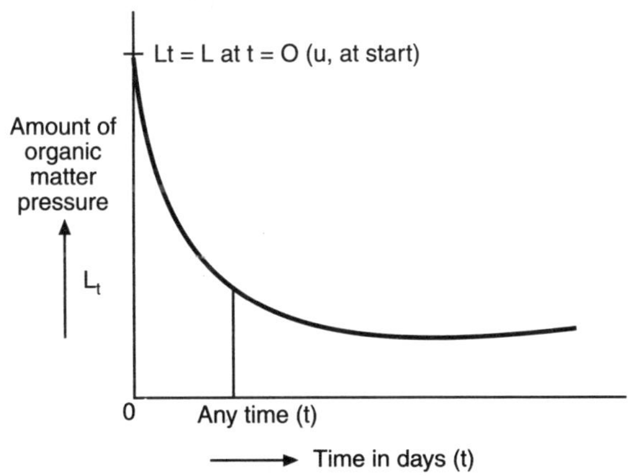

Fig. 5.3. BOD curve-first order

Rate of Deoxygenation: The rate at which BOD is satisfied at any time depends on temperature and also on the amount and nature of organic matter present in the sewage at that time.

At a certain temperature, the rate of deoxygenation is assumed to be directly proportioned to the amount of organic matter present in sewage mathematically.

$$\frac{dL_t}{dt} = -K \cdot L \qquad ...(i)$$

$$\left[\frac{dL_t}{dt} \propto L \text{ or } \frac{dL_t}{dt} = K.L(-)\right]$$

where L_t = Oxygen equivalent of carbonaceous oxidisable organic matter present at any time 't' in mg/1 [minus sign indicates that with passage of time the value of Lt decreases]

t = time in days

K = Rate constant → rate of oxidation of organic matter and it depends upon the nature of organic matter and temperature unit/per day

Integreting eqn. (i)

$$\int \frac{dL_t}{L_t} = -\int K.dt$$

$$\log L_t = -K \cdot t + C \qquad ...(ii)$$

C = Constant of integration and can be evaluated from the boundary conditions at the start

L = oxygen equivalent of organic matter (or start) present in the sewage and in a fixed quantity

When $t = 0$, and at start $L_t = L$

Substituting in eqn. (ii) we have

$$\log_e L = -K(0) + C$$

or $C = \log_e L$ substituting this value of C in (ii)

$$\log_e L_t = -K \cdot t + \text{Log}_e L \text{ or } \text{Log}_e L_t - \text{Log}_e L = -K \cdot t$$

or
$$\log \frac{L_t}{L} = -K \cdot t$$

or
$$2.3 \log_{10} \frac{L_t}{L} = -K \cdot t$$

or
$$\log_{10} \frac{L_t}{L} = \frac{-K \cdot t}{2.3} = 0.434 K \cdot t$$

Using $0.434 K = K_D$ or K_1

where K_D or K_1 = Deoxygeneration constant at the given temp. = 0.434 K

We have

$$\log_{10} \frac{L_t}{L} = -K_1 \cdot t \text{ or } \frac{L_t}{L} = (10)^{-K_1 \cdot t} \qquad ...(iii)$$

Now L is the organic matter at the start, expressed as oxygen equivalent

L_t = Organic matter left after t days

∴ Quality of organic matter oxidised = $L - L_t$

Let Y_t = total amount of organic matter oxidised in t days

= BOD in t days

$$y_t = L - L_t = L\left(1 - \frac{L_t}{L}\right) \text{ or } \frac{Y_t}{L} = 1 - \frac{L_t}{L}$$

$$\frac{L_t}{L} = 1 - \frac{L_t}{L}$$

Substituting this value of L_t in eqn. (*iii*) we have

$$1 - \frac{Y_t}{L} = (10)^{-K_1 \cdot t} \text{ or } \frac{Y_t}{L} = 1 - (10)^{-K_1 \cdot t}$$

or $$Y_t = L[1 - (10)^{-K_1 \cdot t}]$$

The ultimate 1st stage BOD (Y_u) can be obtained when $t = \infty$ days

$$y_t = L[1 - (10)^{-K_1 \cdot \infty}]$$

$$= L\left[1 - \frac{1}{(10)^{\infty}}\right]$$

$$= L[1 - 0] = L \text{ or } Y_t = L$$

Hence Y_u of a given sewage is equal to initial oxygen equivalent of the org. matter (L) present in this sewage. Finally

$$K_{1(T°)} = K_{1(20°C)}\,[1.047]^{T - 20°}$$

The value of $K_{1(20)}$ varies between 0.1 to 0.15

EXAMPLE 5.1. 7 days 25°C BOD of sewage is 250 mg/l. The BOD rate constant is 0.23/day (Base *e*). Calculate 5 days 20°C BOD of the sewage.

Solution: Given BOD, = 250 mg/l (at 25°C)

BOD rate constant = 0.23/day (base *e*) or $K_1 = 0.1$

To find BOD_5 at 20°C

Using $$Y_t(25°) = L[1 - (10)^{-K_1 \cdot t}]$$

$$Y_t = 250 \text{ mg/l (given) } t = 7 \text{ days}$$

$$K_{1(20°)} = 0.1$$

$$250 = L[1 - (10)^{-1258 \times 7}]$$

or $$L\left[1 - \frac{1}{10^{0.78}}\right] = 250$$

or $$L = \frac{250}{0.88} = 283.86 \text{ mg}/l$$

$$K_{t\,(25°)} = K_1\,(20°)\,[1.047]^{T - 20°}$$

$$K_{1\,(25°)} = 0.1\,[1.047]^{25 - 20} = 0.1258$$

$$K_{1\,(20°)} = 0.1$$

$$Y_{t\,(20°)} = L[1 - (10)^{-K_1 \cdot t}]$$

$$= 283.86[1-(10)^{-0.1\times5}] = 283.86\left[1-\frac{1}{10^{+0.5}}\right]$$

$$= 194.1 \text{ mg}/l$$

EXAMPLE 5.2. If 5 days 20°C BOD of a sewage sample is 150 mg/*l*. What will be its 8 days 15°C BOD? Assume deoxygenation constant at 20°C as 0.23 (base *e*).

Solution: Given BOD_5 = 150 mg/l (at 20°C) K = 0.23 base *e* at 20°C

To find BOD_8 at 15°C $K_1 = 0.1$

$$Y_{t\,(20°C)} = L[1-(10)^{-k_1 t}]$$

$$150 = L[1-(10)^{-0.1\times5}]$$

$$= L\left[1-\frac{1}{10^{0.5}}\right], \text{ or } L = \frac{150}{0.684} = 219.37 \text{ mg}/l$$

$$K_{t\,15°} = K_{1\,(20°)}\,[1.047]^{T-20}$$

$$= 0.1\,[1.047]^{15-20} = 0.0795$$

$$\therefore \quad Y_{t\,5°} = 219.37\,[1-(10)^{-0.0795\times8}]$$

$$= 219.37\left[1-\frac{1}{10^{0.64}}\right] = 168.63 \text{ mg}/l \quad \textbf{Ans.}$$

EXAMPLE 5.3. The BOD of a sewage sample incubated for 5 days at 30°C is 105 mg/l. Calculate the BOD at 20°C.

Assume $K_{1\,(20°)} = 0.1$

Solution: Given Y = 105 mg/l

Using equation $K_{1\,(T°)} = K_{1\,(20)} \times [1.047^{(T-20)}]$

$$K_{1\,(30)} = 0.1 \times [1.047^{(30-20)}]$$

$$K_{1\,(30)} = 0.158$$

Now using BOD equation

$$Y = L_0[1-(10)^{-K_1 \cdot t}]$$

$$105 = L_{30}\,(1-10^{-158\times5}) \text{ Here } t = 5 \text{ days}$$

or L_{30} = 58.66 mg/l

Now usir g $L_T = L_{(20)}\,(0.02\,T + 0.6)$

$$L_{20} = 58.66 \times (0.02 \times 30 + 0.6) = 70.39 \text{ mg/l}$$

then BOD at 20°C, $Y_{20} = 70.39\,(1-10^{-0.1\times5}) = 48.13$ mg/l

EXAMPLE 5.4. Calculate the BOD after 2 days at 35°C of a sewage sample whose 5 days 20°C BOD is 120 mg/l.

Assume C = 0.1 at 20°C.

Solution:

$$K_{(37)} = 0.1[1.04^{(35-20)}] = 0.2$$

$$Y_{(20)} = 120 - L_{20}(1 - 10^{-0.1 \times 5})$$

or $$L_{20} = 175.5 \text{ mg/l}$$

$$L_{(35)} = 175.5\,[0.02 \times 35 + 0.6] = 228$$

$$Y_{(35)} = 228\,(1 - 10^{-0.1 \times 2}) = 84.2 \text{ mg/l}$$ **Ans.**

(*d*) Test for Chemical Oxygen Demand (COD): COD of a sewage is the quantity of oxygen required for chemical oxidation of organic matter with the help of strong chemical oxidant in an acidic medium. Potassium dichromate ($K_2Cr_2O_7$) has been found suitable as chemical oxidant.

This test is used to measure the content of organic matter of sewage. The COD test is useful to measure the organic matter in some of the industrial wastes which do not respond to BOD test. It takes only 3 hours to complete. The only disadvantage of this test is that it cannot differentiate between biologically oxidizable and biological inert matter.

(*e*) Test for Relative Stability (Methylene Blue Test): Relative stability is defined as ratio of the amount of oxygen available in sewage to the amount of oxygen required to completly satisfy the first stage of BOD. It is thus a measure of oxygen demand except that the values obtained are relative and not absolute. This tesr gives an idea of when aerobic decomposition will end and anaerobic decomposition starts.

Stability is determined by methylene blue test, in which methylene blue indicator is added to the sample incubated at standard temperature of 20°C or 37°C after inserting the stopper. The point at which methylene blue colour disappears or hydrogen sulphide gas is produced, is taken as an end-point. The number of days taken for decolouration is noted. Then relative stability is calculated by using the formula

$$S = 100\,(1 - 0.794\,t_{20})$$

or $$S = 100\,(1 - 0.630\,t_{37})$$

where t_{20} and t_{37} are the number of days of incubation at 20°C and 37°C respectively and S is relative stability in %.

This test is not suitable for raw sewage as the colloidal solids precipitate the colour of the dye, i.e., methylene blue. It is useful to determine the stability of treated sewage effluents.

(*f*) Test for Nitrogen: Proteins, amino acids and urea are the main sources of nitrogenous compounds in sewage. Nitrogen in sewage may be present in different forms such as : (*i*) albiminoid nitrogen, (*ii*) ammonia nitrogen, (*iii*) organic nitrogen, (*iv*) nitrate and nitrite nitrogen, depending upon the condition of sewage. The importance of the test arises from the fact that nitrogen being an essential constituent of the biological protoplasm, its determination indicates state of sewage and the methods required for treating at the treatment works.

(*g*) **Test of Chlorides and Sulphides:** The sources of chlorides in sewage are urine, common salt from kitchens and other industrial wastes, Chloride determination is used as an index of sewage strength. Very high chloride content indicates infiltration of saline water into the sewers.

The sources of sulphides are industrial waste and decomposition of organic and inorganic matters of sewage. The presence of sulphides in sewage gives bad odour.

(*h*) **Test for Chlorine Demand:** Chlorine is used to disinfect the treated sewage effluents to ensure that all the bacteria are killed. For effective disinfection at least 50 mins of contact time is required. This test is done during disinfection of chlorine. The chlorine demand of sewage is the amount of chlorine in mg/litre that must be added to it so that after a definite contact time, the residual chlorine must be sufficient to kill the bacteria. Chlorine demand gives a rough measure of the strength of sewage.

(*i*) **Test for pH-value:** pH value gives the degree of acidity or alkalinity. It is determined for regulating the various operations of treatment works. Fresh sewage is little alkaline, but stale or septic sewage is acidic in nature. This test is useful in calculating coagulant does and disinfectant does, corrosion control etc.

(*j*) **Test for oil and Grease:** The main source of oil and grease are discharge from animal and vegetable matter or wastes from kitchens, hotels and garages etc. These are undesirable for sewage as cause trouble in the flow of sewage by precipitating and clogging the interstices of filter media. These also form objectionable scum in tanks and streams. Oil and grease are soluble in ether, so these are mixed with ether and digested to covert them into gases.

5.4.3. Biological or Bacteriological Analysis

In sewage, living micro-organisms are present in large quantities. Most of these micro-organisms are beneficial in cleaning polluted water and stabilizing the organic matter. Biological tests of sewage are carried out to determine the types of micro-organisms present in it. This test is done with microscope. In this test bacteria are allowed to develop under favourable conditions in culture media, which may be solid or liquid. The sewage is then added in the culture media and kept in incubator at specified temperature and time. The colonies of descendant cells in solid media are formed by bacteria which is seen by microscope. The kind of bacteria is directly seen by the microscope and its size is measured by stage micrometer.

MODEL QUESTIONS

1. (*i*) What is the nature of composition of sewage ?
   (*ii*) Explain the concept of BOD and its importance.
2. Clearly differentiate between :
   (*i*) Bacterial and plant life.
   (*ii*) Aerobic and anaerobic reactions.
3. Write short notes on the following:
   Strenght of sewage, fresh sewage, stale sewage, pathogenic bacteria, non-pathogenic bacteria, anaerobic and aerobic bacteria.
4. Define sewage characteristics. On what factors do these demand?
5. Explain the terms
   (*i*) BOD
   (*ii*) OD and
   (*iii*) CO
6. Write short notes on:
   (*i*) Smapling of Sewage
   (*ii*) Chlorine demand of sewage
   (*iii*) Strong sewage and weak sewage
7. What is general composition of sanitary sewage?
8. Write short notes on:
   (*i*) Aerobic bacteria and anerobic bacteria
   (*ii*) pH value
   (*iii*) Sampling of Sewage
9. What is BOD? How does it siffer from DO and COD.
10. Write a detailed note on the chemical and bacteriological composition of sewage.
11. Write short note on total solids.
12. Explain the concept of BOD and its importance in connection with sewage Treatment. What do you understand by the term "Sewage Sickness"?
13. Give a brief account of general composition of sewage. What is the purpose and principal involved in its treatment and disposal?
14. What is sewage strength? Also explain briefly characteristics of sewage.

❑❑❑

CHAPTER 6

NATURAL METHODS OF SEWAGE DISPOSAL

Learning Objectives

- General
- Objects of Sewage Disposal
- Disposal Methods
- Nuisance Due to Disposal
- Purification of Sewage
- Disposal by Land Treatment
- Broad Irrigation
- Application of Sewage
- Favourable Conditions for Sub-surface Ace Irrigation
- Self Purification of Streams
- Sewage Sickness
- Preventing Measures for Healthy Conditions

6.1. GENERAL

The methods used for disposal of sewage are called disposal methods. The sewage can be disposed of after suitable treatment or without treatment. Sewage is finally dis-posed off either on land or in large bodies of water like sea, river or lake.

6.2. OBJECTS OF SEWAGE DISPOSAL

- to eliminate or reduce danger to the public health by possible contamination of water supplies.
- to render the sewage inoffensive without causing odour or nuisance.
- to prevent the life of fish or other aquatic life by allowing raw sewage into bodies of water as such.

6.3. DISPOSAL METHODS (FIG. 6.1)

There are two principal methods of sewage disposal by utilizing natural agencies i.e.,

(*i*) Dilution i.e., Disposal of sewage in water.

(*ii*) Land Disposal or Irrigation.

Both methods are very simple. But these may be regulated carefully so that the quantity of sewage put in water or applied to land is such that they are capable of receiving the organic load present in the effluent.

6.3.1. Dilution

Dilution is the disposal of sewage by discharging it into large bodies of water like sea, stream, rivers etc. This method is possible only when the natural water is available in large quantity near the town. Proper care should be taken while discharging sewage in water so that sewage may not pollute natural water and make it unfit for any other purposes like bathing, drinking, irrigation etc.

6.3.1.1. Conditions Favouring Dilution

Following are the conditions which are favourable for dilution :

(*i*) Where sewage is fresh

(*ii*) Where favourable currents exist in a steam.

(*iii*) Where sewage is almost free from floating settleable solids.

(*iv*) Where thorough mixing is possible.

(*v*) Where diluting water has high quantities of dissolved oxygen (DO).

(*vi*) Where the city is situated near river or sea.

The following care should be taken at the site of sewage disposal:

(*i*) It should not cause destruction of aquatic life and contamination of underground water in any way.

(*ii*) The sludge banks should not be formed at the top.

(*iii*) The sewage should not get deposited at the bottom and start giving offensive odour.

(*iv*) The unsightly matter should not float on the surface.

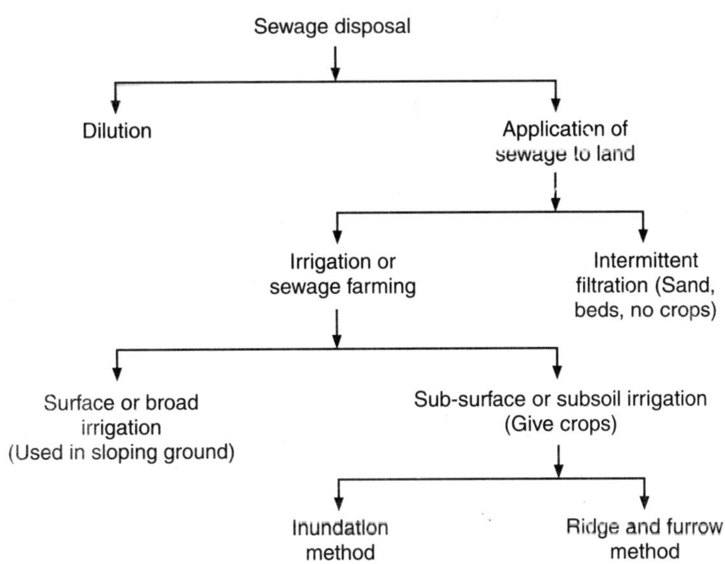

Fig. 6.1. Methods of sewage disposal

> *Note:* The ratio between the amount of diluting water and sewage is called Dilution Factor. Nuisance is experienced when the dilution factor is less than 20.

6.3.1.2. Sources of Sewage Disposal by Dilution

For dilution, sewage can be discharged into the following sources:

(*i*) Lakes

(*ii*) Creeks

(*iii*) Streams and rivers

(*iv*) Groundwater

(*v*) Sea or ocean

(*i*) Lakes: Lakes are sometimes used for the dilution of sewage. The factors which help the process of dilution in lakes are size, shape and nature of area, volume of fresh water flowing into it.

(*ii*) Creeks: Creeks are in the form of an inlet on sea coast. Sewage in creeks are diluted due to large amount of water available. Sewage in creeks should be disposed off very carefully.

(*iii*) Streams and rivers: As there is huge amount of flowing water available in streams and rivers, sewage can be disposed off in rivers and streams. But during summer, when discharge in streams and rivers are less, it is difficult to carryout dilution of sewage. Also temparature during summer is high which results in low solubility of oxygen in river and stream water, therefore sewage should be properly treated before disposal to streams and rivers.

(*iv*) Groundwater: In this method, sewage is disposed on open ground and allowerd to filter through the different layers of soils and finally meets groundwater.

(*v*) Sea or Ocean: This is one of methods of sewage disposal by dilution. Sea or ocean have plenty of water for sewage dilution. The specific gravity of sea water is more than sewage. The sewage discharged into sea water remains at the surface and form a thin film. Due to this reason, following points should be kept in mind while disposing sewage in sea.

(*a*) The sewage should be discharged at a point where the depth of water is more.

(*b*) It should be discharged in such a way that it is carried away from the shore, so that no nuisance occurs to bathing places and recreation centres on the sea shore.

(*c*) Detailed study of ocean currents, velocity and direction of wind should be carried out before dishcarging sewage into sea or ocean.

6.4. NUISANCE DUE TO DISPOSAL

Nuisance from stream disposal can be classified into the following types:

(*i*) Physical nuisance

(*ii*) Chemical nuisance

(*iii*) Bacterial nuisance

(*i*) Physical Nuisance: Physical nuisance is mainly due to the deposit of suspended solids near the bank of stream. These unsightly suspended solids have certain nuisance characteristics. The factors which affect the physical nuisance are volume of water and velocity of flowing water.

(*ii*) Chemical Nuisance: Chemical changes in sewage takes place due to biological oxidation. During oxidation, dissolved oxygen in water is consumed by bacterias and in extreme cases, the stream water without dissolved oxygen may destroy aquatic life.

(*iii*) Bacterial Nuisance: Bacterial nuisance is mainly due to possible presence of pathogenic bacterias in the sewage and thus in the polluted stream. Such harmful bacterias cause various diseases and even death of aquatic life.

6.5. PURIFICATION OF SEWAGE

Purification of sewage by this method is based on the following three actions:

(*i*) Biological action

(*ii*) Chemical action

(*iii*) Physical action

(*i*) Biological action: As soon as sewage is applied on land, the bacterias present in the soil attack on sewage and converted it into plant food which is used by the roots of plants.

While percolating through the soil, the suspended particle of sewage are trapped in soil voids and get oxidised by aerobic process as oxygen is present in different soil layers. Soil gives up carbon dioxide and absorb oxygen from the atmosphere through wind movements, drainage of soil, pressure and temperature difference etc.

(*ii*) Chemical action: The organic matter present in sewage is oxidised by bacterias present in soil and this oxidation takes place by the chemical process.

(*iii*) Physical action: It takes place through the process of filteration. The suspended particles of sewage are caught in the soil voids while percolating. The particles are oxidised by aerobic action if proper aeration of these voids is maintained. Such aeration and aerobic condilions will more likely to prevail in pervious soils like sands than impervious soils.

6.6. DISPOSAL BY LAND TREATMENT

The method in which raw sewage or partly treated sewage is evenly spread on the surface of land is called land treatment. This method is quite suitable when no dilution projects are possible. In this method water of sewage percolates in the ground and suspended solids remain at the surface of the ground. These suspended organic solids are partly oxidised by exposure to atmospheric action of light, air and heat and partly acted upon by the bacteria. This is a very good method of sewage disposal in a country like where lands are available at cheap rates and there is lots of sunshine which helps biological activity. Land treatment can be classified as:

1. Sewage farming or Broad irrigation
2. Filtration

In filtration more depth of soil below the ground level takes part in sewage treatment whereas in broad irrigation only upper 58 mm soil does this work.

6.6.1. Advantages and Disadvantages of Land Disposal

Advantages

(*i*) Adds manure to land.

(*ii*) Pollution of natural water courses is minimised.

(*iii*) Increases fertility of land.

(*iv*) Gives high calorific value to crops grown in sewage farms.

(*v*) Does not require any installation of equipment involving high initial cost.

(*vi*) Crops could be grown and hence a return value is always possible to obtain.

(*vii*) Method specially suitable where large quantity of river water is not available at all times of the year.

Disadvantages

(*i*) Difficult to get land during rainy and harvest seasons.

(*ii*) Additional land is required for 'Reserve'.

(*iii*) Sanitary reasons may not permit growing of crops on sewage farms.

(*vi*) More land area is required if sewage volume is greater since land capacity is limited.

(*v*) If all precautions are not taken, sewage farming results in sewage sickness to land and health hazards to life.

6.7. BROAD IRRIGATION

In this method, sewage is allowea to flow over cultivated lands, from which a part of the sewage evaporates, some percolates and the rest escape into surface drainage channels. Sewage waters the land and adds to its fertilizing value, due to the presence of nitrogen, phosphorous, potsh etc. These fartilizing elements of sewage are consumed by the roots of crops. Crops like cotton, Potatoes, sugarcane, grass etc. can be profitably growing. This is also called sewage farming.

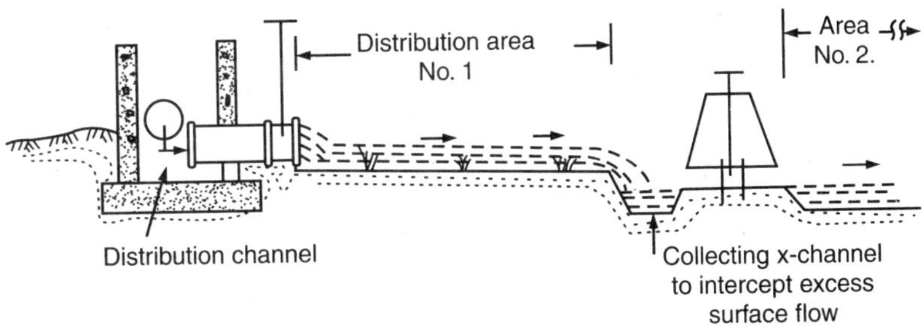

Fig. 6.2. Broad irrigation

Table 6.1. Dose of sewage for sewage farming

S.No.	Nature of soil	Maximum dose of sewage in m^3/hectare/day	
		Raw sewage	Settled sewage
(i)	Loam soil	50–80	110–180
(ii)	Clayey loam soil	40–50	55–110
(iii)	Clayey soil	30–35	33–55
(iv)	Sandy loam soil	80–110	170–225
(v)	Sandy soil	120–150	225–300

6.8. APPLICATION OF SEWAGE

The sewage may be applied to land by the methods given below:

(i) Flooding: The area to be irrigated is divided into various parts surrounded by dykes as shown in Fig. 6.3. The sewage is filled like small ponds in between the dykes. The depth of dose varies from 3.0 cm to 5.0 cm depending on the irrigation requirements.

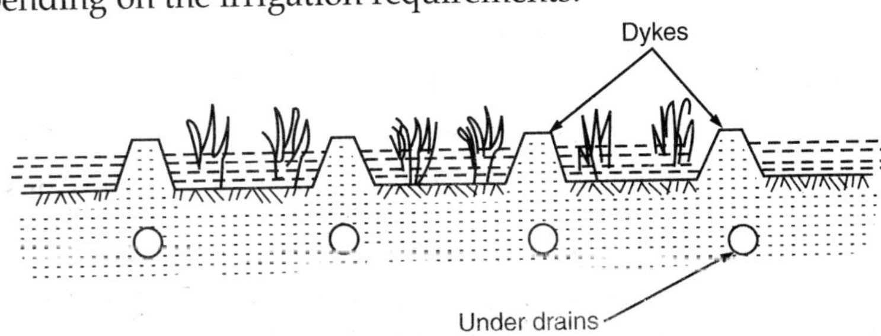

Fig. 6.3. Flooding

***(ii)* Surface Irrigation:** This method is most suited in sloping area. Here, parallel drains are construct in the fields. All these drains are connected to a distributary drain with the help of regulating device so that sewage may flow in the required drain. Here when sewage flows over the fields, its large quantity is absorbed by the fields and only excess quantity reaches another drain.

***(iii)* Spray Irrigation:** Here first the sewage is filled in tanks so that sludge may settle. Then the sewage is sprayed over the fields by pumping it through pipes. The pipes may be fitted with nozzles at other end. The method is not used in India.

***(iv)* Ridge and Furrow:**

(a) **In straight run method:** Here the surface of the land is arranged in the form of ridges and furrows. As shown in the Fig. 6.4 furrows are filled with sewage while crops grow on or between the ridges.

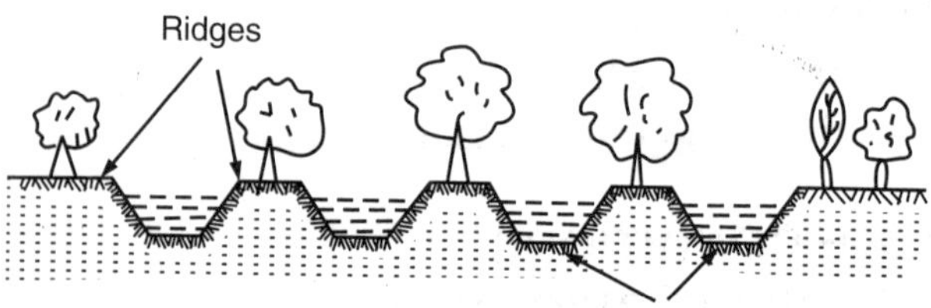

Fig. 6.4. Ridge and furrows

(b) **In the Zig-Zag Run-method:** In this method the ridges are arranged in a zig-zag method with corresponding furrows by their side.

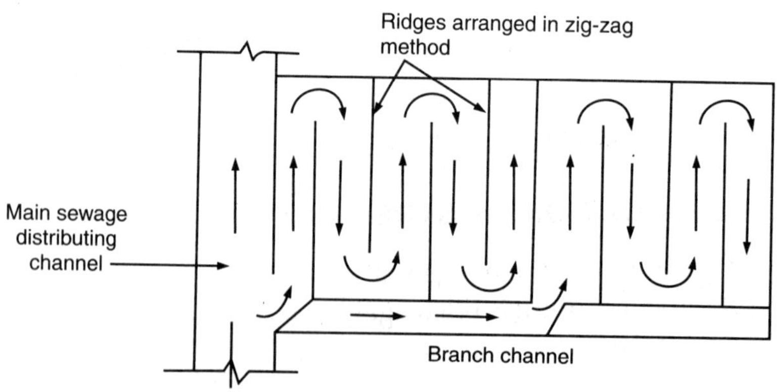

Fig. 6.5. Zig-zag method of irrigation

***(v)* Lagooning (or Sewage Ponds):** These are used cheaply for sewage disposal. In this method the sewage is allowed to settle in a natural depression available or artificial constructed tank. Detention period is about a month. During this period the sludge is stabilized and dried. The purified effluent passes way from an outlet placed at the other end. Lagoons should be shallow and must be constructed away from the town.

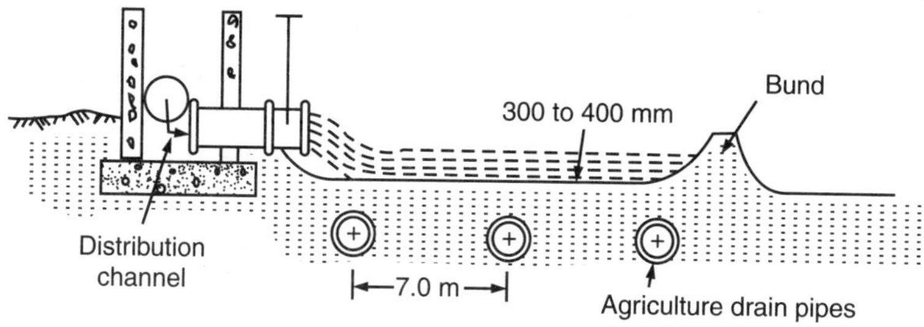

Fig. 6.6. Sub-irrigation-inundation method

***(vi)* Basin Method:** In this method big trees are planted in an isolated manner, basins are formed around each tree. These basins are filled with sewage. This method is suitable for fruit gardens.

***(vii)* Sub-surface Irrigation:** Here sewage is applied at the roots of plants., through the open jointed agricultural drain-pipes. These pipes are laid about 1.0 m below the ground level. The sewage rises up due to capillary action. Here soil takes less load but this is an economical method.

6.9. FAVOURABLE CONDITIONS FOR SUB-SURFACE IRRIGATION

(*i*) Subsoil water level is low.
(*ii*) Land is cheap.
(*iii*) Rainfall is less and irrigation water demand is heavy.
(*iv*) Large areas are available.
(*v*) Where dilution water is not easily available.
(*vi*) Sub-surface strata is porous, favouring infiltration.
(*vii*) Climate is dry favouring drying up conditions.

6.10. SELF-PURIFICATION OF STREAMS

The natural process in which the oxygen of water of stream is consumed by sewage and at the same time it is filled again by the atmosphere is called self-purification of streams.

6.10.1. Factors Helping Self-Purification of Streams

The various natural factors which help in effecting self-purification are:

(*i*) Dilution
(*ii*) Sedimentation
(*iii*) Sun-light
(*iv*) Oxidation
(*v*) Reduction
(*vi*) Temperature
(*vii*) Bacteria

***(i)* Dilution:** Organic matter present in sewage is diluted due to dilution by water and therefore nuisance value is decreased. Surface tributaries and underground streams accelerate the process.

***(ii)* Sedimentation:** Settleable solids drop down to the bottom of the stream in the form of sludge deposits. Hence it is purified to some extent.

***(iii)* Sunlight:** It causes bleaching effects and kills bacteria. It helps certain organisms to give off oxygen. Hence it is effective in self-purification.

***(iv)* Oxidation:** When the organic matter meets the water; it starts getting oxidized due to development of oxidising organisms in water. The process continues till the organic matter has been completely oxidized. The oxygen demand is fully satisfied, the stream becomes pure. Water always contains some dissolved oxygen and it is responsible for oxidation of organic matter.

***(v)* Reduction:** It occurs due to hydrolysis of organic matter either biologically or chemically. It also helps for stabilization by oxidation.

***(vi)* Temperature:** Temperature also affects the rate of biological and chemical activities. These activities are more at higher temperature and less at lower temperature. Also at higher temperature the capacity to maintain dissolved oxygen of water is low, while the rate of biological and chemical activities are high, causing rapid depletion of dissolved oxygen. The larger the amount of dissolved oxygen present in water, the better self-purification will occur.

***(vii)* Bacteria:** Organic solid matter is converted into gasses and liquids by bacteria, therefore it helps the stream to purify itself.

6.11. SEWAGE SICKNESS

Continuous appliction of sewage on land results in the clogging of the pores of the soil thus preventing the free cirulaticn of air. This causes the purification of organic matter and objectionable smell is given out. At this stage the land has its power of receiving the load of sewage.

6.11.1. Preventing Measures for Healthy Conditions

(*i*) Pre-treatment of sewage sedimentation.

(*ii* Resting Periods Sewage Dose is given for 6-9 hours in a day and land is given rest for rest of a period. After 15 to 20 days, land is given continued rest for an equal period before it is brought to service. Land is throughly ploughed during resting period.

(*iii*) Reserve Land—to meet the demand during the resting period.

(*iv*) Rotation of crops—will aerate the soil and utilize the fertilizing elements of sewage.

(*v*) Drainage of land.

(*vi*) Sewage should not be applied in excess.

MODEL QUESTIONS

1. What crops will you cultivate by sewage farming?
2. In what ways will you dispose of the sewage in your town?
3. How does the stream get purified?
4. (*a*) What do you mean by Sewage Irrigation and where is it employed?
   (*b*) Discuss the factors governing the selection of site for sewage disposal from a town?
5. Write a note on sewage farming.
6. Explain the factors governing self-purification of streams.
7. Write short notes on:
   (*i*) Broad irrigation (*ii*) Basin method
   (*iii*) Spray irrigation (*iv*) Intermittent sewage filtration
8. (*a*) What are the aims and objects of sewage disposal.
   (*b*) Explain the land treatment of sewage irrigation.
9. Explain the following:
   (*i*) Ridge and furrow method (*ii*) Inundation method
   (*iii*) Lagooning (*iv*) Flooding
   (*v*) Sewage sickness (*vi*) Dilution
10. Describe briefly the different methods of disposal of sewage.
11. What is sewage sickness? How to cure it? What precautions are to be taken to prevent sewage sickness.
12. What is sewage disposal system? Sketch a layout of sewage disposal system.
13. Write short notes on "Self purification of streams."
14. Explain 'Disposal by Land Treatment".

❑❑❑

CHAPTER 7

SEWAGE TREATMENT

Learning Objectives

- Sewage Treatment
- Objects of Sewage Treatment
- Classification of Sewage Treatment
- Screens
- Grit Chambers
- Skimming Tanks
- Sedimentation
- Septic Tanks
- IMHOFF Tanks
- Sewage Filters
- Trickling Filters

7.1 SEWAGE TREATMENT

The artificial process for changing or removing the objectionable constituents present in sewage to make it less dangerous and offensive is called Sewage Treatment.

7.2. OBJECTS OF SEWAGE TREATMENT

- To reduce the strength of sewage, so that it may not pollute the receiving water.
- To reduce its volume in general.
- To destroy the pathogens i.e., bacteria content.
- To recover its industrial value.
- To enable its safe and satisfactory disposal.

7.3. CLASSIFICATION OF SEWAGE TREATMENT

Sewage treatment plant consists of the following three processes:

(*a*) Preliminary treatment

(*b*) Secondary treatment

(*c*) Final treatment

(*a*) Preliminary Treatment: This treatment is given to make sewage fit for further treatment. Screens, shredder, grit chamber and skimming tank are used in this treatment. Screen is meant for removing floating and suspended matter such as cloth, paper, wood, cork etc. Grit chamber is meant for removing the heavy particles. Skimming tank is used to remove lighter and greasy matter. This treatment removes about 10% of total solids.

(*b*) Secondary Treatment: Preliminary treatment removes solids which are mostly inorganic in nature. In secondary treatment attempts are made to remove fine, very fine and colloidal organic matter. This is done with the help of plain Sedimentation tanks. In this treatment about 40% of total solids gets removed. The BOD reduction varies from 25 to 40%. But the dissolved solids including organic matters escape along with the effluent.

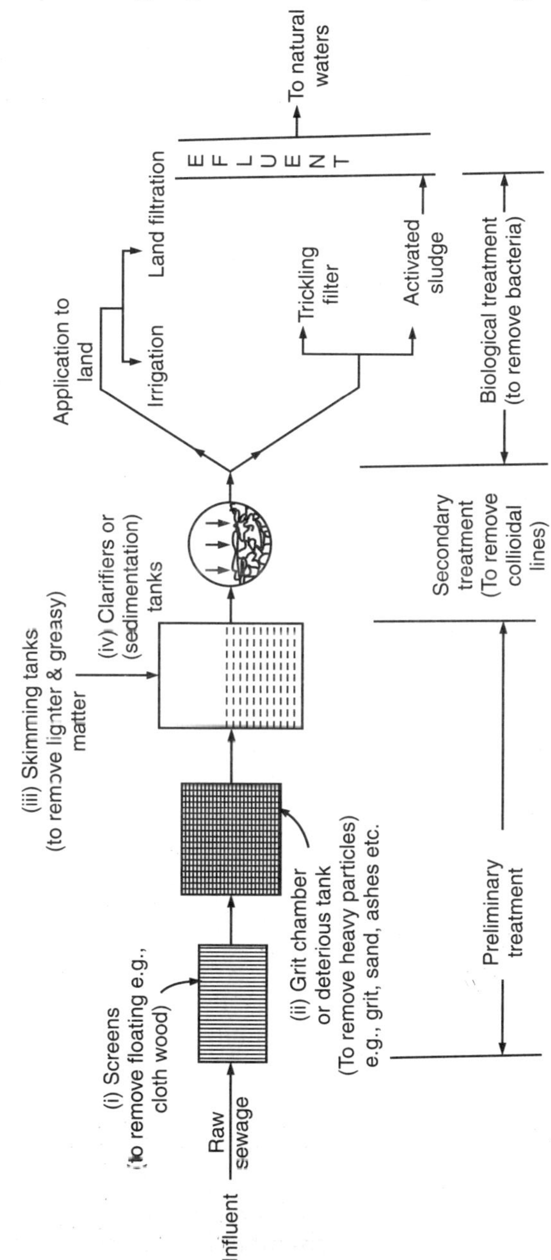

Fig. 7.1. Flow diagram of a typical sewage treatment plant

(c) Final Treatment: This is also called biological treatment. Non-settleable and dissolved organic matters get removed in this treatment. After this treatment about 85% of BOD and 90% of total solids get removed frorn sewage. Intermittent filters, trickling filter, activated sludge process etc., are used in this treatment.

Fig. 7. l shows the flow diagram of a complete sewage treatment plant.

7.4. SCREENS

The process of separating large particles of floating or suspended matter by using screens is called screening. Screens used may be coarse, medium or of fine variety.

7.4.1. Objects of Screens

(*i*) to prevent the formation of unexpected scums in settling and aeration tanks.

(*ii*) to prevent clogging of sprinkling nozzles.

(*iii*) to prevent formation of floating matters in the receiving bodies of water.

(*iv*) to protect pumps at the plant and out-fall.

(*v*) to prevent clogging of trickling filters.

7.4.2. Types of Screens

Screens are classified based on:

(*i*) Size of openings

(*ii*) Mode of setting

(*i*) According to size of openings: Screens are of following types:

(*a*) Coarse screens

(*b*) Medium screens

(*c*) Fine screens

(*a*) Coarse Screens: Coarse screens are also known as *racks*. Their opening (spacing between bars) size is about 50 mm or more. These screens help in removing large floating objects like rags, wood paper etc., from sewage. These screenings will not purify and may be disposed of by incineration, burial or dumping. They collect about 6 litre of solids per million litre of sewage.

(*b*) Medium Screens: Medium screens are similar in construction to coarse screen. But the size of opening is much smaller varying from 6 to 40 mm. These screen contain some quantity of organic matter and may be disposed by incineration or burial. These screens collect about 30 to 90 litres of solids per million litre of sewage.

(c) Fine Screens: Fine screens are usually constructed using non-corrding plates of brass or bronze. The size of perforation ranging from 1.5

mm to 3 mm. Such screens are usually in the form of disc or drum and are normally mechanically operated. They remove about 20% of suspended solids from sewage and need frequent cleaning. These screens are therefore used for treating industrial waste.

Table 7.1. Screens and sizes of openings

S. No.	Types	Class	Size of opening (in mm)	Volume of screenings in l/ml of sewage
(*i*)	Racks or bar screens	Coarse Medium	More than 50 25 to 50	
(*ii*)	Screens	Fine Medium Fine	Less than 25 6 to 10 Less than	6 30-90 20% of suspen -ded matter

***(ii)* According to the mode of setting:** Screens are of following types:

(*a*) Fixed screen

(*b*) Moving screen

(*c*) Movable screen

(*a*) Fixed Screen: The screens which are permanently set in position are known as fixed screens. Bar type coarse or medium screens are fixed screens.

(*b*) Moving Screen: The screen which are moving during its operation are known as moving screen.

(*c*) Movable Screen: The screens which are stationary during operation. But they can be removed from their position for the purpose of cleaning.

Nowadays ractangular type bar screens of coarse and medium types are used in sewage treatment plants. Bar screens are made of steel bars placed perallel to each other at desired spacing on a rectangular steel frame. These screens are kept inclined at about 30 to 60° to the direction of flow, this helps in making the screens more effective by reducing the flow velocity and getting increased opening area.

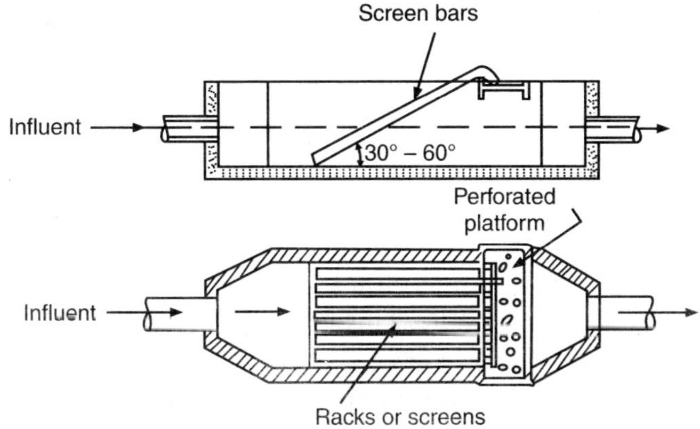

Fig. 7.2. Screening chamber

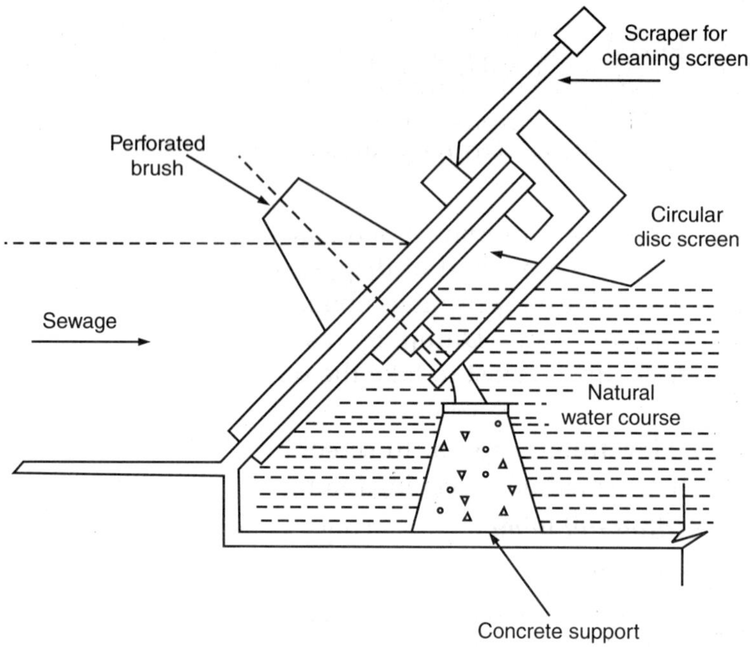

Fig. 7.3. Disc type fine screen

7.4.3. Cleaning of Screen

The art of maintaining the efficiency of screens by removing the material collected in between the screen bars is called **Cleaning of Screen.**

Generally Racks are used for cleaning purposes. They may be hand or mechanically operated. Screen should be cleaned twice or thrice a day so that they may work efficiently.

7.4.4. Disposal of Screenings

The material separated by screens is known as screening. It contains large qunatity of moisture (85 to 90%) and organic matter. These organic matters causing bad smells and nuisance. To avoid such possibilities, the screening must be disposed off properly.

7.4.4.1. Methods of Disposal of Screenings

The following methods are used for disposal of screening:

(*i*) by burning

(*ii*) by burial

(*iii*) by dumping

(*iv*) by digestion

(*i*) By burning: Burning of screenings is done in incinerators. This process is known as incineration. Before burning the moisture content of screenings are reduced to about 60% either by air drying or by compressing through hydraulic prosses. The incineration is carried out at 760 to 815°C temperature to avoid bad smells.

***(ii)* By burial:** This processes is also known as composting. In this method, screenings are buried in a trench 1 m to 1.5 m wide and then covered with 0.3 to 4.45 m of porous earth. Over the period of time, oxidation-reduction of screenings will take place and the contents can be used as manure.

***(iii)* By dumping:** Dumping of screenings in low lying areas or in large water bodies is also a method of disposing screenings. Dumping in low lying areas is suitable for screenings obtained from large screens as they do not contain organic water. Dumping in sea will be suitable in area where strong forward current exist to take the dumped material away from the sea shore.

***(iv)* By digestion:** Disposal to screenings by digestion is not found successful. In this method, screenings along with sewage sludge is digested in a sludge digestion tank.

EXAMPLE. 7.1. Design a medium bar screen for 30 MLD (Million litres per day) sewage flow so that the velocity through the screen does not fall below 0.65 m/sec. Assume the depth of flow as 0.9 m.

Solution: Given discharge, $Q = 30 \text{ MLD} = \dfrac{30 \times 10^6}{24 \times 60 \times 60 \times 1000}$

$$= 0.347 \text{ m}^3/\text{s}$$

Velocity through the screen, $V = 0.75$ m/s

$$\therefore \text{ Area of screen openings} = \frac{Q}{V} = \frac{0.347}{0.75} = 0.463 \text{ m}^2$$

$$\text{Width of clear openings of screen} = \frac{\text{Area of screen openings}}{\text{Depth of flow}}$$

$$= \frac{0.463}{0.9} = 0.57 \text{ m}$$

Let us assume that the screen bars are made of MS flat of size 10 mm × 50 mm with clear openings of 25 mm (being medium bar screen)

$$\text{No. of openings} = \frac{0.57 \times 1000}{25}$$

$$= 22.8 \text{ say } 23 \text{ nos.}$$

$$\text{No. of bars} = 23 - 1 = 22 \text{ nos.}$$

$$\text{Hence cross width of the screen} = 0.57 + \frac{22 \times 10}{1000}$$

$$= 0.79 \text{ m} \simeq 0.8 \text{ m} \quad \textbf{Ans.}$$

Assuming the inclination of the screen is 60° with the horizontal then

Inclined area of the screen = $\frac{0.8 \times 0.9}{\sin 60°}$ = 0.83 m^2 **Ans.**

Length of the Screen channel is given by an emperical formula

$$L = (d_c + 0.3) \cot \alpha + 1.73 (W_c + d)$$

where L = Length of the channel, in m

d_c = Depth of flow in screen chamber, in m

α = Angle of screen bars with the horizontal

W_c = Width of the screen chamber in m

d = diameter of the incoming sewer, in m

Now assuming of the dia. of the incoming sewer = 0.6 m

Then length of screen channel required

$$= (0.9 + 0.3) \cot 60° + 1.73 (0.8 + 0.6)$$

= 3.10 m **Ans.**

7.5. GRIT CHAMBERS

Sewage contains both inorganic and organic solids. The inorganic content of sewage is termed as grit. The size varies from 0.2 mm diameter and above. It consists of sand, gravel, ash, egg shells, bones, clinkers and inert inorganic materials.

A grit chamber is a long basin or an enlarged channel in which cross-section is increased to reduce the velocity of flowing sewage to permit the settlement of grit only. The velocity of flow is maintained between 0.15 to 0.30 m/s depending on grit specific gravity and diameter. This can be achieved by installing velocity control devices at outlet. At least two separate chambers should be provided, one to take care of low flow and the other for the high flow. A detention period of 1 minute is commonly employed which is just sufficient to remove grit. The length to breadth ratio of chamber varies from 6 to 15 and length to depth ratio varies from 10 to 30.

Grit chambers are cleaned by hand, mechanically or hydraulically after every one or two weeks. Hand cleaning is done only in case of smaller plants. Adequate capacity is provided below sill of weir for storage of grit. Mechanical cleaning is preferred for flows over 10 ml/d. Grit is extracted from suction pump or air lift pump from the grit storage pit, where the slopes of the channel are towards the pit. In hydraulic cleaning, the deposited material is flushed out under fire streams directed from a central point and removed through pipes in the bottom of channel or side walls.

Table 7.2. Settling velocities and overflow rates for grit removal devices at 10°C

S.No.	Diameter of particles (mm)	Settling velocity (m/s) for sp. gr. 2.65	Overflow rate in the grit chamber ($m^3/d/m^2$)
1.	020	0.02	1700
2.	0.18	0.018	1600
3.	0.15	0.015	1300

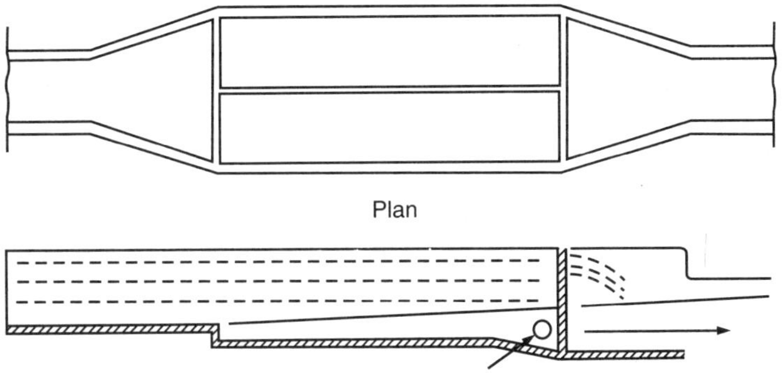

Fig. 7.4. Frit chamber (multiple parallel channel)

7.6. SKIMMING TANKS

Sewage contains floating matters such as oil, grease, fat etc. These can be removed from sewage by means of floatation or settling as a scum or sludge. The scum formation can be encouraged by passing sewage through diffused air aeration tanks called the skimming tanks.

A skimming tank is a chamber so constructed that the floating matters like oil, grease, fat etc., rise and remain on the surface of the sewage until removed, while the liquid flows out continuously under baffles. Removal of floating matter from the sewage is necessary due to the following difficulties:

(*a*) They clog the fine screens.

(*b*) They reduce considerably the efficiency of the activated sludge process.

(*c*) They form scum when discharged into stream, the scum retards reaeration of streams which leads to anaerobic action.

(*d*) They adversely affect biological growth due to which various bacteria and protozoa cannot perform their work.

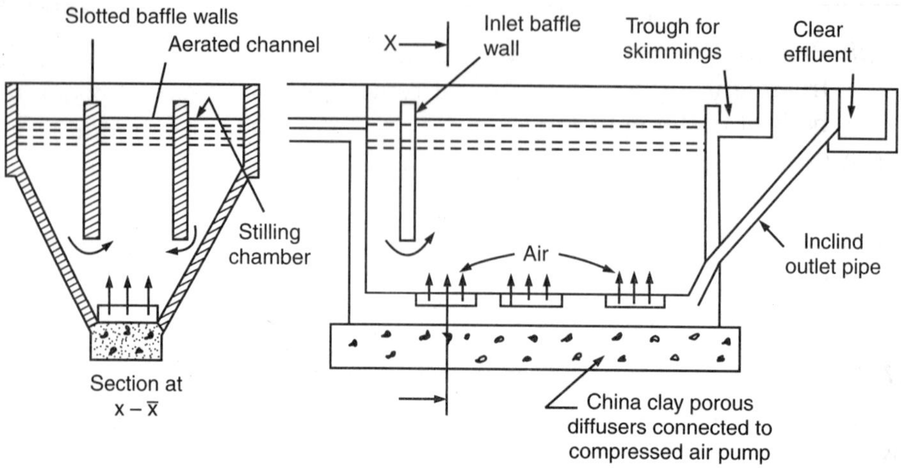

Fig. 7.5. Aerated skimming tank

Skimming tanks are in the form of narrow rectangular tanks with longitudinal baffle walls having slots dividing the tank into three compartments. Perforated plates/diffusers are fixed on the floor through which compressed air is blown upward which help in rising of floating matter. Sewage enters the tank from one end, flows longitudinally and leaves out through a narrow inclined outlet. A detention period of 3 mins is usually sufficient and the amount of air required is about 0.2 m^3 per cu.m of sewage.

Skimming are usually disposed off by burning or burying in the ground. Sometimes these may be used for the manufacture of soap, wax, lubricant etc.

7.7. SEDIMENTATION

The operation in which suspended particles heavier than water settles under the action of gravity is called sedimentation. In waste water treatment sedimentation is the most useful operation to remove fine suspended matters from the sewage.

When the suspended particles are separated from the suspending fluid by gravitation and natural segregation of the settling particles, the operation is called *Plain Sedimentation.*

When chemicals are added to hasten the settling process, the operation is called *chemical precipitation.*

In sewage treatment plants, Sedimentation is carried out twice, once before the biological treatment and then after the biological treatment.

7.7.1. Aims of Sedimentation

(*i*) To remove gross solids from sewage and finally the BOD.

(*ii*) Primary Sedimentation prepared sewage for the biological treatment. If the solids are not removed then they will chock the beds of trickling filter.

(*iii*) Secondary Sedimentation removes precipitated solids from the sewage.

(*iv*) To increase the efficiency of disinfection of sewage.

(*v*) Chemical precipitation may be done when intermediate treatment is required.

7.7.2. Theory of Sewage Sedimentation

The sedimentation of particles takes place according to the following laws:

(*i*) Stoke's Law

$$V_S = \frac{g}{18}(\rho_S - \rho)\frac{d^2}{\mu}$$

(*ii*) Newton 's Law

$$V_S = 1.8\sqrt{g(\rho_S - \rho)d}$$

where V_s = Settling velocity in cm/s

g = Acceleration due to gravity

ρ_s = Specific gravity of particles

ρ = Specific gravity of fluid

d = Diameter of particles in cm

μ = Kinematic viscosity of fluid in cm^2/s

Sedimentation of particles in settling tank is governed by the following factors:

(*i*) Specific gravity of particles

(*ii*) Size of particles

(*iii*) Temperature.

7.7.3. Types of Tanks

Horizontal-flow and vertical-flow sedimentation tanks are commonly adopted in sewage treatment. Fig. 7.6 to 7.8 shows various types of sedimentation tanks. Horizontal-flow tanks are usually rectangular in plan and vertical-flow tanks are circular. The diameter of circular tank ranges from 10 m to 30 m depending upon the available size of scraper mechanism and its stability. The breadth of rectangular tank ranges from 10 m to 30 m and length to breadth ratio is 2 to 5. The length to depth ratio is 5 to 10. Bottom slope is 1 % towards the inlet end.

In rectangular tanks, sewage enters continuously from one end and passes at the other end, generally over a weir, sludge is removed manually or mechanically. Mechanical cleaning is more popular and is done by using scrappers.

In circular tanks, sewage enters at the centre, rises vertically to be drawn off by flowing over a peripheral weir arranged at the surface. The effluent flowing over the outlet weir is collected in an outlet pipe for further treatment.

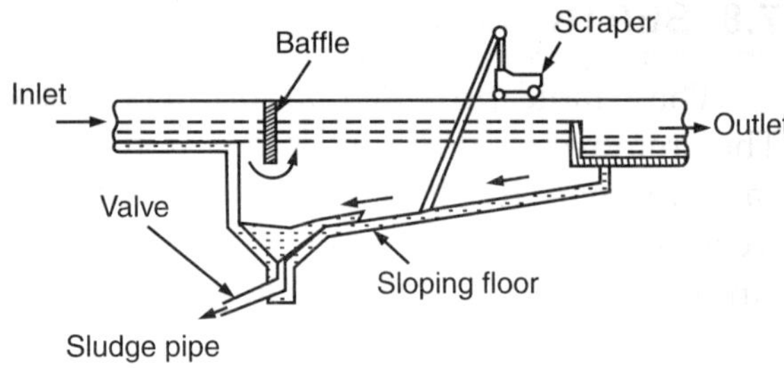

Fig. 7.6. Rectangular sedimentation tank

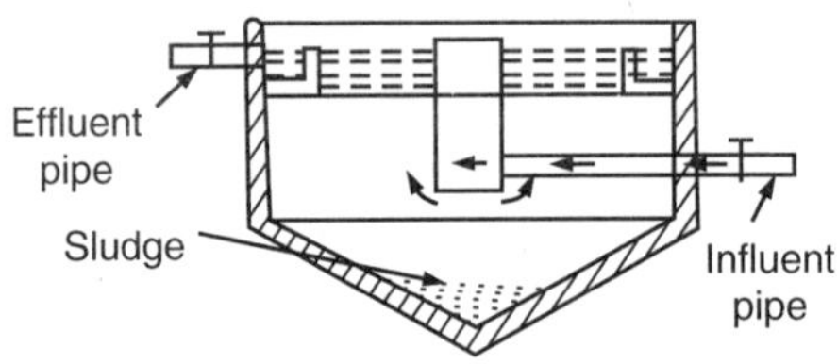

Fig. 7.7. Sedimentation tank (Circular bottom)

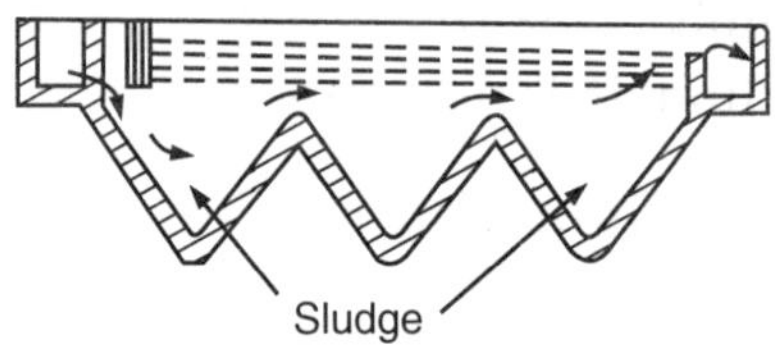

Fig. 7.8. Sedimentation tank (Hopper bottom)

7.7.4. Design Criteria

(*a*) Tank is designed for average flow conditions.

(*b*) Rate of flow – 225 to 300 $m^3/m^2/day$

(*c*) Surface loading – 40 to 50 $m^3/m^2/day$

(*d*) Detention time – 1 to 3 hours.

(*e*) Depth – 2.4 to 3.6 m

(*f*) Velocity of flow- 0.3 to 0.6 m/min.

7.7.5. Clarifiers

Clarifiers is a type of sedimentation tank and the design is similar to plain sedimentation tank. In this tank, the floc produced in the sewage is allowed to settle and is separated from the sewage. Some clarifiers are also fitted with continuous sludge removing arrangements.

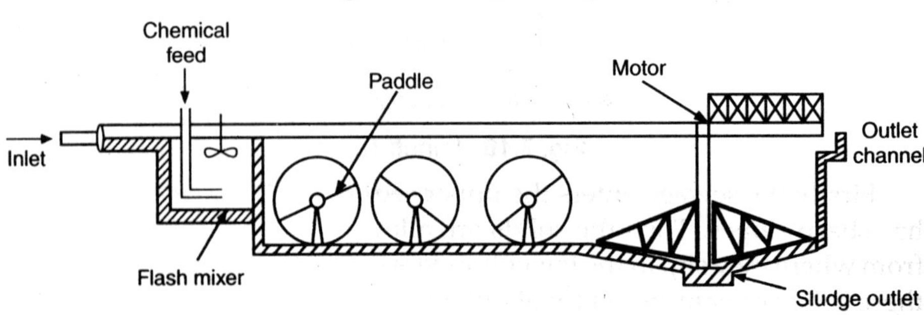

Fig. 7.9. Clarifier of M/s. Dor Olivr & Co. USA

7.8. SEPTIC TANKS

These tanks provide both primary and secondary treatment to sewage. They remove solids by sedimentation which is a primary treatment. At the same moment these solids are digested by bacteria which is the secondary treatment. Tanks have one chamber in which both the treatments occur simultaneously. For construction and design details.

7.9. IMHOFF TANKS

These tanks are the improvements over the Septic Tanks. It combines sedimentation and sludge digestion as in a septic tank but there are different chambers for each action. They are good for small colonies of towns. Name of "Imhoff Tank" is after its inventor Karl Imhoff, a German Engineer.

7.9.1. Construction Details and Working

Fig. 7.10 shows the essentials of the tank. Imhoff Tank is a double chamber tank. The upper chamber is called the Sedimentation chamber through which sewage flows at a very low velocity. The lower chamber is known as digestion chamber in which septic (or anaerobic) decomposition occures. Number of inter-connected hoppers are provided below settiing tanks.

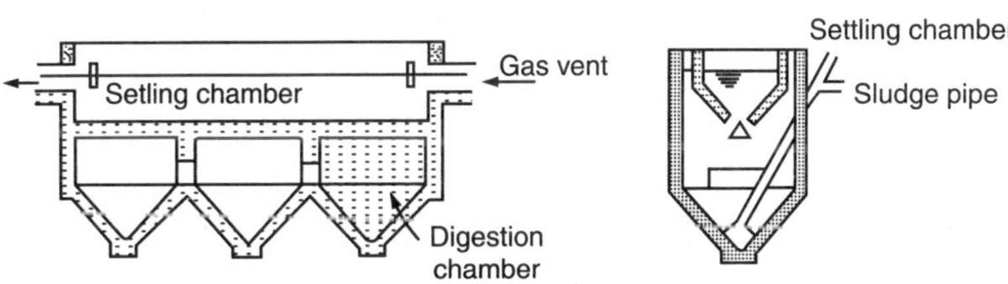

(a) With rectangular plan for small communities

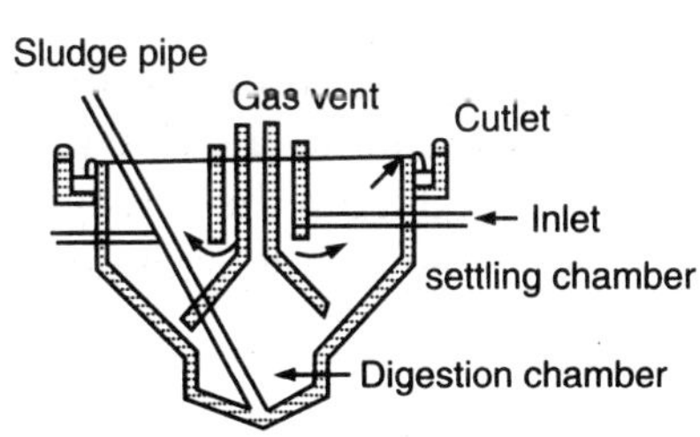

(b) With circular plan for single residence

Fig. 7.10. Imhoff Tanks

Firstly the sewage enters the upper sedimentation tank whose bottom has sharp slopes. Then the solids are allowed to settle in the upper tank from where they slip in the lower hoppers through the slots. In the hoppers these solids remain stored for about one month. During this duration they are acted upon by the bacterias and are converted into the stable solids, gases etc. Gases of digestion escape through the gas vents. The sludge is removed periodically through sludgepipe under hydrostatic pressure of 1.0 to 2.0. The flow of solids in the lower hoppers is regulated with the help of triangular beam. Filled hopper is closed with triangular beam and in that hopper digestion starts.

To prevent sludge or scum from penetrating into the sedimentation chamber, the sludge and scum must be at 50 cm distance below and above the slots. This clear zone is known as *Neutral Zone.*

The tanks may have different shapes of x-section either rectangular or sometimes circular tanks are used. Scum must be cleaned every day from the Imhoff tank. Cleaning period of the tank is generally one to two months.

7.9.2. Advantages and Disadvantages of Imhoff Tanks

Advantages

(*i*) Simplicity of operation—no mechanical appliances used.

(*ii*) Have better economy.

(*iii*) Gives good results without skilled attention.

(*iv*) There is automatic, uniform and continuous feeding of organic solids to be digested, into the digestion compartments, hence gives better efficiency of digestion.

(*v*) There is less problem of sludge disposal.

Disadvantages

(*i*) As tanks are two storeyed and are very deep (9 to 10 m) due to steeply sloping hoppers, hence uneconomical.

(*ii*) Difficult to construct in quick sand or solid rock or where ground water levels are high, therefore costly.

(*iii*) Foul smelling gases of digestion escape out, thereby polluting the atmosphere.

(*iv*) Unsuitable where acidic conditions exist.

(*v*) No proper control over their operation because there is no mechanical mixing or any heating arrangement.

(*vi*) Not suitable for large treatments.

7.10. SEWAGE FILTERS

Sewage filters are of following types :

(*a*) Contact beds

(*b*) Intermittents and filters

(*c*) Trickling filters

(*a*) **Contact Beds:** A contact bed is a water tight tank with masonry walls in which contact media like broken stone of size 25-75 mm is filled over underdrains placed at the bottom of the tank. The tank is filled with sewage and allowed to stand full over a period of two hours and then emptied. The tank is then left empty for 3 to 4 hours. This period is called "left over period". During this period air penetrates into the bed, permitting aerobic bacteria to oxidize the deposited organic matter. The tank is again filled with sewage after the left over period and the cycle of operation is continued. This method of biological treatment is no longer in use now a days.

(*b*) Intermittent Sand Filters: Intermittent sand filters are the early forms of biological units of treatment. It is a sand filter where sewage is applied at regular intervals. Fig. 7.11 shows an interniittent sand filter. It consists of layers of sands having effective size of 0.20 to 0.50 mm and uniformity coefficient between 2 to 5. The depth of bed is kept 450 mm to 750 mm. The size of bed is l/10th to 2/5th of a hectare, generally rectangular in shape. Filter loading is normally kept between 1000 to 2500 litres per square meter per day. About 90 to 95 percent of BOD is expected to be removed by this method.

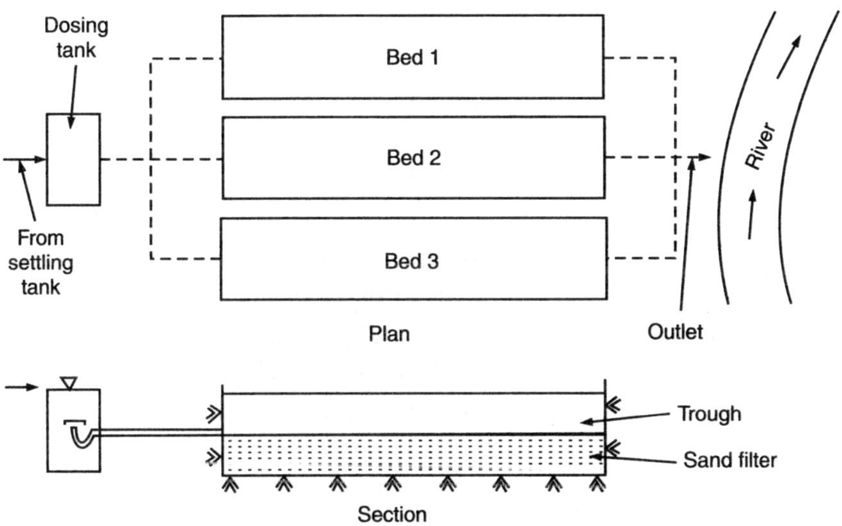

Fig. 7.11. Intermittent sand filter.

The sewage is applied evenly on the surface of the sand bed by means of a dosing tank. It is then allowed to flow over the sand bed and after an interval, sewage is applied over a second bed while the first bed is kept in rest. The aerobic condition is maintained due to entrance of air into the sand bed.

Advantages

1. Its operation is simple.
2. Less head of flow is required.
3. Secondary sludge disposal is not required.
4. Effluent obtained is of good quality.

Disadvantages

1. Due to slow rate of loading it requires large area.
2. Due to frequent clogging of sand bed it gives trouble regularly.

7.11. TRICKLING FILTERS

The fllters's over which sewage is allowed to trickle over or sprinkle are called *trickling filters.*

These are mostly used in the sewage treatment works. Here the sewage is not flooded over the filter as in case of contact beds etc., but it trickles over the filter media. Many times it is sprayed over the filter media at a uniform rate. Hence these filters are continuous in operation and also called as percolating or sprinkling filters.

The filter media is crushed rock, but large gravel, blast furnace, broken bricks, cinder etc., can also be used.

Trickling filter has the ability to recover from shock loads and to provide good performance with a minimum of skilled technical supervision. A relatively low operating cost, ability to work under extreme weather conditions particularly in low temperatures and a satisfactory efficiency in reduction of BOD are the advantages. See Fig. 7.12 below for construction details.

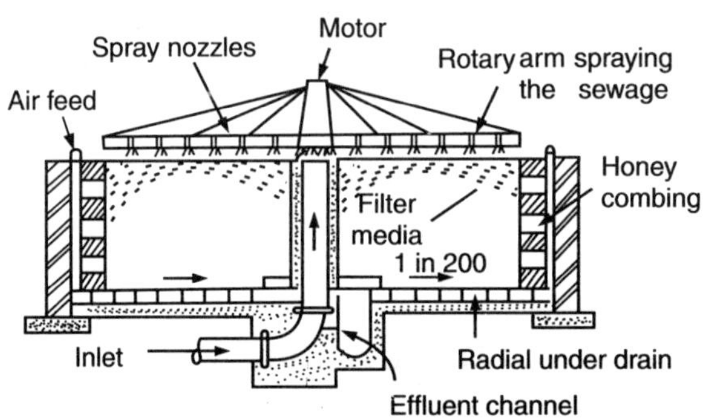

Fig. 7.12. Trickling filter

The disadvantages of the trickling filter at many plants is the head that is lost through the filter between 1.5 to 3 mm in addition to the depth of the filter. Other disadvantages include odour and fly nuisance (specially in case of Standard rate filter), the large area required and relatively high construction cost. Sedimentation of the effluent is usually necessary to remove the settleable solids.

7.11.1. Classification of Trickling Filter

The filters are mainly divided into two classes.

(*a*) Standard rate filter.

(*b*) High rate filter.

The comparison of these filters is indicated below:

Table 7.3. Comparison between standard rate and high rate trickling filters

S.No.	Characteristic	SRTF	HRTF
1.	Filter Loading (*i*) Volumetric kg/Hectare/day (*ii*) Hydraulic ML/Hac./day	 925 – 2200 22 – 45	 7400 – 18500 110 – 330 (Recirculation included)
2.	Depth of contact media	1.8 – 2.4 m	1.2 – 1.8 m
3.	Preliminary and final Treatment	Necessary for proper functioning	Also necessary for the same reason
4.	Method of operation	Continuous application Less flexible requiring less skill and supervision	Continuous application More flexible needs more skill and supervision
5.	Recirculation	May or may not be addopted	It is a must
6.	Type of effluent	More stable, highly nitrified	Less stable, Deflicient in nitrate content
7.	BOD removal	56 – 98%	63 – 78%
8.	Ploughing of film	Intermittent	Continuous
9.	Secondary sludge	Black, highly oxidised, light fine particles	Brown, not fully oxidised fine, particles tending to septicity
10.	Operational Difficulty	Odour and fly nuisance	Relative very less nuisance
11.	Operating cost	More	Less for equal performance
12.	Construction cost	High	Low

7.11.2. Advantages and Disadvantages of Trickling Filters

Advantages

(*i*) These increase water content of sludge upto 99%.

(*ii*) These are flexible in operation.

(*iii*) These reduce BOD upto 75%.

(*iv*) These reduce 80% of colloidal matter.

(*v*) Because their working is simple, hence do not require any skilled super-vision.

(*vi*) These give highly stabilised effluent which require less quantity of water for dilution.

Disadvantages

(*i*) They need primary treatment.

(*ii*) Large area is required.

(*iii*) Construction cost is high.

(*iv*) They develop bad smell.
(*v*) They develop fly nuisance.

EXAMPLE 7.2. Design a low rate trickling filter bed for treating 3 × 10⁶ litres/day of sewage. The BOD of the raw sewage entering in the TF is 130 mg/l and the required BOD of the effluent is 10 mg/l. An extra TF is required to work as standby.

Solution: Required efficiency of the *TF*, $\eta = \dfrac{130-10}{130} \times 100 = 92.3\%$

Total BOD applied filter, $W = \dfrac{130 \times 3 \times 10^6}{1000} = 39 \times 10^4$ gm

Efficiency of the TF is also given by $\eta = \dfrac{100}{1 + 0.01398\,(W/V)^{1/2}}$

where V = Volume of the filter bed
Using the value of η, and *w* as calculated above we get

$$92.3 = \frac{100}{1 + 0.01398\left(\dfrac{39 \times 10^4}{V}\right)^{1/2}}$$

or $V = 10952.3\ \text{m}^3$
Assuming the depth of filter $d = 2.6$ m

$$\text{Surface area of } TF = \frac{V}{d} = \frac{109.52}{2.6} = 4212.4\ \text{m}^2 \text{ (For twoTFs)}$$

$$\therefore \text{Surface area of one filter} = \frac{4212.4}{2} = 2106.2$$

Hence diameter of the *TF* = 91.78 m say 92 m.

$$\text{Organic/BOD loading} = \frac{39 \times 10^4}{10952.3} = 35.6\ \text{gm/m}^3$$

[It is well within the permissible limit of 80 to 400 g/day/m³]

$$\text{Surface/Hydraulic loading} = \frac{3 \times 10^6}{2106.2} = 1425\ \text{l/m}^2\text{/day}$$

[It is also within the permissible limits 1250 to 5000 l/m²/day]

EXAMPLE 7.3. Design a low rate trickling filter for a town from the data given below:

Town Population = 1,00,000
Quantity of Sewage = 125 l/day
BOD of the Influent = 220 mg/l

It is proposed to construct 5 filters each with an efficiency of 30%. The filteration is to be done after primary treatment.

Solution:

$$\text{Daily flow in each filter} = \frac{1,00,000 \times 12^5}{5} = 2.5 \times 10^6 \text{ litres}$$

$$\text{BOD/day in each} \quad TF = \frac{2.5 \times 10^6 \times 220}{10^6} = 550 \text{ kg/day}$$

Allowing BOD loading 275 g/m^3

$$\text{Volume of } TF \text{ bed,} \quad V = \frac{550 \times 10^3}{275} = 2000 \text{ m}^3$$

Again allowing surface loading 200 l/m^2

$$\text{Surface area of the } TF \text{ Bed} = \frac{2000 \times 10^3}{2000} = 1000 \text{ m}^2$$

$$\therefore \text{ Depth of bed} = \frac{\text{Volume}}{\text{Area}} = \frac{2000}{1000} = 2 \text{ m}$$

$$\text{Check for efficiency of the filter, } \eta = \frac{100}{1 + 0.443\left(\frac{5.50}{2000}\right)^{1/2}} = 89.1\,\% \quad \textbf{Ans.}$$

7.11.3. Bio-Filters (High Rate Trickling Filters)

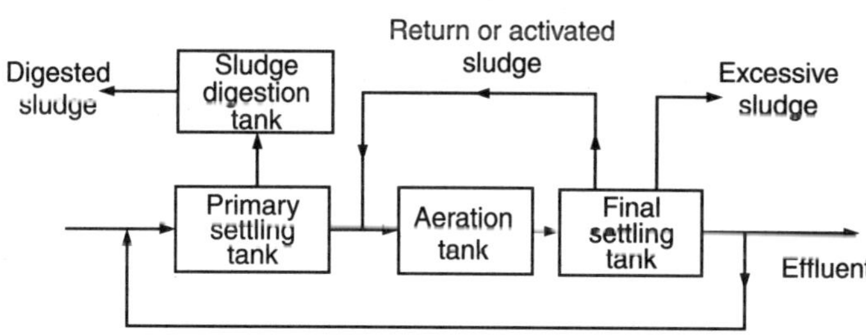

Fig. 7.13. Details of activated sludge process (One stage)

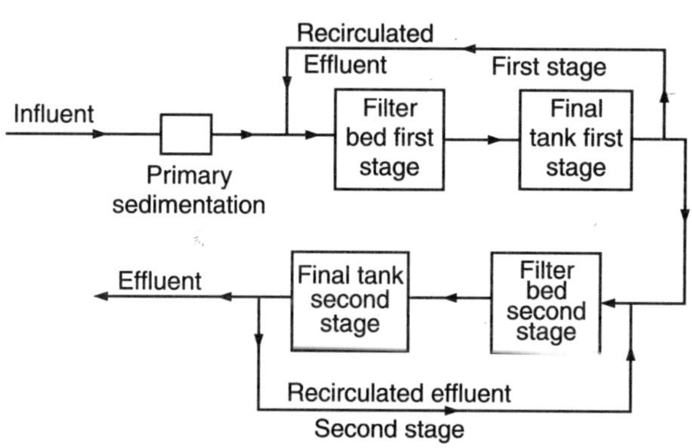

Fig. 7.14. Activated sludge process (Two stage)

The Bio-filter is a high rate trickling filter working on the system of recirculated flow. In the low rate trickling filter humunes like sludge formed in the filter beds, which comes out with the effluent, requires to be separated in the settling-tanks. Sometimes intermediate sedimentation or clarifier tanks are also provided. But in the case of bio-filters the operation may be either single recirculation (See Fig. 7.13) or two stage recirculation (See Fig. 7.14) depending upon the degree of treatment required.

Recirculation consists in pumping back part of the filter effluent to the pri-mary settling tank thus letting flow again pass through it. For treating strong sewages and to obtain effluent of comparable quality as in a standard rate filter, it is usual to employ two stage bio-filter treatment. As the main action is involved in the upper surface layers of the filter, it is considered expedient to reduce the depth of filtration in the bio-filters to 1 to 1 .5 m as against 2 to 3.5 m in the case of slow rate trickling filters. Hence they are economical.

7.11.3.1. Advantages and Disadvantages of Biofilters

Advantages

(*i*) They reduce the cost of filter by increasing the rate of loading.

(*ii*) They reduce the filter size.

(*iii*) The capital cost is less.

(*iv*) They are not effected by the strength and character of sewage.

(*v*) They do not develop bad smell.

Disadvantages

(*i*) They require primary treatment of sewage.

(*ii*) As the effluent is not highly stabilised, hence require more quantity of water for dilution.

MODEL QUESTIONS

1. (*a*) What is sludge? When is it necessary to treat the sludge?

(*b*) What are the methods of disposal of sludge?

2. Describe the detritus tank and screens used for sewage treatment. The capacity of detritus tank and screens should be given in MKS units.

3. Draw a neat sketch of the following and explain their working:

(*i*) Grit chamber

(*ii*) Screening chamber.

4. Write short notes on the following:

(*i*) Aeration tanks

5. What is the composition of sludge? How is the sludge treated and finally disposed off?

6. What is necessity of Sewage Treatment.
7. Explain Sewage Treatment Plant with the help of a neat flow diagram.
8. (*a*) Sketch and explain the working of a primary sedimentation tank in sewage treatment.

   (*b*) What is chemical sedimentation, when it is adopted in sewage treatment?
9. Explain briefly the following:

   (*i*) Screening chamber

   (*ii*) Skimming tanks
10. Draw a line plan showing the location of the various units required to be installed in connection with the treatment and disposal of sewage from a town. State the principles involved in the function of each unit provided there in.
11. Why is it necessary to treat sewage? Describe principles of primary treatment.
12. Write short nots on:

    (*i*) Grit chamber

    (*ii*) Trickling filters
13. Explain the following:

    (*a*) Meaning and principles of Primary treatment and Secondary treatment.

    (*b*) Trickling Filters (Sketch is essential)

❑❑❑

CHAPTER 8

ACTIVATED SLUDGE PROCESS

Learning Objectives

- Introduction
- Activated Sludge Process
- Methods of Aeration
- Sludge Digestion
- Sludge Treatment
- Sludge Drying
- Sludge Disposal
- Oxidation Ponds
- Chlorination of Sewage

8.1 INTRODUCTION

After primary treatment the secondary treatment is carried out in order to reduce the BOD of the effluent. Activated sludge process is a biological treatment process employing a sludge which by aeration and agitation becomes floculent and accumulates a very large population of aerobic bacteria. This sludge is known as Activated Sludge. When this sludge is added to sewage and the mixture is agitated in the presence of oxygen taken from the atmosphere for a sufficient time, the mixed liquor is then allowed to settle, the clear liquid is the effluent with BOD considerably reduced.

8.2 ACTIVATED SLUDGE PROCESS

Activated sludge is the sludge settled from the sewage previously agitated in the presence of abundant oxygen.

It is a biological treatment process and may be used in place of trickling filter. In this process biologically active mass, called activated sludge, is continuously mixed with the sewage and the mixture, known as mixed liquor, is agitated in the aeration tank in the presence of oxygen for a considerable period. The oxygen is supplied to the mixed liquor either by diffused compressed air or by mechanical agitation. The mixed liquor is then allowed to settle. The clear liquid obtained is effluent and the activated sludge which has settled is recirculated to the aeration tank. The flow diagram is shown in Fig. 8.1.

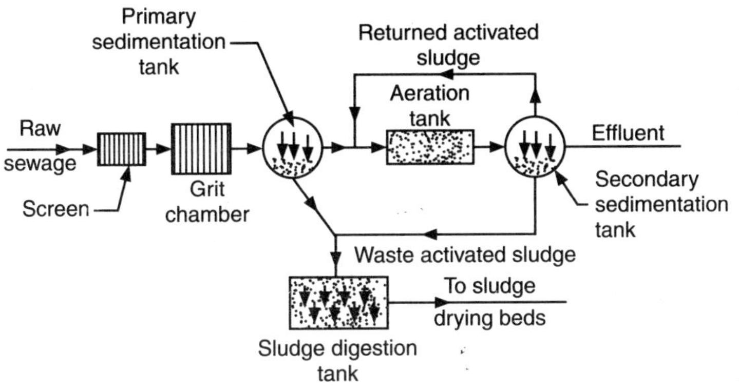

Fig. 8.1. Flow diagram of activated sludge process

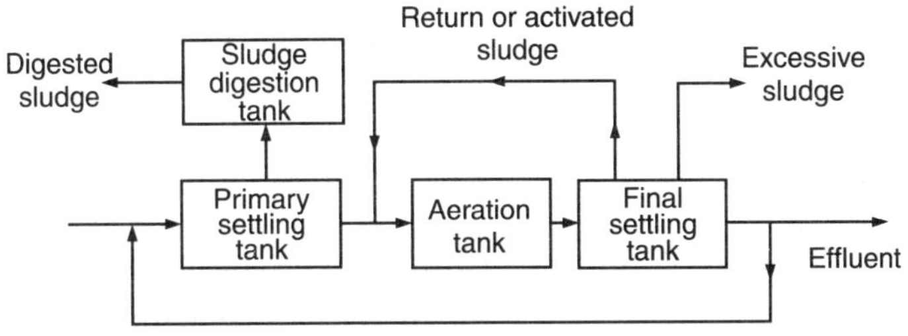

Fig. 8.2. Activated sludge process (one stage)

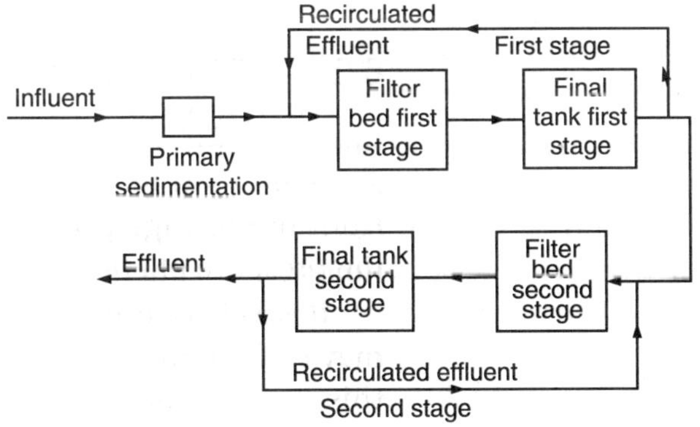

Fig. 8.3. Activated sludge process (two stage)

8.2.1. Theory

When activated sludge is added to the sewage, containing organic matters, flowing into the aeration tank, the active biomass (sludge) metabolizes the organic matter due to aeration by taking dissolved oxygen and produce new cells. Protozoa utilizes these bacterial cells as their food and grows rapidly. Some of the cells die due to endogeneous respiration. The biological floc, formed by synthesis, is transferred to a settling tank for settling. A part of settled floc is returned continuously to the aeration tank and the effluent is discharged. The recycling to activated sludge helps in the initial build up of high concentration of micro-organism in the mixed liquor, which accelerates BOD removal.

8.2.2. Advantages

(*a*) The operating head required is less and hence no pumping is needed.

(*b*) The effluent produced is clear and non-putrescible.

(*c*) Limited land area is required in this process.

(*d*) There is no offensive odour or fly nuisance during operation.

(*e*) There is a control over the quality of effluent desired due to greater flexibility treatment.

(*f*) The sludge obtained from this process has commercial value.

8.2.3. Disadvantages

(*a*) Initial investment is more.

(*b*) Skilled personnel and careful supervision is required for operation and maintenance.

(*c*) The operation plant is more sensitive to the change in quality of influent.

(*d*) It produces large volume of sludge which is difficult to handle.

8.3. METHODS OF AERATION

Methods employed for aerating the mixed liquor are generally of two types.

(*a*) Diffused aeration or diffused air system

(*b*) Mechanical aeration

(*a*) Diffused Aeration: In this method, mixed liquor (sewage mixed with activated sludge) is aerated by introducing compressed air through submerged porous diffusers or air nozzles. The diffusers are in the form of porous tubes or porous plates of aluminium oxide or quartz, from which the air is released in the form of fine bubbles. Tube diffusers are suspended along one side of the tank, whereas the plate diffusers are placed at the bottom of the tank. The plate diffusers are commonly used in practice and are either flat tiles of 300 mm × 300 mm × 250 mm with pores of about 0.3 mm diameter or dome-shapped tiles of 100 mm to 175 mm in diameter. Dome-shaped tiles are more popular these days. Generally two types of arrangement of diffuser tiles are adopted.

(*i*) Ridge and furrow types

(*ii*) Spiral flow type

In ridge and furrow type, ridge and furrows are constructed in the floor of the tanks, as shown in Fig. 8.4. The diffusers are placed on the furrows and are connected with air mains.

The diffusers are placed on one side of the tank in spiral flow method. The tank corners are chamfered so that the air bubbles rising at one side of the tank are deflected inducing rotation of the sewage.

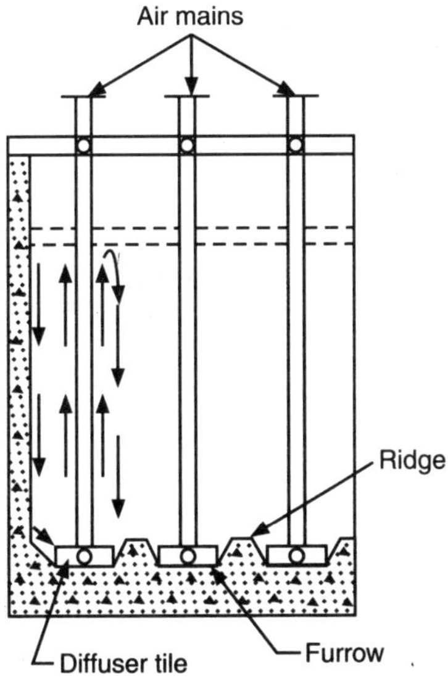

Fig. 8.4. Ridge and furrow tank

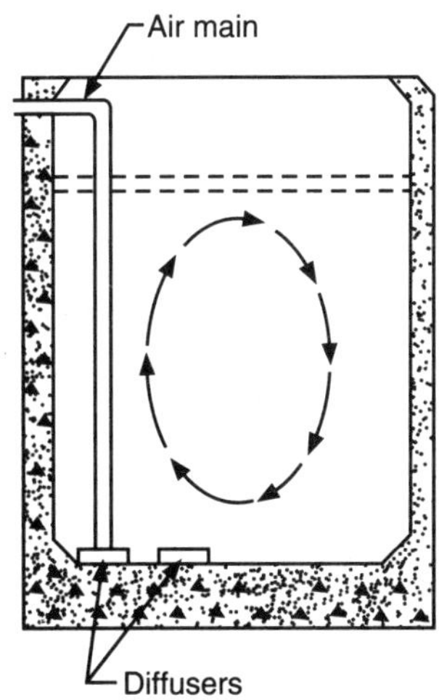

Fig. 8.5. Spiral flow tank

***(b)* Mechanical Aeration:** In mechanical aeration method, the sewage is stirred constantly by mechanical means to bring it in contact with the atmospheric air. This method requires lesser detention period and is simple in operation and maintenance. Generally two types of aerators are used.

(*i*) Simplex aerator (Surface aerator)

(*ii*) Dorr co aerator (Imhoff aerator)

***(i)* Simplex Aerator:** In this method, the aeration tank is either circular or square in shape with hopper at the bottom and is vertical flow type. It is shown in Fig. 8.6. The tank is fitted with a central uptake tube which is widened out at the bottom and has an inverted cone suspended centrally

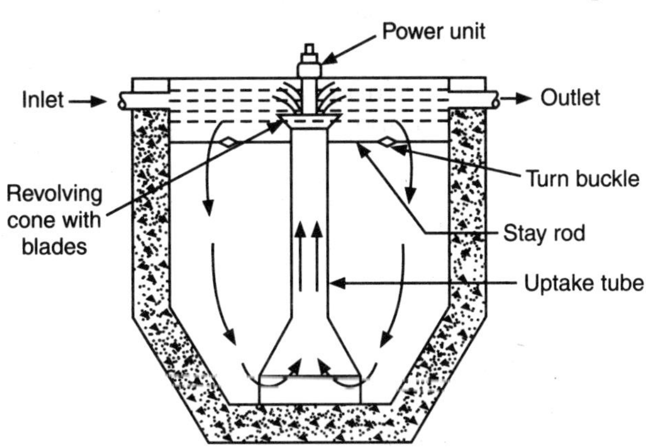

Fig. 8.6. Simplex aerator.

at the top of the tube. The cone has vanes on the upper side and is driven by motor. During its revolution, it sucks up sewage from the bottom and sprays at the top, thereby setting a circular motion in the sewage. The detention period is 8 hours or more for complete treatment.

(*ii*) Dorr Co Aerator: In this aerator, diffusion and mechanical aeration are combined in one unit, as shown in Fig. 8.7. It consists of a tank of 3 to 4.5 m deep and of equal width. Diffusers are fixed at one side in the bottom of the tank and submerged paddle wheel is fixed at mid-depth of the tank. Paddle has two blades mounted on a horizontal shaft which is rotated by motor with a speed of 10–12 rpm in the direction opposite to that the rising air bubbles. The detention period is 3 to 4 hours.

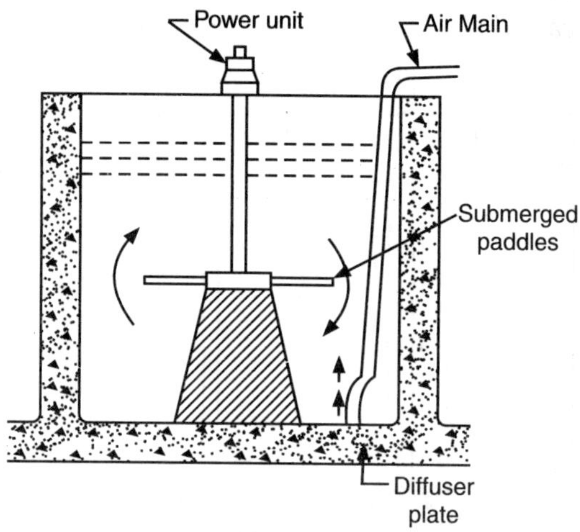

Fig. 8.7. Door Co Aerator

8.4. SLUDGE DIGESTION

This is an effective method for the treatment of highly concentrated organic waste. Sludge digestion is the process of decomposing highly organic matter of sewage-sludge anaerobically under conditions of adequate operational control. By digestion sludge is broken up into three different forms:

(*a*) humus like solid matter

(*b*) supernatant liquor containing liquified and finely divided solid matter

(*c*) gases like methane (CH_4), carbondioxide (CO_2), nitrogen (N_2) etc.

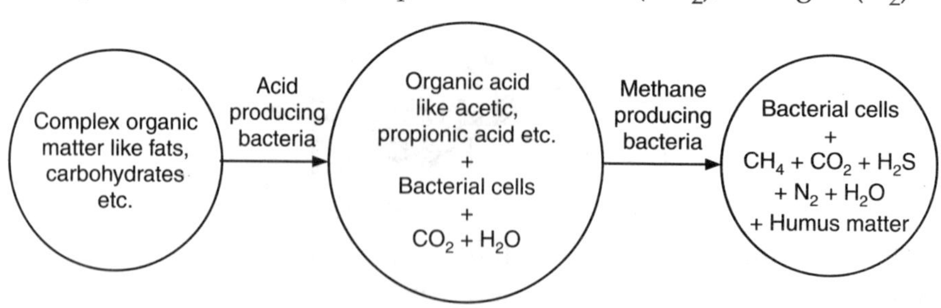

Fig. 8.8. Stages of sludge digestion

Advantages

(*a*) Digested sludge is used as fertiliser and the gas produced is used as fuel.

(*b*) It reduces the volume of the sludge.

(*c*) It breaks down the organic matter into simple compounds

The digestion of sludge is temperature dependent and the period of digestion varies with temperature. Sludge digestion is carried out in a tank called sludge digestion tank.

Table 8.1. Period of digestion at different temperatures

S.No.	Temperature in °C	Digestion period in days
1.	10	75
2.	15	56
3.	20	42
4.	25	33
5.	30	27
6.	35	24
7.	40	25
8.	45	23
9.	50	16
10.	55	14
11.	60	18

8.4.1. Sludge Digestion Tank (Digestor)

A sludge digestion tank is a RCC or steel tank with hopper bottom. It is covered with fixed or floating type of roof, as shown in Fig. 8.9.

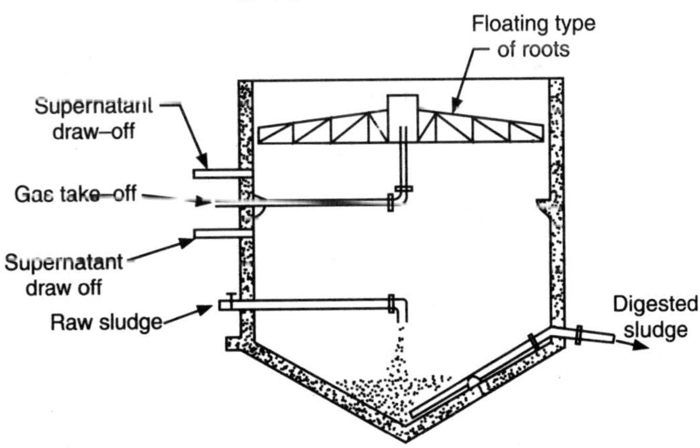

Fig. 8.9. Floating cover sludge digestion tank

The floating type roof makes the operation much more effective. The diameter of the tank varies from 6 m to 55 m and the depth of liquid in the tank should not be more than 9 m. The bottom slope should not be less than 1 in 12.

Table 8.2. Design capacities of digestion tanks on per capita basis

S.No.	Type of raw sewage to be digested	Gms of volatile solids present per litre of sludge per month	Per capita capacity in litres/capita
1.	Primary	8	21
2.	Mixture of primary sludge and secondary sludge from trickling filters	7.36	36
3.	Mixture of primary sludge and secondary activated sludge	5.76	61
4.	Chemically coagulated sludge	–	56

Raw sewage is pumped into the digestor through pipes where it is mixed with digested sludge. Due to digestion gases are formed which are collected in the gas dome. The digested sludge which settles down to the bottom of the tank is removed periodically under hydrostatic pressure. The tank is provided with a heating coil to maintain the optimum temperature. The liquid between the scum and the sludge called *supernatant liquid* is withdrawn through number of pipes located at different elevations of the tank. The scum formed is either broken up automatically due to recirculation of flow or through mechanical rackers called scum-breakers.

The volume of sludge gas produced varies from 0.014 to 0.028 m^3 per capita, normal value of 0.017 m^3 per capita. The sludge gas produced contains 65% of methane (CH_4), 30% of carbon dioxide (CO_2) and 5% of nitrogen and other inert gases. Methane has a calorific value of 5400-5850 k.cal/m^3.

8.5. SLUDGE TREATMENT

The process of treating sludge before disposing off finally is called sludge treatment.

8.5.1. Objects of Sludge Treatment

- To reduce the water content for final disposal.
- To stabilize the organic matter contained in the sludge.
- To destroy the pathogenic organisms.

8.5.2. Methods of Sludge Treatment

The most commonly used method of sludge treatment is sludge digestion. The sludge produced in the settling tanks contains more than 90% moisture and 10% solid. For effective and quick treatment of sludge, the solid concentration should be more. Solid concentration of sludge is increased by giving some pre-treatment. They are :

(*a*) Conditioning (*b*) Elutriation

(*c*) Gravity thickening (*d*) Vaccum filteration.

(*a*) Conditioning of Sludge: This method is also called sludge thickening. In this method certain chemicals called conditioners, are added to the sludge to increase the rate of dewatering. The most commonly used conditioners are iron salts like $FeCl_3$, lime, Alum and Polyelectrolytes. These conditioners act by binding sludge particles together and permitting water to run-out.

(*b*) Elutriation of Sludge: Washing the sludge with water or effluent is known as elutriation. In this, the sludge is washed with water to remove ammonia compounds which interfere with coagulants and they are difficult to dewater. Washing is done by gentle mixing of water with sludge in the ratio of (1.5 :1 to 4.5 : 1) for about 10 mins and then allowing the mixture to settle.

(*c*) Gravity Thickening of Sludge: This is the most common method for concentration of sludge. Gravity thickening is useful pre-treatment prior to other mechanical methods of dewatering.

(*d*) Vacuum Filteration of Sludge: This is the most widely used mechanical sludge dewatering. By this method, activated sludge can be concentrated upto 25% of solids and primary sludge upto 32% of solids.

8.6. SLUDGE DRYING

When sludge is to be used as fertiliser, its moisture content is to be reduced to less than 10%. So the filtered sludge will require drying. The drying of sludge is done by the following methods:

(*a*) Air drying

(*b*) Heat drying

(*a*) Air Drying: This is one of the most common method of sludge drying. In this, under favourable climatic condition well digested sludge is run on to a sand bed of thickness 200–300 mm. It will dry within a week or two without any odour. Drying of fresh sewage sludge is not suitable by this method as it gives foul odour while drying and does not lose moisture satisfactorily. Air-drying of sewage can be done in open-air or under glass covered roof. In air drying, moisture is lost to the atmosphere by evaporation and to the drying bed by percolation.

The drying bed consists of a layer of 300 to 450 mm depth of graded gravel or stone ballast over which 150 to 300 mm depth of coarse sand is provided. Below the gravel layer underdrains are provided.

(*b*) Heat Drying: In this, the sludge is dried in rotary drier. Two types of rotary driers are in use—the kiln drier and cage mill flash drier. The drying can be done more efficiently if it is mixed with previously dried sludge to reduce the moisture contents.

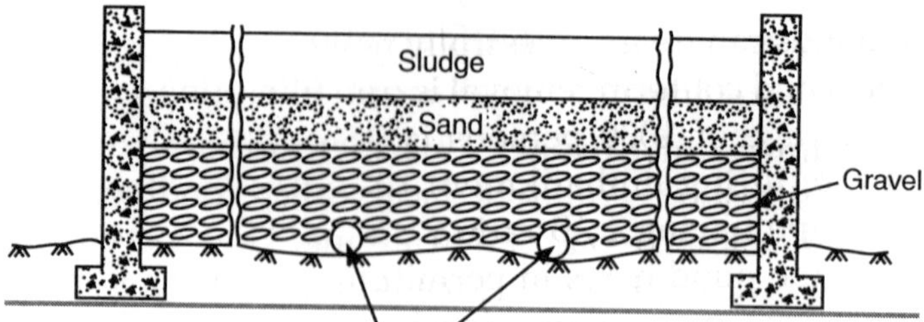

Fig. 8.10. Sludge drying bed.

8.7. SLUDGE DISPOSAL

Sludge disposal is carried out by the following methods:

(*a*) **Land Dumping:** At place where sufficient lands are available, sludges can be disposed off on open land. But the land should not cause contamination of surface and groundwater.

(*b*) **Shallow Burial into the Ground:** Sludge is disposed off by running into trenches of cross-section 0.9 × 0.6 m and spaced at 0.9 to 1.5 m apart. When the sludge dried up in the trenches, it is covered with a thin layer of soil. The land is then ploughed with powered lime and planted with crops after a month.

(*c*) **Lagooning:** In this method, earth tanks or ponds of depth 0.6 to 1.2 m are made and are provided 100 mm dia tile underdrains spaced at 2.7 m apart. Bottom of pond is covered with 150 mm thick layer of clinker or ashes. Sludge is then run into it and allowed to remain for 2 to 6 months. When the sludge dried up, the residue is used as manure.

(*d*) **Sea Dumping:** This method is suitable for coastal sites. The sludges are discharged into deep waters. Action is through the process of dilution.

8.8. OXIDATION PONDS

Oxidation ponds or lagoons or waste stabilization ponds are shallow body of water (ponds) in which sewage can be retained for a sufficient time to satisfy the bio-chemical oxygen demand (BOD) and thereby make the sewage non-putrescrible.

Treatment in a oxidation pond depends on the conditions such as sunlight, algae, evaporation, percolation, sedimentation and oxidation. The bacteria metabolize organic matter releasing nitrogen, carbondioxide etc. Algae use these compounds alongwith sunlight for synthesis releasing oxygen. This oxygen is taken up by bacteria for metabolism, thus closing the cycle. This type of relatinship between bacteria and algae is called *symbiosis*.

Oxidation ponds can treat successfully raw sewage. The BOD removal is as much as 90% and coliform removal is upto 99%. The area required for pond to treat sewage of 1000 persons is 0.2 to 0.4 hectares. These are constructed by building embankment of earth. These are generally rectangular in shape having L/B ratio 2 : 1 and side slopes 1 : 1.5. The shallow depth of the pond helps in permitting penetration of sunlight to all parts of the sewage encouraging algael growth.

8.8.1. Advantages and Disadvantages of Oxidation Ponds

Advantages

(*i*) The efficiency of BOD removal is high.

(*ii*) Maintenance of oxidation ponds is easy.

(*iii*) Ponds suitable in small cities where large land areas are cheaply and easily available.

(*iv*) Ponds quite suitable in Tropical countries having warm temperatures and dry climate.

(*v*) These provide suitable contents in industrial use.

Disadvantages

(*i*) They create nuisance due to mosquitoes breeding.

(*ii*) They require more land area.

(*iii*) They are costly to maintain.

(*iv*) They tend to pollute underground water.

8.9. CHLORINATION OF SEWAGE

It is done in sewage-treatment works as a disinfectant. It is used to reduce its BOD value, to suit the ability of receiving waters. Action and methods of application of chlorine are the same as explained in the Author's Water Supply Engineering.

8.9.1. Aims of Chlorination

(*i*) It prevents corrosion of sewers.

(*ii*) It controls sewage nuisance.

(*iii*) To disinfect sewage especially when the effluent is to be discharged into body of water to be used for water supply.

(*iv*) It reduces BOD (10 to 35%).

(*v*) It acts as a general disinfectant.

(*vi*) To prevent algae growth and plant life in sewers.

(*vii*) To reduce ponding in case of trickling filters,

(*viii*) Chlorine is used for grease removal.

(*ix*) To kill the filter flies

(*x*) It forms floc in coagulation in combination with other chemicals.

(*xi*) To make treatment units more effective.

8.9.2. Chlorine Dose

Here the amount of chlorine dose required is much larger as compared to water purification. It is due to great amount of organic matter in sewage which tries to neutralize the chlorine. Refer Table below for recommended dose values:

Table 8.3. Chlorine dose values

Type of effluent	Dose (ppm)
(*i*) Fresh sewage	10
(*ii*) Septic sewage	20
(*iii*) Intermediate sand filter effluent	2
(*iv*) Trickling filter effluent	3 to 6
(*v*) Activated sludge effluent	5
(*vi*) Imhoff Tank effluent	8 to 10

8.9.3. Application of Chlorine

Action and methods of application of chlorine to sewage are the same as used for water. Here the addition of chlorine has an immediate effect upon the oxidation of the organic matter while the bacteria are subsequently attacked. It may be noted that 1.5 to 2.0 ppm. Chlorine is required for p.p.m. deduction in BOD.

MODEL QUESTIONS

1. (*i*) Sludge digestion is _______ process. (aerobic/anaerobic)
   (*ii*) Activated sludge process has _______ efficiency.
   (*iii*) More land is required in case of _______ .
   (Activated Sludge Process/Trickling Filter Process)
2. How is capacity of senice reservoir determined?
3. Describe an oxidation ditch.
4. (*a*) What is sludge? When is it necessary to treat the sludge.
   (*b*) What are the methods of disposal of sludge?
5. Write explanatory notes on:
   (*a*) Aeration. (*b*) Activated Sludge.
6. Describe the detritus tank and screens used for sewage treatment. The capacity detritus tank and screens should be given in S.I. units.
7. What is activated sludge and how does it contribute to efficient treatment of sewage?
8. Describe the activated sludge process. Compare it with trickling filter.
9. What is the difference between a contact bed and Trickling filter and how do these help in the purification of the sewage?
10. Define the following :
    (*i*) Pre-chlorination and Post-chlorination.
    (*ii*) Plain chlorination and Super-chlorination.
11. Explain "Activated sludge and sludge digestion".

CHAPTER 9

BUILDING DRAINAGE

Learning Objectives

- Building Drainage
- Aims of Building Drainage
- Requirements of Building Drainage
- Terms used in Building Drainage
- Flushing Storage Tanks
- Traps
- Layout of House Drainage
- Sanitary Fittings
- Plumbing
- Inspection and Testing
- Fixtures in Modern Bathrooms

9.1. BUILDING DRAINAGE

The arrangement provided in a building for collecting and conveying the waste water through drain pipes by gravity to join a sewer is known as *Building Drainage.*

9.2. AIMS OF BUILDING DRAINAGE

(*i*) To dispose off liquid waste as early as possible.

(*ii*) To prevent foul air entering the building from the sewer.

(*iii*) To dispose off storm water into open surface drains.

(*iv*) To facilitate quick removal of foul matter i.e., human excreta etc.

(*v*) To collect and dispose the water matter systematically.

(*vi*) To provide healthy conditions in a building.

9.3. REQUIREMENTS OF BUILDING DRAINAGE

(*i*) The lay out of building drainage must be simple.

(*ii*) The lay out of building drainage must allow quick removal of foul waste matter.

(*iii*) Levels of building, sewer, outlets must be determined accurately.

(*iv*) The drains must be laid in straight lines.

(*v*) The drains should provide self cleaning velocity.

(*vi*) The drains should not pass under buildings as far as possible.

(*vii*) The diameter of drains must be sufficient for carrying domestic sewage and sullage.

(*viii*) The drainage pipes must be of non-absorbent materials.

(*ix*) To prevent entry of foul gases from the sewer into the building, intercepting trap should be provided in house drain.

(x) The branch drains should be short as far as possible.

(xi) All the pipes should be laid with water-tight joints.

(xii) The drainage system must be vantilated properly.

(xiii) There should be proper means of access for inspection of drainage system.

(xiv) Proper arrangement must be ensured for the quick disposal or rain- water from buildings.

9.4. TERMS USED IN BUILDING DRAINAGE

(*As per I.S. Code of Practice for Building Drainage*) Terms associated with code of practice are as under:

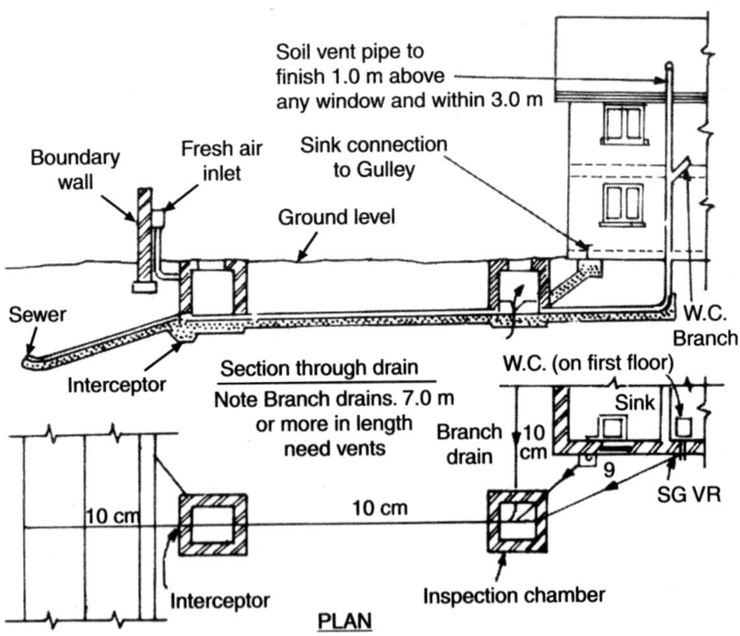

Fig. 9.1. House drainage

(i) Flushing Tank: It is a cistern with discharge arrangements for Flushing & W.C. (water closet), urinal drain or sink.

(*ii*) Stack: It is a general term used for any vertical line of drainage. Pipe may be installed in vertical, horizontal or inclined position.

(*iii*) **Branch:** It is a part of the Piping system discharge into a stack.

(*iv*) **Cross Connection:** It is an arrangement connecting two piping systems. One containing potable and other polluted water.

(*v*) **Back Siphonage:** It is the flowing back of used or polluted water from a plumbing fixture or vessel into a water supply pipe due to a lowering of pressure in such a pipe.

(*vi*) **Water seal:** It is a vertical distance between the dip and crown weir of a trap (See Fig. 9.2).

(*vii*) **Waste Pipe:** A pipe which carries liquid waste that does not include night soil.

(*viii*) **Vent Pipe:** It is a pipe line installed to provide flow of air to drainage system in order to protect the water seal of traps against back pressure and siphonage.

(*ix*) **Ventilating Pipe:** It is a pipe which enables the foul gases of the sewers to be safely discharged into the atmosphere.

(*x*) **Anti Siphonage Pipe:** It is a pipe to preserve the water seal or traps through proper ventilation.

(*xi*) **Self Siphonage:** It is the breaking up of the trap-seal as a result of removing the water therefrom by the discharge of the fixture to which the trap is connected.

(*xii*) **Induced Siphonage:** It is the breaking up of trap seal as a result of removing the water therefrom by the sudden flush of water from a fixture on the upper floor.

(*xiii*) **House connection:** It is the connection of a drain with a public sewer laid under the street.

(*xiv*) **Storage Tank:** A tank for storage of water which is connected to the water main by means of a supply pipe.

(*xv*) **Geyser:** It is an appliance for heating of water with a water control on the inlet side and free at the outlet.

9.5. FLUSHING STORAGE TANKS

These tanks are required to supply water to all flushing cisterns through down-take-pipes. It is desired that supply pipes should not feed these cistern directly. Flushing storage is based on the number of W.C. seats in case of residential building. In case of public building it is based on number of urinals and water closet seat. Refer Table 9.1 and 9.2 for Flushing storage capacities for different types of Buildings.

Table 9.1. Flushing Storage Capacities (As per I.S. 2065-1963)

Type of buildings	Storage capacities
1. For tenements having common conveniences.	900 litres net per W.C. Seat.
2. For residential premises other than tenements having common conveniences.	270 litres net for one WC Seat and 180 litres for each additional seat in the flat.
3. For Public Assembly Halls, Cinemas.	900 litres per WC seat and 360 litres per urinal seat.
4. For Workshop and Factories.	900 litres per WC seat and 180 litres per urinal seat.

Table 9.2. Capacity of various fixture units

Description of Fixture Unit	Capacity of Storage in litres/day/unit
Wash Hand Basin	70
Kitchen Sinks	100
Shower	135
Urinal	160
Bath Tub	200
Flushing Cistern for WC	270

9.6. TRAPS

Trap is a bend or depression in a Sanitary fittings which is always full of water. It does not allow air connection between the fitting and outside drain or sewer pipe. Foul gases thereby cannot enter the Building to cause nuisance. Efficiency of the trap depends on depth of water seal. Its minimum value is 50 m.m. Greater the depth of seal, more effective is the trap. Depth of seal in practice varies from 25 mm to 75 mm.

9.6.1. Requirement of a Good Trap

(*i*) It should be made of non-absorbent material.

(*ii*) It should provide good depth of water seal all the times having large surface area.

(*iii*) It should be self-cleaning having no projections inside which may obstruct the flow of sewage.

(*iv*) It must be provided with suitable access for cleaning.

9.6.2. Seal in Traps

The vertical distance between inside lowest point (or dip) and inside highest point (or crown weir) is known as water seal or Seal in Trap.

Depth of water seal represents strength of a trap. Depth of seal varies from 25 mm. to 75 mm.

9.6.3. Causes of Breaking of a Seal

(*i*) If partial vaccum is created in the sewer fittings, it will suck up the seal water.

(*ii*) It may break due to any crack in the bottom of seal.

(*iii*) It may be due to faulty joint.

(*iv*) If due to blockage, there is increase in the pressure of the sewer gases, it will pass through the seal water.

(*v*) Water of seal may evaporate when the seal is not in use for a long time.

9.6.4. Precautions to be Taken

(*i*) To avoid breakage of water seal due to creation of partial vaccum, the portion between the trap and the soil-pipe must be connected to the vent pipe.

(*ii*) An anti-siphonage pipe is used in the buildings to prevent breakage of trap seal.

9.6.5. Classification of Traps

Traps may be classified into following ways:

(*A*) According to their Shape

(*i*) P-Trap

(*ii*) Q-Trap

(*iii*) S-Trap.

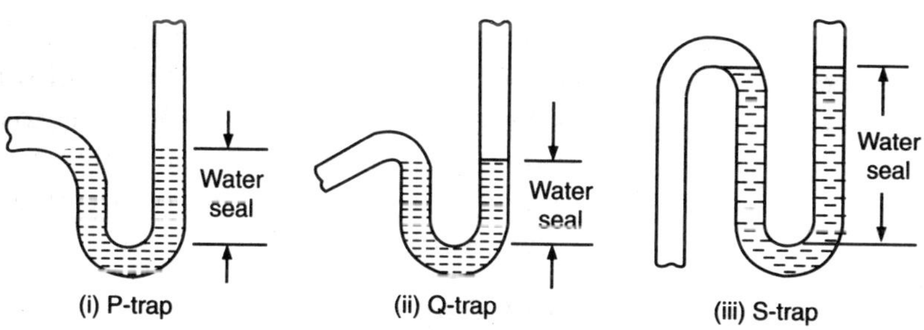

Fig. 9.2. Trap

These traps essentially consists of a U-Tube which retain water acting as a seal between foul gases and atmosphere.

(*B*) According to their Uses:

(*i*) Floor-traps

(*ii*) Gully-traps

(*iii*) Grease-traps

(*iv*) Intercepting Traps

(*i*) Floor Traps: The traps which are used for receiving surface waste water from the floor of bath and kitchens are called Floor Traps. A form of floor trap much used in practice is the Nahani Trap.

A floor trap is used for admitting waste water from floors of kitchens, baths, courtyards or washing places. As shown in the Fig 9.3, it is provided with CI grating at the top to avoid passing of coarse solid matter to house drain.

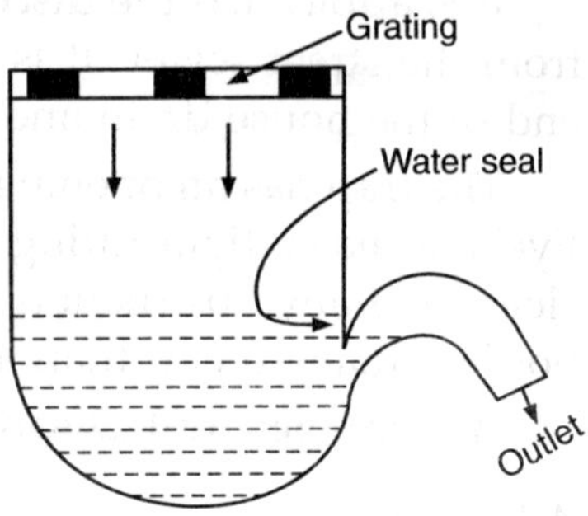

Fig. 9.3. Floor Trap

(ii) Gully Traps: The traps which are used to receive waste water from wash basins, sinks, baths etc., are called Gully traps.

These traps are provided at different places in the drain pipes. Sullage from baths, sinks, basin etc., enters in through back inlet and rain or surface water from house tops etc., enters from top. A coarse screen grating is provided at the top to prevent entry of solid matters in the tap. Fig. 9.4 shows gully trap.

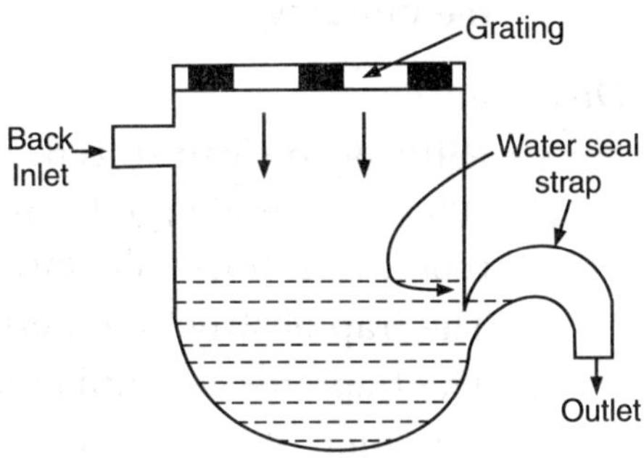

Fig. 9.4. Gully trap

(iii) Grease Traps: The traps provided in the plumbing system to remove grease oil etc., are called grease traps.

It is a variety of gully trap. It is fixed outside Kitches, Hotels Canteens and wash up rooms near motor garrages to intercept grease from entering the drainage system.

(vi) Intercepting Traps: The traps which are provided at the junction of a house drain and sewer are called an Intercepting Traps.

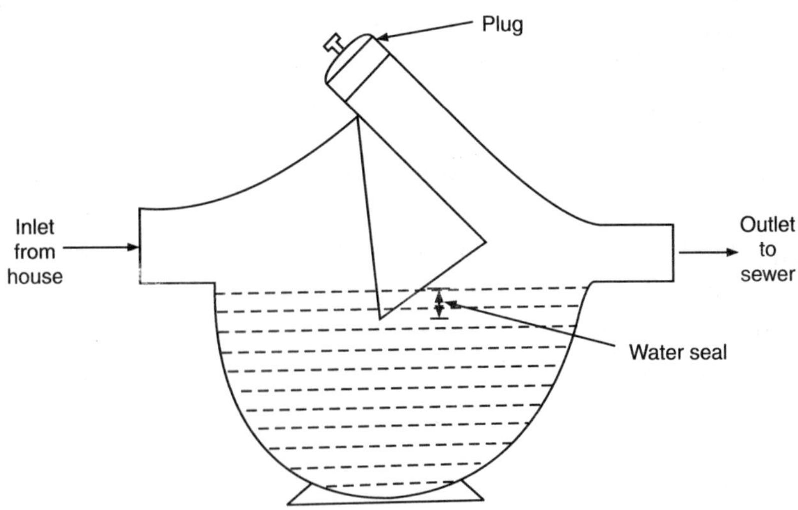

Fig. 9.5. Intercepting Trap

It is also called the disconnecting trap. It disconnects the house drain from the street sewer. It is placed in small chamber between the lowest end of the house drain and the street sewer.

The trap has an opening at the top called 'Rodding Arm' or 'Cleaning Eye' and has a tight fitting plug. This enables the trap to be periodically cleaned of any obstruction present inside. It has a deeper water seal (i.e., not less than 10 cm) than an ordinary trap.

Advantages and Disadvantages of Intercepting Traps

Advantages

(*i*) Helps in quick removal of foul matter received from building drains into the sewer.

(*ii*) Helps in preventing entry of foul gases from sewer into the building.

(*iii*) Helps in preventing entry of harmful bacterias from sewer into the building.

Disadvantages

(*i*) Difficult to clean trap through inspection arm.

(*ii*) When the discharge from building drain is less then solid matter retained in trap will start decomposing.

(*iii*) The trap obstruct the normal flow of sewage.

(*iv*) The trap effects ventilation of sewers.

(*v*) Provision of trap increases the cost of building drainage.

9.7. LAYOUT OF HOUSE DRAINAGE

The following points should be kept in mind during layout of house drainage:

(*i*) Layout should be direct and simple as far as possible. All pipes (horizontal or vertical) should be laid in straight lines.

(*ii*) Pipe should be laid in such a way that any abrupt changes in the direction of flow should be avoided.

(*iii*) The drains should be laid at such a slope that self-cleaning velocity is developed in them. (Ref. Table 9.3).

(*iv*) At the pipe intersections and bends, manholes or inspection chambers should be provided.

(*v*) For gradual and smooth change in direction of flow the contained angle between the intersecting pipes should be less than 45°.

(*vi*) Laying of sewers under a building should be avoided for fear of undermining its foundation, when it is unavoidable for want of sufficient gradient in the sewer, then only cast iron pipes should be used.

Table 9.3. Recommended gradient, discharge for pipe sewer

S.No.	Dia. of pipe (mm)	Gradient	Discharge (m^3/minute)
1.	100	1 in 35	0.18
2.	150	1 in 65	0.42
3.	230	1 in 120	0.93
4.	300	1 in 200	1.66

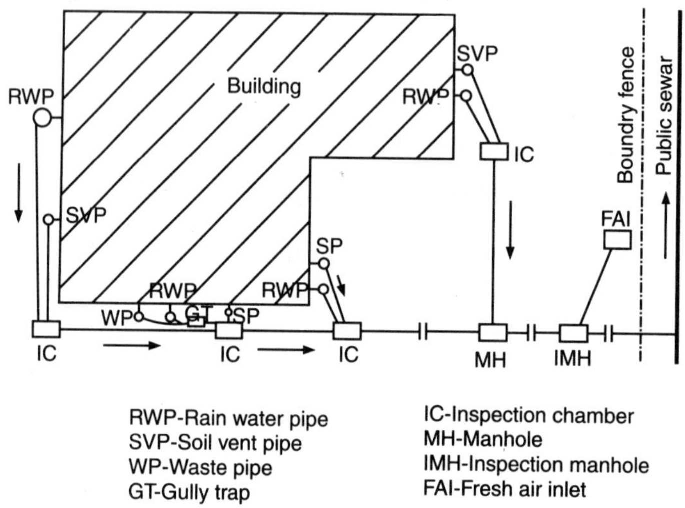

Fig. 9.6. Typical layout of house drainage system

9.8. SANITARY FITTINGS

The following sanitary fittings are used in the drainage system of buildings:

1. Sinks
2. Wash basin
3. Water closets
4. Urinals
5. Baths
6. Flushing cisterns

9.8.1. Sink

Sink is a rectangular receptacle used in kitchen or laboratory for draining off water. It is commonly made of glazed earthen ware. Now-a-days kitchen sinks made of steel are mostly in use. The sink is provided with circular waste hole. The floor of sink is given a slope towards the waste hole. The kitchen sink is provided with a draining board on the right of the user. Usual size of sink is 600 mm × 450 mm × 250 mm. Other sizes of sink are also available which are given in Table 9.4. The sink is so installed that the height of the top of the sink from the floor is about 900 mm. It should be located as to directly receive light through window. Fig. 9.7 shows kitchen sink.

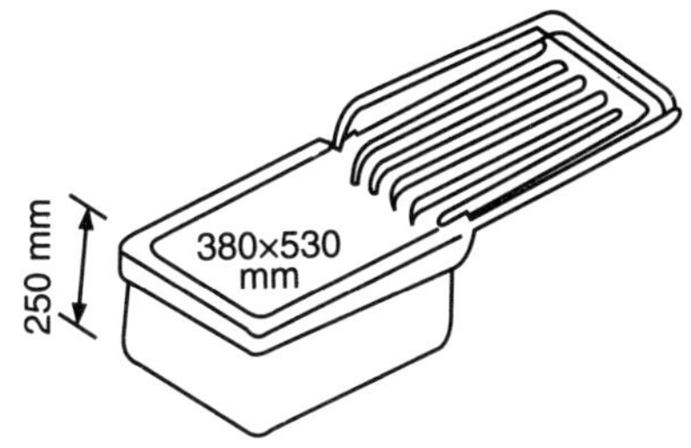

Fig. 9.7. Kitchen sink

Table 9.4. Different sizes of sink

Type of sinks	Dimensions (mm)
Kitchen sinks	750 × 450 × 250
	600 × 450 × 250
	600 × 450 × 150
Laboratory sinks	600 × 400 × 200
	500 × 350 × 150
	450 × 300 × 150
	400 × 250 × 150

9.8.2. Wash Basin

It is made of vitreous china clay, stoneware etc. It is generally of two types:

(*a*) Flat back

(*b*) Angle back

Flat back type is for fixing on walls and angle back type is for fixing at the junction of two walls. Both the types are either fixed on brackets secured into the wall or on pedestals rising from the floor. It is provided with a circular waste hole in the bottom and an internal soap holder recess which drains into the bowl. The wash basin is fixed at a height of 750 to 800 mm from the floor. Table 9.5 gives different sizes of wash basin. Fig. 9.8 shows wash basin.

Table 9.5. Different sizes of wash basin

Type	Size (mm × mm)
Flat back	550 × 400
	630 × 450
Angle back	400 × 400
	600 × 480

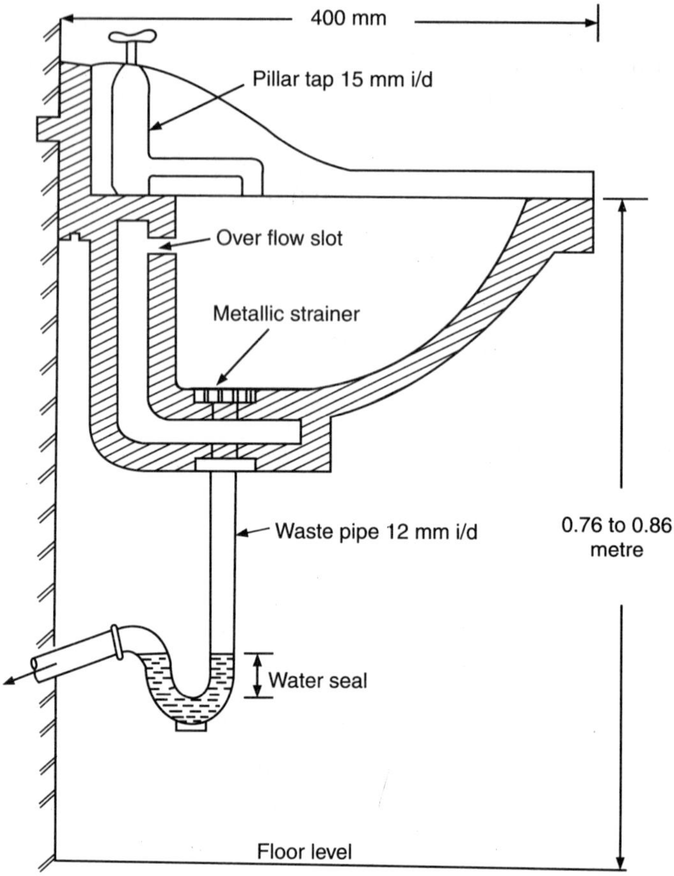

Fig. 9.8. Wash basin

9.8.3. Water Closets (WC)

Water closet may be defined as a water flushed plumbing fixture designed to receive human excreta directly from the user. It is connected to the soil pipe by means of a trap.

In residential buildings they should be located keeping in view the prevailing winds, so as not to cause any nuisance.

They are made of either of Glazed Earthware or Porcelain and are available in several varieties in the market.

9.8.3.1. Requirements of Good Water Closet

(*i*) It should be such that night soil do not stick to it and flow down easily.

(*ii*) It must provide easy flushing with less quantity of water.

(*iii*) It must be suitable in use for all persons.

(*iv*) When night soil falls into the trap, its water should not be splashed.

9.8.3.2. Types of Water Closets: Water closets are of following types

1. Indian or squatting type
2. European or Pedestal type

1. Indian or Squatting Type: This is also called orissa pattern of Indian WC. It is made of vitreous China clay having its inner portion glazed to make it easy in cleaning. The pan is 450 mm to 630 mm in overall length and 450 mm to 500 mm height. The pan is connected with P-trap or s-trap and with flushing cistern. The trap is connected to the soil pipe. Footrests are constructed integratedly with pan or are separately fixed on the floor and finished with mortar. Fig. 9.9 shows Indian type WC.

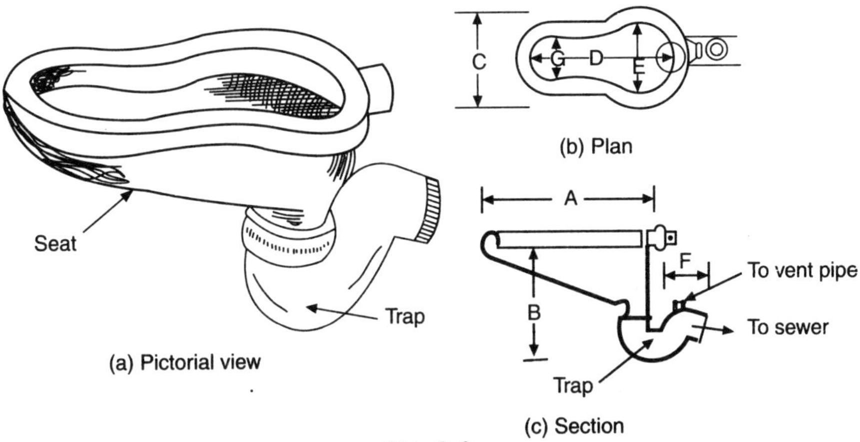

Fig. 9.9.

2. European or Pedestal Type: This is a wash down water closet provided with a seat and cover. It is fitted with P-trap or S-trap. European type WC is made of vitreous china clay with its inner portion glazed. The overall length is 630 mm and height 340 mm to 400 mm. This type of WC requires less space and is flushed by a low level cistern. Fig. 9.10 shows European type of WC.

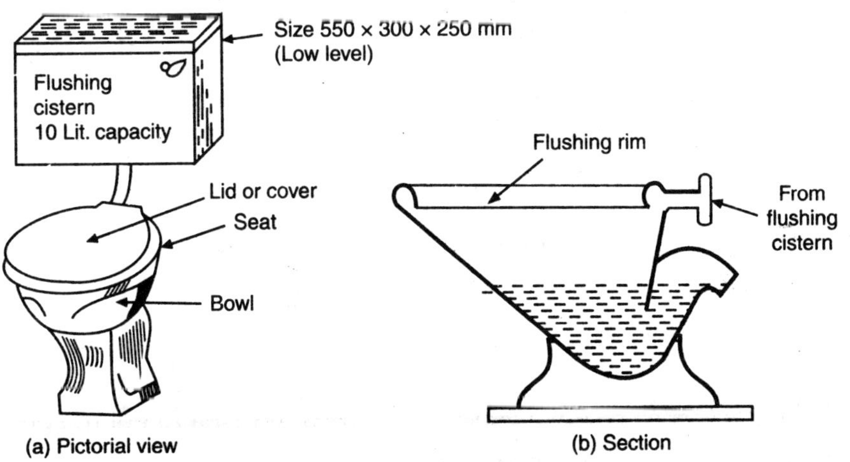

Fig. 9.10. European type

9.8.4. Urinals

The arrangement provided for receiving human urine is called urinals. Urinals are available in different sizes and patterns. Table 9.6 shows sizes and patterns of urinals.

Table 9.6. Different patterns of urinals

Pattern	Sizes
(*i*) Bowl type	Mini 450 × 250 × 350 mm.
(*a*) Flat back	Mini 350 × 425 × 275 mm.
(*b*) Angle back	Single 1150 × 450 × 400 mm.
(*ii*) Stall type	Single 450 × 1000 mm.
(*iii*) Slabtype	100 × 100 mm.
(*iv*) Squatting plate	350 × 450 mm. to 350 × 600 mm.

Bowl type urinals are of one piece construction and are made generally of porcelain, vitreous china clay. Each urinal is provided with two fixing holes on the side for fixing it on the wall. The height of the bowl above the floor should not be more than 500 to 600 mm. In a battery of urnials, the centre-to-centre distance between two bowls should not be less than 750 mm. At the bottom of the bowl an outlet horn is provided for connecting it to the trap. The bottom of urinal is given a proper slope from the front for efficient drainage of the urinal. These urinals are also provided with flushing arrangement. Fig. 9.11 shows bowl type urinals.

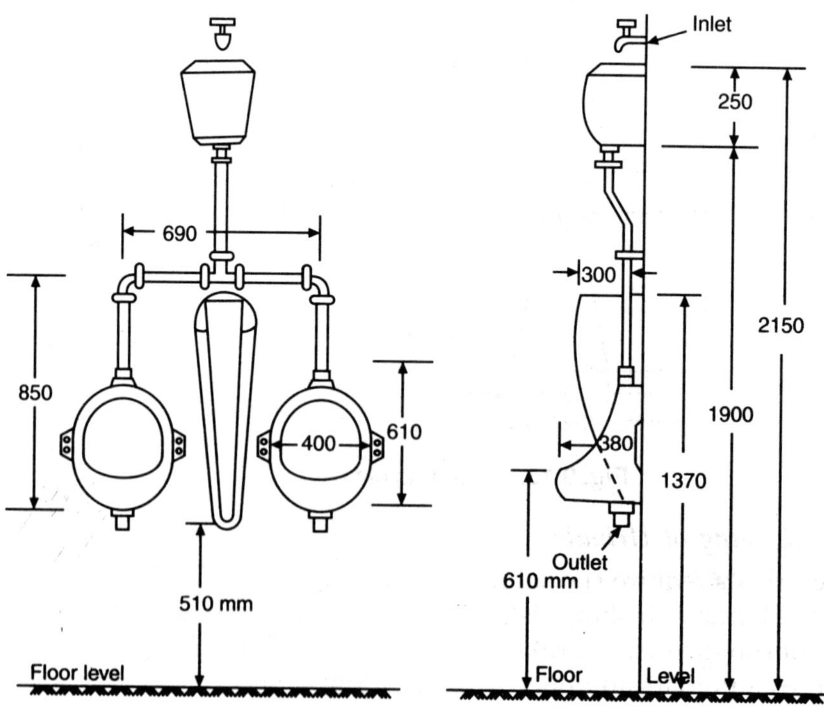

Fig. 9.11. Bowl type urinals

The slab and stall type urinal consists of a flat wall slab with partitions on sides and floor channel to drain off the discharge through a trapped outlet (See Fig. 9.12). These are generally provided in public places like restaurants, cinema houses, railway stations etc.

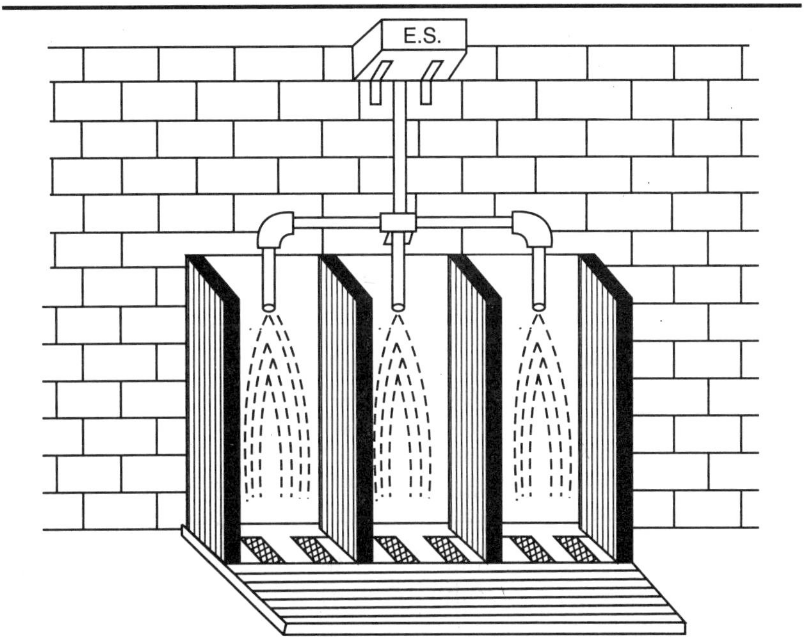

Fig. 9.12. Slab and stall type urinals

The squatting plate urinals are mostly installed in ladies lavatories. These are one piece construction as shown in Fig. 9.13.

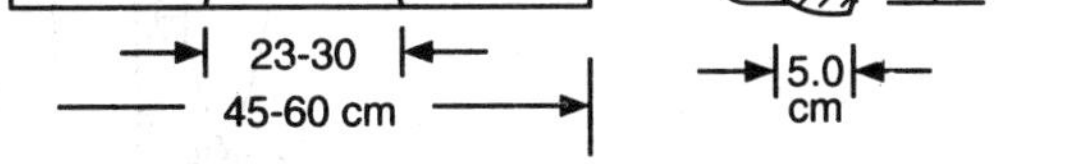

Fig. 9.13. Squatting plate urinals

9.8.4.1. Flushing of Urinals

The urinals require special flushing to prevent their offensive odour. A small automatic flushing cistern of 2.50 to 10 litre. capacity is provided for the purpose to each urinal or a set of urinals. This flushing cistern discharges automatically once in every 15 minutes or so. Sometimes in case of bowl type the hand operated flushing cistern of 5 litre. capacity is used.

Table. 9.7. Sanitary fitting for office building

Fittings	For male personnel	For female personnel
Wash Basins:	1 for 30	1 for 30
Water closets	1 for 25	1 for 20
Urinals	2 upto 20	1 for 15
	2 for 21-45	1 for 15
	3 for 46-75	-do-
	4 for 76- 100	-do-
Baths	1 for each floor	-do-
Drinking fountain	1 for every 100	-do-
Ablution taps	1 in each WC	-do-
Cleaners sinks	1 for each floor	-do-

9.8.5. Flushing Cistern

The arrangement provided for flushing out water closets or urinals is called a Flushing Cistern.

It is used for flushing urinals and water closets after use. It is made of cast iron of minimum thickness 5 mm. The internal parts are of corrosion-resisting material. Cistern for a urinal or group of urinals may be of 4.5 litre. capacity. Nowadays plastic cisterns are also available in the market.

9.8.5.1. Types of Cisterns

Cisterns are of following two types:

(*a*) Bell type without valve

(*b*) Flat bottom type with valve.

(*a*) Bell Type: This type of cistern is not in use now-a-days. It is made of cast iron having a suitable cover with a sump formed in the centre through which a flush pipe of 32 mm bore passes. The upper end of flush pipe is projecting little above the maximum water level in the cistern, while the lower end is fitted with the water closet pan. The flush pipe is covered by a C.I. bell to the top of which a lever arm with a chain is fitted. When the chain is pulled, the bell is lifted and water enters the flushing pipe and siphonic action is started. The entire water of the cistern then rushes down into the WC Fig. 9.14 shows bell type cistern.

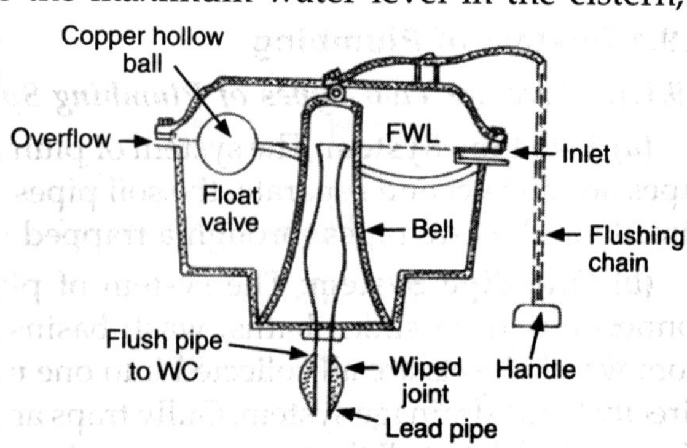

Fig. 9.14. Bell type cistern

(*b*) **Flat Bottom Type:** Refer Fig. 9.15 for construction details. It is the valveless syphonic type flushing cistern. Here when the chain is pulled, disc is lifted which also suddenly lifts the water above it and starts the syphonic action. The valve allows the water to rush in the outlet pipe. The cistern is provided with inlet, overflow pipe, automatic closing float valve etc. Its use is limited in practice.

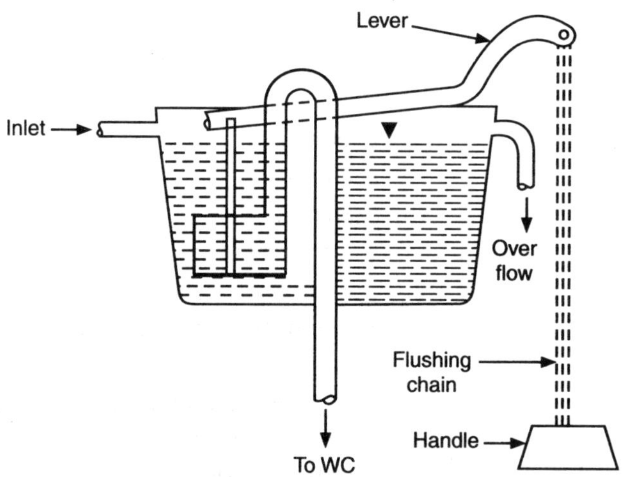

Fig. 9.15. Flat bottom type cistern

9.9. PLUMBING

It is one of the important links of the chain of sewage disposal in the house drainage system. House drainage covers the entire sanitary arrangements above or below the ground.

Leakage of plumbing system will cause nuisance. Insects and rats may enter the house through a faulty plumbing. Therefore it is recommended that only authorised plumbers should be employed for plumbing.

9.9.1 System of Plumbing

9.9.1.1. There are Two Types of Plumbing System

(*a*) **Two Pipe System:** The system of plumbing in which soil and waste pipes are distinct and separate, the soil pipes being connected to the drain directly and waste pipes through a trapped gully.

(*b*) **One Pipe System:** The system of plumbing in which the waste connections from sinks, baths, wash basins and the soil pipe branches from water closets are all collected into one main pipe which is connected directly to the drainage system. Gully traps and water pipes are completely dispersed with but all the traps of water closets, basins, etc., are completely ventilated to preserve the water seal.

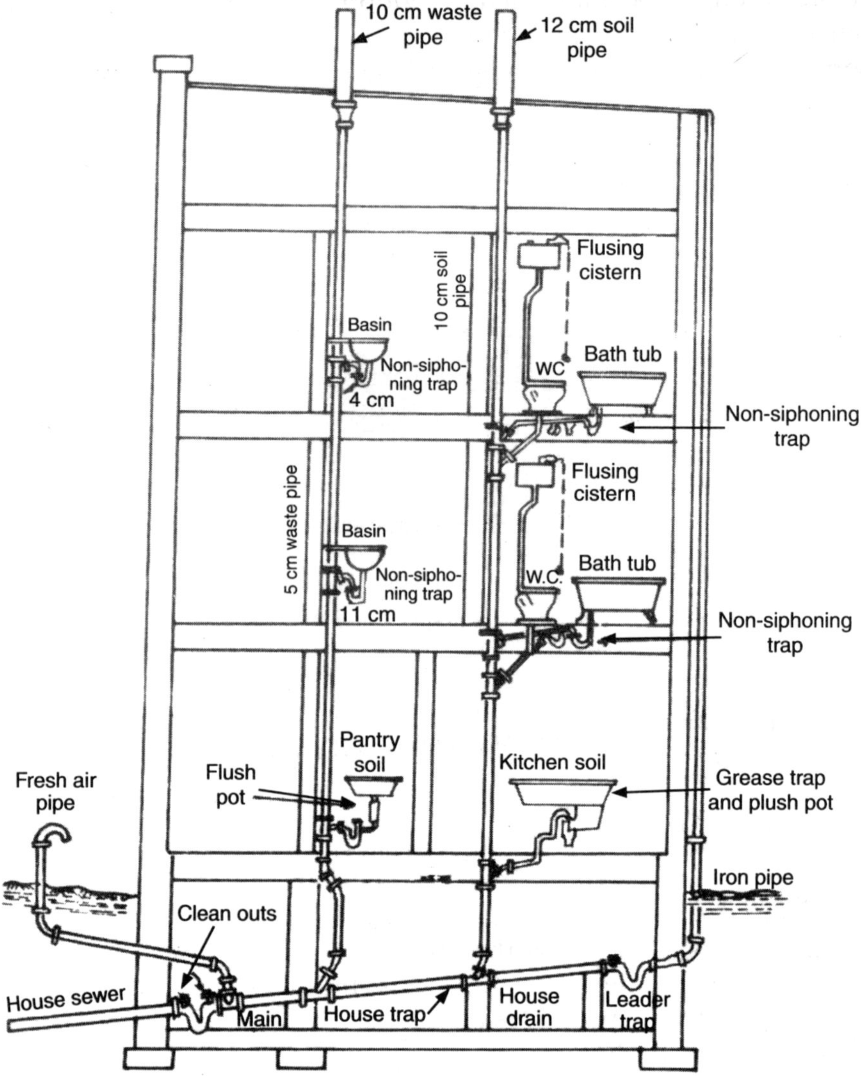

Fig. 9.16. House drainage arranged on one pipe system

Single-Stack System: It is a one pipe system without trap ventilation pipe work. The traps of water closets, sinks, basins, etc., are directly connected to the single stack.

One-Pipe System, Partially Ventilated (Single Stack, Partially Ventilated: A via media between the one pipe system and the single-stack system. There is one soil pipe into which all water closets, baths, sinks and basins discharge. In addition, there is a relief vent which ventilates only the traps of water closets.

9.9.1.2. Choice of Plumbing System

For the disposal of sanitary wastes from a building, the two-pipe system of plumbing is safer than other systems of plumbing and is, therefore, strongly recommended for adoption however, where one-pipe system or partially ventilated one-pipe system is to be used for any special reason and in such cases the safeguards given below be observed.

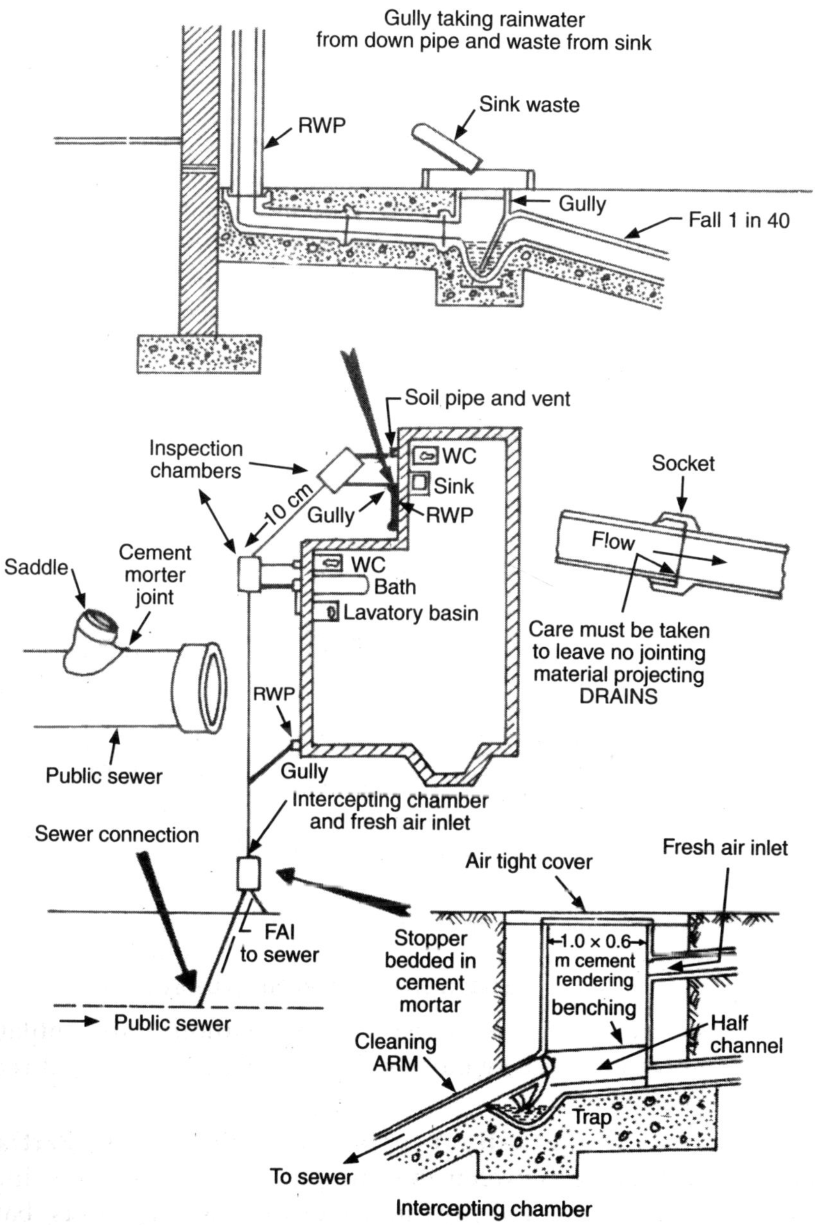

Fig. 9.17. House drainage details

(*i*) Safeguards for one pipe system:

(*a*) The waste pipe shall join the stack above the soil branch at each floor;

(*b*) Cast iron fittings and branches for waste pipes shall be of the same quality as for soil pipe joints shall be made gas tight;

(*c*) Each waste pipe shall be connected directly to the common stack;

(*d*) All traps shall be of the deep seal type with 7.5 cm water seal; and

(*e*) All traps shall be fitted by means of pipe of not less than 50 mm diameter for preserving their seals.

(*ii*) The safeguards for partially ventilated one-pipe system shall be as follows:

(*a*) Water pipe shall join the stack above the soil branch at each floor;

(*b*) All traps shall be of deep seal type with 75 mm water seal;

(*c*) Waste pipe shall fall gradually and continuously in the direction of flow; waste pipe shall have a slope of 1:12 to 1:48; and

(*d*) Water closets shall be ventilated by a pipe of not less than 50 mm diameter for preserving their seals.

In designing a drainage system for individual building or a housing colony, the aim shall be to provide a system of self-cleansing conditions and without risk of nuisance.

9.10. INSPECTION AND TESTING

(*i*) Inspection: All sanitary appliances shall be carefully examined for defects before installing them and also on completion of the work.

Pipes are liable to damage in transit and, not withstanding tests that may have been made before despatch, each pipe shall be struck with a mallet and those that do not ring true and clear, rejected. Sound pipes shall be carefully stored, to prevent damage. Any defective pipes shall be segregated, marked in a conspicuous manner and their use is re-examined for soundness before laying.

(*ii*) Close Inspection: Cast iron pipes shall be carefully examined for a damage to the protective coating. Minor damage shall be made good by painting over with hot tar, preferably bitumen. But if major defects in the coating exist, the pipe shall not be used unless recoated. Each pipe shall be carefully re-examined for soundness before laying.

Close inspection shall be maintained at every stage in the work, particularly as to the adequacy of timber supports used in excavation and the case thoroughness exercised in filling.

Carefully note shall be kept of the condition of any sewer, man-hole or other existing work which may be uncovered and any defects evident shall be pointed out immediately to the appropriate authority's representative.

No work shall be covered over or surrounded with concrete unit it has been inspected and approved by the authorized representative of the appropriate authority.

9.11. FIXTURES IN MODERN BATHROOMS

According to European system, the modern bathroom consists of the following fixtures:

(*i*) Water-closet (WC)
(*ii*) Urinals
(*iii*) Bath tub with shower arrangement
(*iv*) Baby bath
(*v*) Waste-hand basin
(*vi*) Mirror
(*vii*) Tooth brush holder
(*viii*) Tower rack or towel rail
(*ix*) Medicine chest
(*x*) Soap case
(*xi*) Soap-holder for bath tub
(*xii*) Toilet paper holder
(*xiii*) Geyser for hot water supply
(*xiv*) Drinking fountain etc.

The floor of bathroom should be of a non-absorbant material with curved angles at the junction with walls for sanitary point of view.

MODEL QUESTIONS

1. (*i*) What is the necessity of trap?
   (*ii*) What is depth of seal in the trap ?
   (*iii*) What are the causes of breakage of seal in the trap ?
   (*iv*) Where will you use Gulley Trap ?
2. (*a*) How will you test the home drains ?
   (*b*) Where will you use floor-trap and intercepting trap ?
3. Explain the 'one' and two pipe system of plumbing and state the conditions under which one is suitable to the other.
4. Draw a layout of drainage for a given building. Find the number of fittings and estimate the cost.
5. What are 'traps'? Why is it necessary that all sanitary fittings should be provided with individual traps? State the different traps in use.
6. Sketch and describe:
   (*i*) An intercepting trap.
   (*ii*) An inspection chamber.
7. In a four-storeyed building, the water closets are fitted in rooms one over the other and are connected to a common soil-pipe. Draw sketches of the water closet and all its fittings on the second floor. Also draw detailed sketches showing the different methods of connecting the water closet to the soil pipe.

8. Write short notes on the following:
   (*i*) House drain (*ii*) Traps
   (*iii*) Water-closets (*iv*) Vent pipe.
9. (*a*) Enumerate the usual sanitary fittings that you would proceed to fix in a modern bath room.
   (*b*) In a European WC in the Vice-chancellor's bungalow, it was observed that a water trap was provided. Is the trap was necessary ? If so, explain its utility.
10. What do you mean by single stack system in sanitary installation of a building? How does it differ from two stack system?
11. Explain the location and functioning of the following:
    (*i*) Gulley trap
    (*ii*) Grease trap
    (*iii*) Intercepting trap.
12. Write short notes on:
    (*a*) "Purpose of anit-syphonage Pipe".
    (*b*) P, Q and S Type of Traps.
13. Write short note on Automatic Flushing tank.
14. What is water closet. Explain briefly various types of water closets
15. Write short notes on:
    (*i*) Anti-syphonage pipe (*ii*) Soil pipe
    (*iii*) Intercepting trap (*iv*) Grease trap and gully trap
    (*v*) Automatic flushing tank (*vi*) Geyser
16. Sketch and explain the following:
    (*i*) Joint of a house drain with street sewer.
    (*ii*) Public Latrines and Urinals.
    (*iii*) Testing of house drains.
    (*iv*) Flushing of sewers and drains.
17. Explain various types of urinals used in buildings.
18. Write a short note on "Working of WC Flushing Cistern."
19. What are the requirements of a good trap? Under what conditions the water seal of the trap can be broken?

CHAPTER 10

RURAL SANITATION

Learning Objectives

- General
- Refuse
- Classification of Refuse
- Collection of Refuse
- Disposal of Refuse
- Excreta Disposal without water Carriage
- Types of Conservancy Closets
- Septic Tank
- Design Problem of a Septic Tank
- Sanitation of Camps
- Sanitation of Fairs and Melas
- Gobar Gas Plant

10.1. GENERAL

Rural sanitation is very important problem specially in India where majority of people live in villages. Rural areas are characterized by stagnant ponds or pools, bad drainage, cattle dung and human excreta almost every where, refuse dumps, unprotected water supply etc. The age-long habits and ignorance of villagers have also contributed largely to the problem of proper sanitation and drainage of villages, Health authorities should help the people to know the principles of Hygiene and Public Health. Documentary films may be arranged frequently showing advantages of light, open air, proper ventilation cleanliness etc., and on the disadvantages of dark, ill ventilated, dirty surroundings etc.

Moreover our masses are not still fully sanitation-minded. It is generally seen that in the best of buildings constructed by citizens, the worst-portion is the bathroom and the latrine. A proper and efficient drainage system of a house although comparatively costly in the first instance, is cheaper in its running charges.

10.2. REFUSE

Refuse is all solid and semi-solid waste matters of a community except night soil.

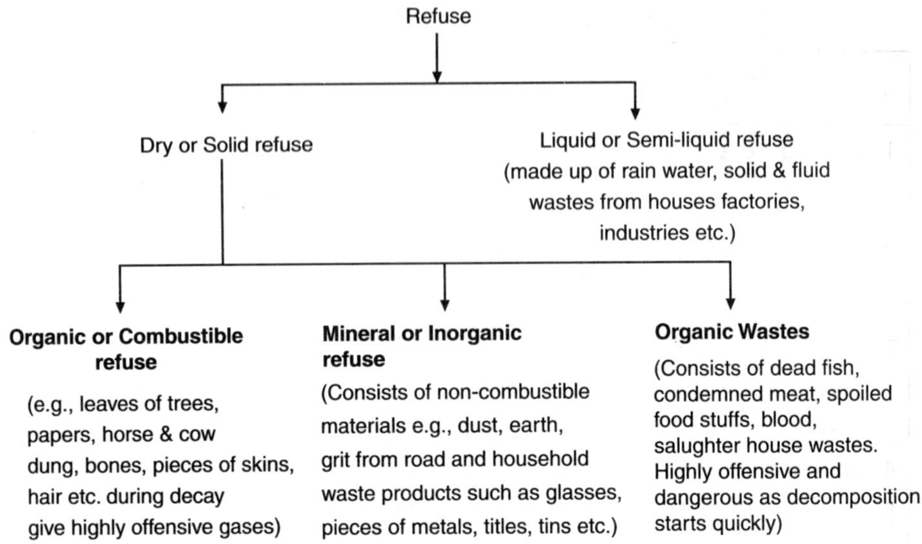

10.3. CLASSIFICATION OF REFUSE

1. Based on type of wastes

(*i*) Garbage

(*ii*) Rubbish

(*iii*) Ashes

(*iv*) Street washings

(*v*) Industrial waste.

Garbage is the decomposing organic wastes from kitchens, restaurants, hotels etc. This includes all waste food articles, vegetable peelings, fruit peelings etc. If it is properly composted and processed, products like grease, fertiliser, hog food etc., may be produced. The quantity of garbage varies from 450 to 900 kg/m^3.

Rubbish includes all non-decomposing organic wastes except ashes, i.e., paper, glass, crockery, rags etc. Density of rubbish is between 50 to 400 kg/m^3.

Ashes are the waste products from hearths and furnaces at houses or industries. The quantity of ashes depends on the geographical location and season. The density of ashes is between 700 to 850 kg/m^3. Ashes result in the creation of dust nuisance.

Street washings are waste materials from street surfaces, side walks etc. Its quantity varies from 250 to 500 kg/capita/year.

2. Based on its sources

(a) House refuse

(b) Street refuse

(c) Trade refuse

Table 10.1. Average composition of refuse

Constituent in India	Percentage in Europe	Percentage	Remarks
1. Fine dust e.g., ashes.	25%	10 to 15%	In Europe more use of electricity.
2 Unburnt fuel e.g., sticks, pieces of wood, charcoal cinders including all volatile debris.	25%	10%	
3. Putrescible matter e.g., kitchen wastes, spoiled vegetables, surplus food stuffs etc.	15%	20%	
4. Papers, rags etc.	15%	10%	
5. Metal, grit, sand etc.	15%	20%	
6. Broken glass bottles, pieces of tiles, broken crockery etc.	5%	25%	In Europe people drink wine, tea coffee etc.

10.4. COLLECTION OF REFUSE

The frequency of refuse collection depends mainly on the quantity of refuse and the season. In India, refuse is generally collected in small containers, called dust bins, kept at suitable position in the rear or front of individual houses. From there, it is transported and dumped into the road side dust bins or refuse depots. The refuse is finally transported to the place of final disposal by municipal trucks.

The collection and carriage of refuse is done under proper sanitary regulation with minimum of nuisance and spreading of dust while loading into vehicles. The trucks transporting the refuse should be covered up.

10.5. DISPOSAL OF REFUSE

Following are the methods of refuse disposal:

(*i*) Dumping, (*ii*) Sanitary fill, (*iii*) Ploughing in Fields, (*iv*) Incineration, (*v*) Compositing, (*vi*) Pulverizing or grinding, (*vii*) Reduction, (*viii*) Salvaging, (*ix*) Barging in to the sea.

(*i*) Dumping: It is open throwing of refuse so as to fill up low-lying areas. The refuse may be ashes, street washings, rubbish and exclude garbage. The method is least satisfactory.

(*ii*) Sanitary Fill: It is the modern method of disposal of refuse. It is economical than incineration method but it requires ample land. About 1.0 m deep trenches are dug out and the refuse is dumped in layers. It is then properly compacted and covered up with 15 cm layer of earth, ashes or sand spread at the end of each working day. Then it is left for sometime for decomposition after which it is used as manure. The method requires regular supervision so that the land fill may not deteriorate into an ordinary dump.

***(iii)* Ploughing in Fields:** This method of disposal of refuse is used only on small scale and is suitable for summer camps etc.

***(iv)* Incineration:** It involves the burning of refuse in an incinerator (See Fig. 10.1). This is the most effective sanitary method of disposal of refuse.

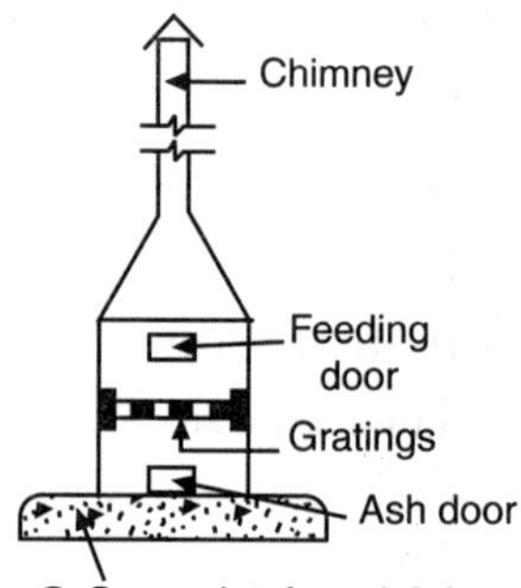

Fig. 10.1. Incinerator

The incinerator consists of a furnace provided with a grating and a chimney charging doors are provided for feeding refuse into the furnace. The burnt ash is removed through ash door. It is an effective method for the disposal of combustible refuse. Bacteria and insects are also destroyed in the process. Incinerators may be:

(*a*) Low temperature Incinerator—which use chimney draughts.

(*b*) High temperature Incinerator—which is provided with furnace and reduce the contents to ashes.

***(v)* Compositing:** The method is good and ideally suited to Indian conditions with plenty of open lands, sun and air.

It is the process of decomposing organic wastes mostly in presence of air. The organic refuse in this case is mixed with some nitrogenous matter like cowdung, sewage, night soil etc., and then it is subjected to aerobic decomposition in trenches. Decomposition may take 4 to 6 months to complete. Resulting matter is used as fertilizer, which is odourless and powdery humus.

***(vi)* Pulverizing or Grinding:** It is the mechanical method of reducing garbage out of refuse in machines into fine powder. The powder thus obtained will have some fertilizing power.

***(vii)* Reduction:** It is the process of recovering grease and other solid residue from garbage. Here solid residue used as a manure. Grease can be used in the manufacture of soap, candles, glycerines etc.

***(viii)* Salvaging:** It is the process of removing certain elements having market value from the refuse e.g., glass, paper, metals, rags etc. This method is much suitable due to its high nuisance value.

***(ix)* Bargings into the sea:** The towns situated on the coast of sea having seaward currents under suitable conditions can dispose of their refuse into the sea. The refuse should be disposed at a safe distance from shore so that it may not come back due to shore currents. It is very economical method of refuse disposal.

10.6. EXCRETA DISPOSAL WITHOUT WATER CARRIAGE

In rural areas of small towns there is no sewerage or water carriage system. Hence some methods must be developed for the safe collection and disposal of human excreta from such areas. It is important that the human excreta should be removed hygienically and in an efficient manner.

10.6.1. Objects to be achieved

Methods employed should fulfil the following objects:

(*i*) All night soil must be removed to an isolated area.

(*ii*) It should not contaminate any surface or ground water supply.

(*iii*) It should not be accessible to flies, insects or other animals.

(*iv*) The methods should be simple and economical both in construction and operation.

(*v*) The methods should ensure privacy and convenience.

10.7. TYPES OF CONSERVANCY CLOSETS

These are:

(*i*) Earth closets or the earth privy

(*ii*) Trench Latrines

(*iii*) Chemical Toilets

(*iv*) Pit Privy

(*v*) Bore Hole Latrines

(*vi*) Soak Pit

(*vii*) Cesspools

(*viii*) Dug-Well Privy

(*ix*) Aqua Privy or village Septic Tank

(*x*) Septic Tanks.

(*i*) Earth closets or the earth privy: It is a pit dug below ground. A latrine structure may be constructed temporarily over it. The night soil falls into the pit which may be 2.0 to 2.5 metres deep and about 1.25 m square. It is kept ventilated by vent pipe to allow the gases to escape. Each time the privy is brought into use, a small quantity of dry earth should be thrown on the excreta. The required quantity of earth is kept in the privy room in a bucket or in the Hopper.

When finally the closet is almost full, it is covered with earth and left out of use for a few weeks. The night soil is converted into manure and can be removed from the pit.

(*ii*) Trench Latrines: Portable latrines of a single seated type are kept on a trench excavated to receive the excreta. Working of the closet is similar to that of earth closet. This is a cheap type of privy.

(*iii*) Chemical Toilets: This is the most satisfactory method of disposal of excreta without water carriage. In this type of privy, powerful chemicals are used in place of earth. It consists of a metal tank containing the solution of caustic soda. The tank capacity is generally 500 litre for each person served. A seat is placed directly over the tank. When the tank is filled up, it is emptied and cleaned.

(*iv*) Pit Privy: This is very economical and requires no operation. It consists of a pit 1.2 m × 1.2 m in plan and 1.0 to 2.0 m in depth. It is covered with squatting slab with a hole or squatting plate. The superstructure is of temporary nature. A capacity of 40-60 litres/person/year is provided. Storage period may be 4 to 10 years.

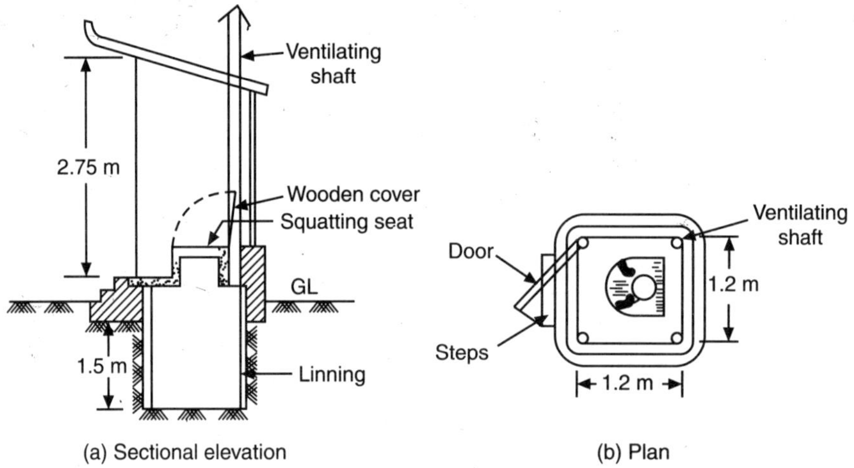

Fig. 10.2. The pit privy

Once the pit is full, new pit is excavated by its side and compartment along with the seat is shifted to the new pit. Its use is restricted to places where the ground water table is sufficiently below the ground surface so that no pollution of water supply can occur.

(*v*) Bore hole Latrines: It is similar to pit privy, the only difference is that in place of pit, it has 400 mm dia. bore hole, normally 4 to 6 m deep. The depth of bore hole should always be 1 m above the ground water table to avoid pollution of ground water. The hole may be lined or unlined from inside. Squatting plate is provided at the top of the bore hole. Either an ordinary RCC seat is used or a pan and a trap arrangement may be used to make things more asthetic. Fig. 10.3 (a) and (b) shows both type of arrangement. In this type of latrine, excreta directly goes into the pit and gets digested, while the liquid portion gets absorbed.

It is simple and economical in construction and maintenance. For a family of 5 persons, such latrine will give good service for 5 to 8 years. The more depth ensures freedom from fly and odour nuisance and this makes it very hygienic. Hence it is universally used.

(*vi*) Soak Pit: The covered pit through which the effluent from Septic Tank is allowed to be soaked into the surrounding soils is called a Soak Pit.

Soak Trench or Soak Pit is constructed for small family Septic Tanks, are installed, in case of subsoil, layers are of porous nature e.g., gravelly or sandy earth. Moreover there should be no well within 15 m from the Soak Trench. Its size is 1.5 m × 1.5 m at the bottom. It is lined on the sides and the bottom with boulders.

Effluent from the septic Tank is led in a pipe. It pours its contents on the top of the filter in the centre at about 22.5 cm below the ground level 5 cm thick layer of flag stones or straw is spread on the top of the filter. Here bottom and sides of the pit absorb the water. Sometimes ventilating pipe (size 7.5 cm.) with its bottom and resting on the loose material at about 30 cm below the surface is provided.

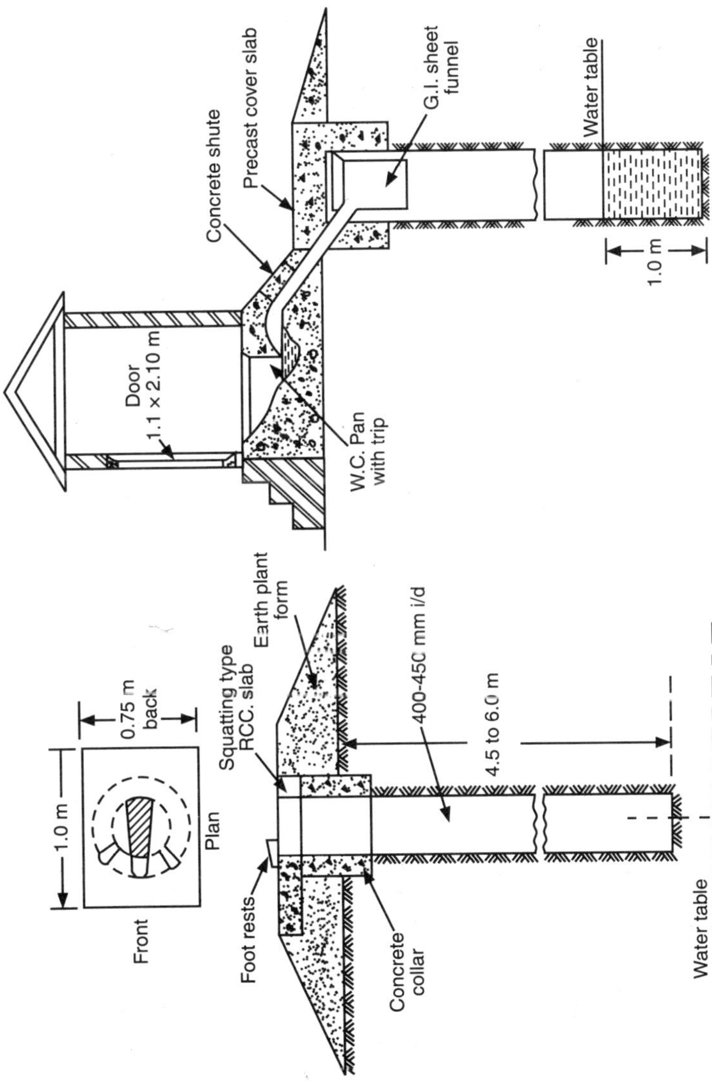

Fig. 10.3. Bore hole latrine

(*vii*) **Cesspools:** It is a circular or rectangular underground chamber, in which the top portion acts as absorption field and bottom portion acts as septic tank. The chamber is lined from inside. The top 600-900 mm of lining or the depth from ground level up to the inlet level is made water tight. The sewage effluent will pass through these linings and disperse easily into the surrounding soils. When the bottom of the cesspool is above the ground water table, the bottom portion is made water tight which retains the heavier sludge from the effluent for septic action. But when the bottom of the cess-pool is below the groundwater table, it is not made water tight, except for side walls in brick masonry. The water tight cess-pools are cleaned every 6 months and their capacity must not be less than 70 l/person/month.

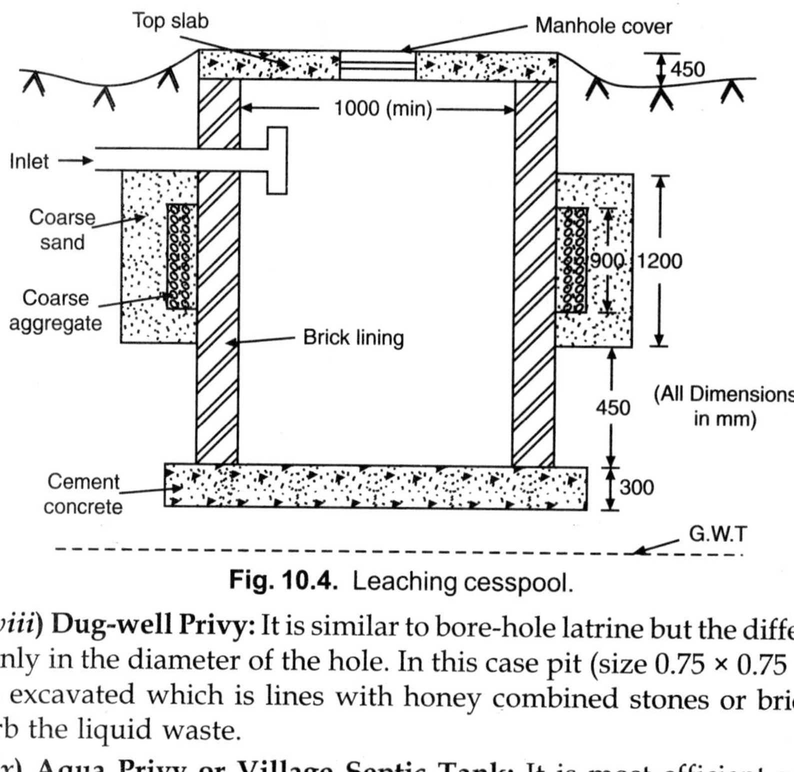

Fig. 10.4. Leaching cesspool.

(*viii*) **Dug-well Privy:** It is similar to bore-hole latrine but the difference lies only in the diameter of the hole. In this case pit (size 0.75 × 0.75 × 3.50 m) is excavated which is lines with honey combined stones or bricks to absorb the liquid waste.

(*ix*) **Aqua Privy or Village Septic Tank:** It is most efficient way of conservacy for a village or small town. It is designed to meet the customs of eastern people. Where there is no sewerage system and pipes water supply and where people use only one lota of water for ablution, this aqua privy is substitute for the ordinary hand removal of excreta.

Fig. 10.5 shows two seated Aqua Privy and it consists of three chambers i.e.,

(*i*) **Anaerobic:** This chamber filled with water to start with Anaerobic bacteria digest. And liquify the solid organic matter in night soil dropping in from the user through funnel shaped seat.

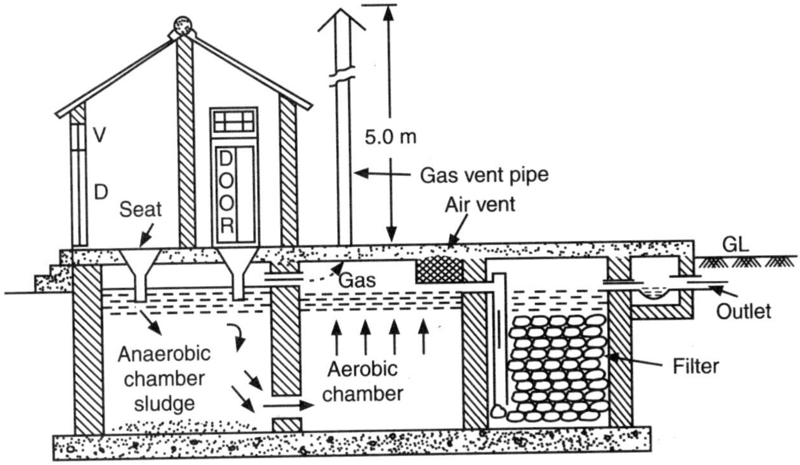

Fig. 10.5. Aqua-privy or village septic tank

(*ii*) Aerobic: Here aerobic bacteria further digest the liquified matter and clarify the resulting liquid.

(*iii*) Filter: In this chamber the effluent issuing from the 2nd chamber is passed through filiter made of gravel or coke pieces in downward and upward flow. The effluent coming from the filter is smell-less and clear. It can be used in small kitchen garden.

Advantages of Providing Aqua Privy

(*i*) No nuisance from flies, hence contamination of water and food is eliminated.

(*ii*) Nuisance from smell and unsightly appearance of night soil is avoided.

(*iii*) Effluent is clear and odourless.

(*vi*) No danger of sweepers strike.

(*v*) Hook-worm infection eliminated.

(*vi*) Very small quantity of water is required for using it.

(*vii*) Maintenance cost much less than that of ordinary pail latrines.

10.8. SEPTIC TANK

A masonry or concrete structure, built below the ground level, where biochemical reactions take-place due to an aerobic bacteria is called a Septic Tank. Septic Tank was first used in INDIA in the year 1806.

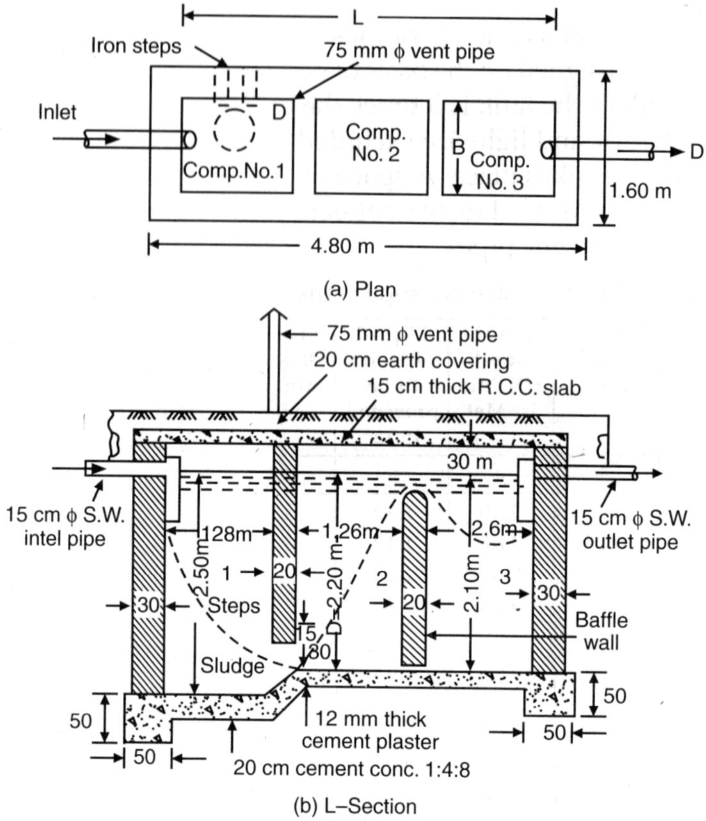

Fig. 10.6. Septic Tank

10.8.1. Principle of Working

These tanks work on the principle of biological actions which is brought about by bacteria which converts the solid matter of sewage into liquids and gases. Gases like carbon-di-oxide, hydrogen sulphide etc., are evolved. Mainly anaerobic bacteria (which thrive in absence of oxygen) takes part in the liquification.

10.8.2. Construction and Operational Features

A Septic tank is a horizontal continuous flow sedimentation tank. Hence sewage moves very slowly so as to be retained for a period sufficient to allow 60 to 70% of suspended matter to settler in the form of sludge at the bottom of the tank.

In Septic tanks the sedimentation takes place in the upper part and sludge accumulation occurs in the lower part. The effluent flows between skum and sludge. Due to decomposition of sludge, the gases are produced and are allowed to escape in the air. Due to liquification, sludge produced in these tanks may be about $1/4^{th}$ of that in sedimentation tanks. Hence they require clearance at longer intervals.

Septic tank is either a concrete or masonry structure and is built below ground level. It is designed on the basis of continuous flow. It is rectangular in size. The length of the tank is 3 times the breadth. It is divided into 2 to 3 compartments. Air and light are excluded from the tank and due to this the anaerobic action takes place. A space of 15 to 30 cm is provided above the scum and bottom of slab for collection of gases which go to the atmosphere through vent pipes.

Table 10.2. Showing sizes of septic tanks [I.S. 2470-1963 (Part I)]

No. of users	Length (L) in Met.	Breadth (B) in Met.	Liquid depth (D) Min in Met.	Liquid capacity to be provided in m^3	Free Board min. in cm.	Sludge to be removed	Recommended ended Interval of cleaning
1	2	3	4	5	6	7	8
			11.0	1.12	↑	0.18	6 months
5	1.5	0.75	11.0	1.12	|	0.36	1 year
			11.05	1.18	|	0.72	2 years
			11.0	1.80	|	0.36	6 months
10	2	0.9	11.0	1.80	|	0.72	1 year
			11.4	2.52	|	1.44	2 years
			11.0	1.10		0.14	6 months
15	2	0.9	11.3	2.35	30 cms	1.08	1 year
			12.0	3.60		2.16	2 years
			11.0	2.50	|	0.72	6 months
20	2.3	1.1	11.3	3.30	|	1.44	1 year
			11.8	4.55	|	2.88	2 years
			11.0	5.60	|	1.80	6 months
50	4	1.4	11.3	7.30	|	3.60	1 year
			12.0	11.20	↓	7.20	2 years

Table 10.3. Period of cleaning and storage capacity

Period of cleaning	Storage capacity
6 Months	0.028 m^3
1 Year	0.050 m^3
2 Years	0.001 m^3
3 Years	0.085 m^3

10.8.3. Detention Period

A minimum period of detention of 12 to 24 hours is essential for raw sewage. Smaller value may be taken when effluent is to be used for sewage forming, while larger one is taken when effluent is allowed to percolate in the subsoil below and depends upon the type of the subsoil. These are not designed for a capacity less than 15 cu. m. As Septic Tank acts as sludge digestion pits, a rational approach to their design must be made for larger tanks.

10.8.4. Performance

The effluent from a tank is foul. It contains large quantities of dissolved and suspended organic matter, the suspended solids amount to 125 to 150 ppm. BOD about 100-150 ppm while all the load of dissolved organic matter present in the sewage, passes on to the effluent. Hence effluent requires to be diluted in large volumes of surface water or by sewage forming or by underground percolation if porous soil is available i.e., in soak pits as discussed earlier.

Table 10.4. Capacity of septic tanks for day schools and camps

Maximum number of persons served		*Nominal sewage capacity of tank in litres*	*Recommended dimensions*			
Camps	*Day school*		*Width m*	*Length m*	*Sewage depth, m*	*Total depth m*
40	60	3800	1.20	2.50	1.20	1.50
80	120	7600	1.50	3.30	1.50	1.80
120	180	11400	1.80	4.00	1.50	1.80
160	240	15200	1.80	4.40	1.50	1.80
200	300	19000	2.20	5.40	1.50	1.90
240	360	22800	2.40	6.00	1.50	1.90
280	420	26600	2.50	6.00	1.60	2.10
320	480	30400	2.50	6.90	1.60	2.10

10.8.5 Advantages and Disadvantages

The main advantages are:

(*i*) The production of a sludge which is lighter in weight and contains less water.

(*ii*) No skilled attendance required.

(*iii*) No mechanical moving parts.

(*iv*) They are very cheap and convenient on small scale.

Septic tanks are unsuitable for large populations due to following:

Disadvantages

(*i*) The size becomes too large.

(*ii*) BOD and bacteria removal efficiency goes low.

(*iii*) Initial expenditure is high.

(*iv*) The effluent is with high BOD and foul gases

10.9. DESIGN PROBLEM OF A SEPTIC TANK

Design and Draw a Septic Tank for one hundred users,

Solution: The tank will be designed on per-capita basis of 12 mm

Effective capacity = 100 × 0.12 = 12 cu m

Sludge capaeity= 100 × 0.01 = 1 cu m
(10%)
Add for future extension at 25% = 3 cu m
Total capacity = 16 cu m

Adopt one No. unit of I6 cu m. If the depth of the tank is 2.0 m including the depth required for sludge, the area of tank will be $\frac{16}{2}$ = 8 sq m

Adopt total lengths as 2 times the breadth, we have:
Length = 4.0 m Breadth = 2.00 m

Longitudinally the tank will be divided into three compartments of 1.5 m (approximately) each. A free board of 0.75 m will be kept for scum etc., in addition to 2.0 m (See Fig. 10.6).

10.10. SANITATION OF CAMPS

The problems met in the sanitation of camps are similar to those of rural sanitation. Care must be taken to preserve the health of workmen and officers in the party. Keep in view the following points:

(*i*) Adequate accommodation with a list of labourers
(*ii*) Supply of good drinking water
(*iii*) Disposal of garbage
(*iv*) Surface drainage
(*v*) Disposal of night soil
(*iv*) Supply of good food stuff
(*vii*) Weekly inspections and disinfection.

In addition to above points a complete plan of the camp must be kept in the office of the XEN for reference.

10.11. SANITATION OF FAIRS AND MELAS

Their sanitation is also similar to that of rural sanitation. Arrangement must be made for supply of good drinking water, proper surface drainage and refuse disposal, night soil and animal dung disposal. It is desired that the proper layout must be done of :

(*i*) Roads and lanes
(*ii*) wells
(*iii*) Fire brigade stations
(*iv*) Police stations
(*v*) Temporary dispensaries
(*vi*) Volunteer camps
(*vii*) Earth closets or cespools.

To avoid fire, special attention must be given on proper garbage collection and disposal.The site for the mela must be carefully selected from the drainage point of view.

10.12. GOBAR GAS PLANT

Nowadays Gobar Gas Plants are becoming very popular in the villages for the disposal of gobar and night soil. These are also used for the production of Gobar Gas.

The villagers keep buffaloes, cows, goats, bullocks etc., animals for their fields Gobar of these animals is mostly dried as used as fuel for kitchens. But it is not possible to dry Gobar in rainy season.

10.12.1 Construction and Working

Ref. Fig. 10.7 for constructional details for Gobar Gas Plant.

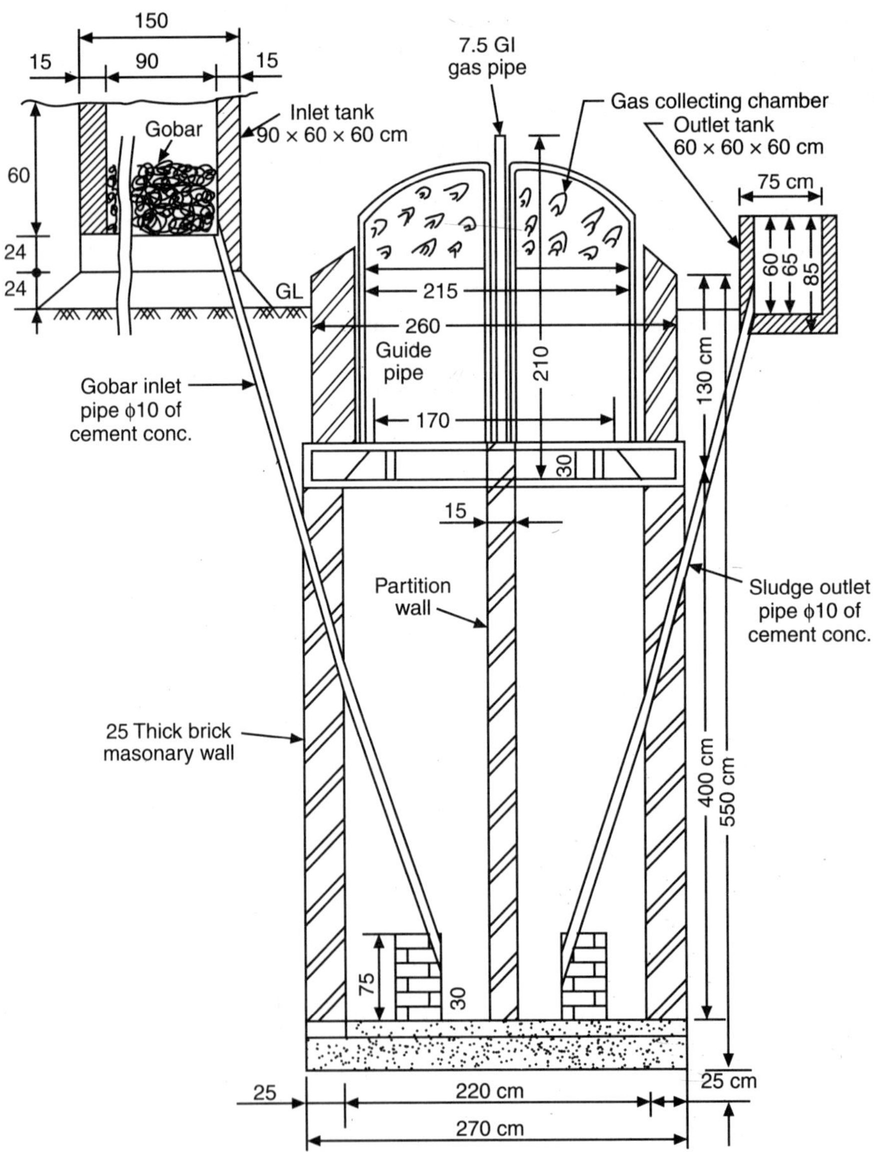

Fig. 10.7. Gobar gas plant

It consists of cylindrical shape divided into two compartments by one partition wall. A cylindrical dome is placed over the top which collects the gas. One inlet is provided for feeding the gobar. An outlet is provided for discharging the rich manure out.

The night soil from the wc reaches directly in the digestion tank through pipes. To form slurry of 10% solids, water is added. This slurry is led into the digestion sludge through the inlet pipe. The gobar gas is collected in the cylinderical dome. Persons who have 50 or more animals, can construct a Gobar gas plant to meet the need of manure and kitchen's fuel gas.

Use of Gobar Gas: From the dome the gas reached into kitchnes and other uses through pipes. These pipes are coloured to differentiate from water supply pipes. Now bulbs, ceiling fans, pumping sets, kitchen furnaces etc., use the gobar gas for their operations.

MODEL QUESTIONS

1. (*i*) What are different types of refuse collected in the Rural Sanitation?
   (*ii*) Septic Tank work on the principle of __________
   (*iii*) Septic Tank is horizontal continuous flow __________ Tank.
   (*iv*) Minimum Detention period for raw sewage in case of Septic Tank is __________
   (*v*) Quantity of street sweepings varies from __________ to __________ kg. capita/day.
   (*vi*) Garbage consists of __________
   (*vii*) Septic Tanks are __________ for large population, (Suitable/unsuitable).
   (*viii* __________ is the process of decomposing organic, wastes mostly in presence of air. (Salvaging/Composition/Pulverizing)
2. What sort of temporary latrine would you propose for a Mela in a rural area of Haryana ? Explain your proposal in detail. Give a sketch and describe the working of latrine proposed by you.
3. Explain with the help of a neat sketch the working of a Septic Tank for a small community of 100 persons.
4. (*a*) Sketch different types of privies construction in rural areas and describe their merits and demerits.
   (*b*) Describe an Incinerator.
5. Mention the various methods commonly used for the disposal of refuse. Describe in detail one of them.
6. Write notes on Soak Pits.
7. Explain the necessity of systematic collection and disposal of waste from a city.
8. Compare and contrast cesspools, privy pit and soak pit in a rural sanitation scheme.
9. Explain working of the following with neat sketches : (i) Privy pit, (ii) Bore-hole Latrine.
10. Explain the following terms:
    [Garbage, Sullage, Sewage, Industrial waste, Storm water flow.]

11. Write how the following types of refuse are collected and disposed off?
 (*i*) Garbage, (*ii*) Sullage, (*iii*) Storm water flow, (*iv*) Night soil
12. What are various ways of collecting and removing the waste of a town? Describe them in brief. For a town having a rich municipality, which method you would recommend and why?
13. Explain briefly the following:
 (*i*) Soak pit and Cesspools
 (*ii*) Earth closets and Chemical closets
 (*iii*) Composing and Salvaging.
14. Write short notes on the following:
 (*i*) Sanitation of camps
 (*ii*) Water Closets
 (*iii*) Septic Tank
 (*iv*) Aqua Privy
15. What sort of temporary latrines would you propose for the Solar Eclipse Fair to be held at Kurukshetra? The number of persons expected to use these latrines would be about 1.5 lakhs. Propose the number of seats and justify the same. Also give a neat sketch of the latrine proposed by you. The note should be detailed and self-explanatory.
16. What is the Principle of Septic Tank Treatment of Sewage ? Comment if the Septic Tank effluent can be let off in the road side drains.
17. Give a brief account of a Septic Tank, a cesspool, a privy pit and a soak pit and explain their utility with the help of sketches in connection with rural sanitation.
18. Write short notes on "Principle and Working of Septic Tanks".
19. Write short notes on:
 (*i*) Biogas Plant
 (*ii*) Cess pools
20. Design a Septic Tank of 200 population colony and illustrate your design with a neat sketch. The rate of water supply may be assumed to be 140 litres per head per day.

❑❑❑

CHAPTER 11

MAINTENANCE

Learning Objectives

- General
- Causes of Sewer Breakage or Damage
- Inspection of Mains
- Cleaning of Sewers
- Flusing of Sewers
- Precautions to be Kept in Mind while Entering Sewers
- Cleaning of House Drainage Lines
- Cleaning of Catch-Pits
- Maintenance of Traps
- Tools and Equipment for Maintenance
- Causes of Explosions in Sewer
- Sewer Repairs

11.1. GENERAL

The purpose of sewer maintenance is to keep them free of any obstruction and to carry out any repairs to damaged portions so as to prolong their life. Hence maintenance must be done on a continuous basis. Generally there should be a proper schedule by which every section of the sewer line is covered and preferably at least once every year.

11.1.1. Maintenance

The art of keeping the entire sewerage system clean and free from observations is known as Maintenance.

11.2. CAUSES OF SEWER BREAKAGE OR DAMAGE

Sewers get damaged or broken due to:

(*i*) Poor foundation on bad soil.

(*ii*) Faulty design or workmanship.

(*iii*) Small cover on top with no provision of RCC slab to protect.

(*iv*) Greater depth causing excessive super imposed load.

(*v*) Understanding due to some reason.

(*vi*) Change in traffic nature, i.e., bringing heavier loads on top.

(*vii*) Deterioration due to corrosive gases like H_2S.

(*viii*) Explosion due to gases like methane when proper ventilation is not done.

(*ix*) Explosion may be when industrial wastes of detrimental nature are ad-mitted without proper pre-treatment.

(*x*) Sometimes tree roots and biological agencies cause the clogging in sewers.

11.3. INSPECTION OF MAINS

It is the first step in a maintenance programme. The purpose of inspection is:

(*i*) To find the extent of deposition and clogging.

(*ii*) To examine the condition of structure.

(*iii*) To check the function of various appurtenances.

(*iv*) To find out that all connections from the house-drains are regular and good.

(*v*) To measure the rate of flow in the sewer.

(*vi*) To check the condition of control mechanisms.

Bigger size sewer should be inspected by getting down through the man-holes and by floating down through the manholes. The most important job in inspection is the detection of cracks and leakages.

Smaller size sewers should be inspected from manholes only by reflecting sunlight by using a mirror or by means of dry cell torches. This will also help to examing the clogging. Record of inspection must be maintained which should include:

(*i*) Name of the inspector

(*ii*) Date of inspection

(*iii*) Portion or length of sewer line inspected.

(*iv*) Defects observed

(*v*) Remedies suggested

(*vi*) Possible causes of defects.

Table 11.1. Periodical inspection of sewers in cities

Items	Period of inspection
Sewer on flat grades	3 months
Sewers troubled by roots	3 months
Sewers having no trouble	6 to 12 months
Intercepting sewers	10 to 30 days
Flushing tanks	45 days
Inverted syphon	10 to 30 days
Storm water overflows	During rains

11.4. CLEANING OF SEWERS

This has normally to be done as many times deposition occurs inside the pipe-line reducing its carrying capacity. Sewers are largely self-cleaning but may get clogged particularly in case of small sewers. It may be due to deposit of sand, grit, grease, fungus growth or roots etc., cleaning of sewers is, hence, an important item in their maintenance. Cleaning is achieved in practice by the following methods:

(*i*) For sewers of bigger diameter, cleaning is done by hand, by going on a boat, shoveling materials into buckets may be carried manually through manholes.

(*ii*) Medium size sewers can be cleaned by using plows and other scraping instruments (i.e., scrapers) which are dragged through sewers as shown in Fig. 11.1.

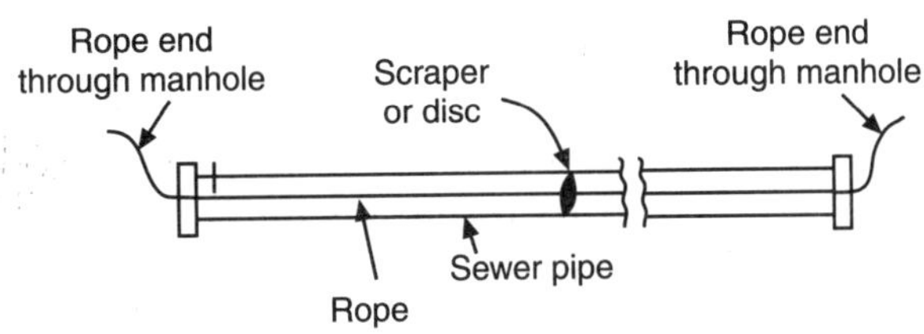

Fig. 11.1. Cleaning of sewers by scraper

(*iii*) Small size sewers are cleaned by thrusting cane-rodding or flexible wooden rods (See Fig. 11.2) into the sewer and dragging to and fro from the open end e.g., manholes.

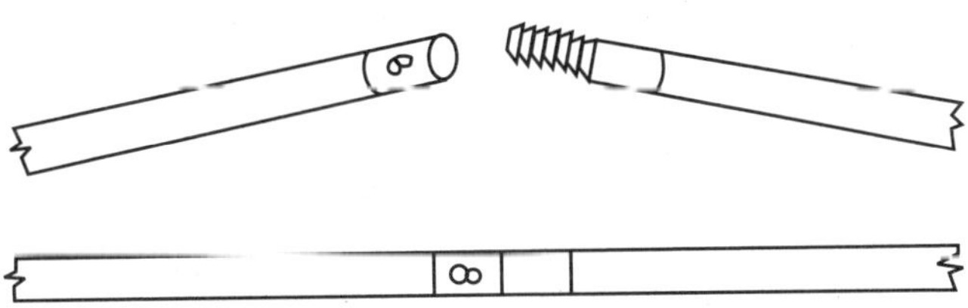

Fig. 11.2. Sewer cleaning rods

The common sewer rods are made of stiff cane 1 to 1.25 m long and are fitted with a steel eye at one end and a hook at the other. The rods can be easily jointed or unjointed in a manhole but cannot be separated in the sewer while being pushed or pulled. They are effective upto 50 to 70 m distance which is normal spacing of manholes in the city areas.

(*iv*) Obstructions in sewer may also be removed by discharging water under high pressure through a small nozzle.

(*v*) Flushing is an efficient method which ensures self-cleaning velocity in the sewers. Flushing tanks are constructed for this purpose and flushing is done automatically.

(*vi*) Another simple method is the use of "Pills" which are light floating wooden or rubber balls of varying sizes. This method has been success-fully used in USA for cleaning badly clogged sewers.

When such a ball is let into the sewer it gets stuck up where heavy deposi-tion has occurred, here area of sewer is considerably reduced resulting in the increase of sewage velocity. This causes vigorous scouring action in the bottom of the sewer due to which all deposits are removed. The Pills are inserted in good number in the sewers at one hole, which are received at the next man hole after the cleaning of the sewers.

11.5. FLUSHING OF SEWERS

Flushing at regular intervals and flushing after the scraping of the sewers, are recommended to maintain the sewers in a good working order. The time chosen for flushing is during the period when there is minimum flow.

Flushing is an efficient method which ensures self-cleaning velocity in the sewers i.e., the velocity of flow at which the floating solid matter in the sewage is carried away and not allowed to settle in the sewers.

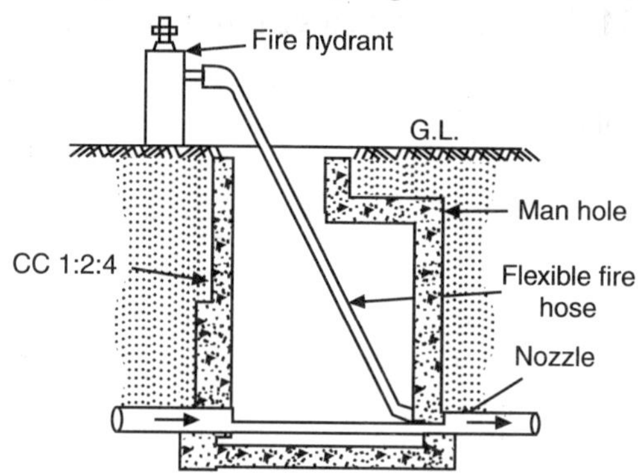

Fig. 11.3. Flushing of sewers

Flushing tanks are constructed for this purpose and flushing is done automatically or flushing of partially clogged small sewer is done as shown in Fig. 12.3 above. Hence one end of a flexible fire hose is connected to the fire hydrant and the other end with nozzle is inserted down inside the sewer and water is discharged under pressure. Hence high velocity of water helps to flush the sewer and keep it in good working order.

11.6. PRECAUTIONS TO BE KEPT IN MIND WHILE ENTERING SEWERS

The following precautions should be taken for the safety of workmen employed for the maintenance of sewers, before they are permitted to enter a manhole:

(*i*) The manhole cover by which a man enters should be thrown open at least one hour in advance.

(*ii*) In addition to this, covers of two adjacent manholes, one on the up-stream and other on the down stream side be removed. This would ensure necessary ventilation.

(*iii*) Test the atmosphere for noxious gases and oxygen deficiencies.

(*iv*) Allow no smoking, open flames or naked lights inside the sewer.

(*v*) Errect warning signs.

(*vi*) Use only safety gas-proof electric lighting equipment.

(*vii*) If the atmosphere is normal, workmen may enter with safety-belts at-tached and with two men available at the top.

(*viii*) First aid should be available for workmen in case of necessity. Specially the breathing apparatus should be kept at hand.

(*ix*) Only permissible rubber boots, non-sparking tools and safety lamps should be used.

(*x*) In case of bad ventilation inside the sewer, fresh air may be supplied by mechanical means e.g., portable air blowers.

11.6.1 Test for Presence of Gases in Sewer

The presence of gases in a sewer may be detected by performing following Tests:

(*i*) The presence of carbon-dioxide (CO_3) is checked by lowering a miners safety lamp near level of sewage. If the flame extinguishes within 5 minutes, then the gas is present.

(*ii*) The presence of hydrogen sulphide (H_2S) is detected by a wet led acetate paper lowered in the manhole for 5 minutes. If the sheet turns, black this shows gas is present.

(*iii*) For methane (CH_4), by lowering the safety lamp in the upper portion of sewer. As methane is lighter than air. Hence when the gas is present, it forms an explosive mixture with air and the gauze-cylinder of the safety lamp is filled with the flame. The lamp must be withdrawn immediately.

Note: When the presence of methane is found, then workmen should be permitted to enter a manhole, with special breathing apparatus fitted with hose marks.

11.7. CLEANING OF HOUSE DRAINAGE LINES

Domestic drainage system shall be inspected at regular intervals. The system shall be thoroughly cleaned out at the same time and any defects discovered shall be made good.

It is required that all the house drainage pipe should be water proof to avoid nuisance due to leakage. Through cleaning of the whole house drainage system must be done every three months or so. All the pipes, drains, gutters etc., should be cleaned by flushing and rodding. Keep in mind the following points for unkeep of house drainage lines:

(*i*) Leaky joints should be repaired.

(*ii*) Traps and inspection chambers should be specially attended to and clogging must be removed.

(*iii*) Examine all the fittings in a building occasionally.

(*iv*) Grease Traps should be cleaned.

(*v*) Solid waste should not be indiscriminately thrown into sewers and house drains.

11.8. CLEANING OF CATCH-PITS

Catch-pits must be cleaned of all silt and debris by means of shovel shortly after every rain. The organic matter, will be removed if there is no slit deposit, otherwise mosquitoes will bread in the basins.

In certain localities, the cleaning of the catch-basins has to be done alongwith the sweeping of the streets. Moreover catch-pits near markets may be cleaned twice or thrice during the day and their walls washed with hose.

11.9. MAINTENANCE OF TRAPS

The function of a trap is to provide a barrier in the way of foul gases in a drain, soil pipe or a waste pipe, so that they cannot escape to the atmosphere either inside or outside a building. Hence regular attention must be given for their maintenance. Refer chapter 10 i.e., "Building Drainage" of this book for detailed study.

11.10. TOOLS AND EQUIPMENT FOR MAINTENANCE

Removal of obstruction and blockages is normally done by using cleaning rods in case of small sized sewers (See Fig. 11.2).

A blockage may be due to either rags, sticks, large objects like brooms, brushes, tins etc., which get entangled amidst depcsits of greasy, fibrous or grity material or even to roots of trees. The latter first enter through pores in the joining material in the form of thin thread like hair. Then it gradually develop into hairy growth of root system, which may often clog the sewer. Special cutting tools attached to cleaning rods are useful. In practice obstructions are removed by steel bands having special tools fitted to their leading ends (Refer Fig. 11.4). The steel tape consists of a flexible band 3 to 5 cm, wide and 3 mm. thick. This is pushed and vigorously twisted by electric motor revolving at high speed i.e., more than 750 RPM. The tools fixed at the advancing end remove the blockage.

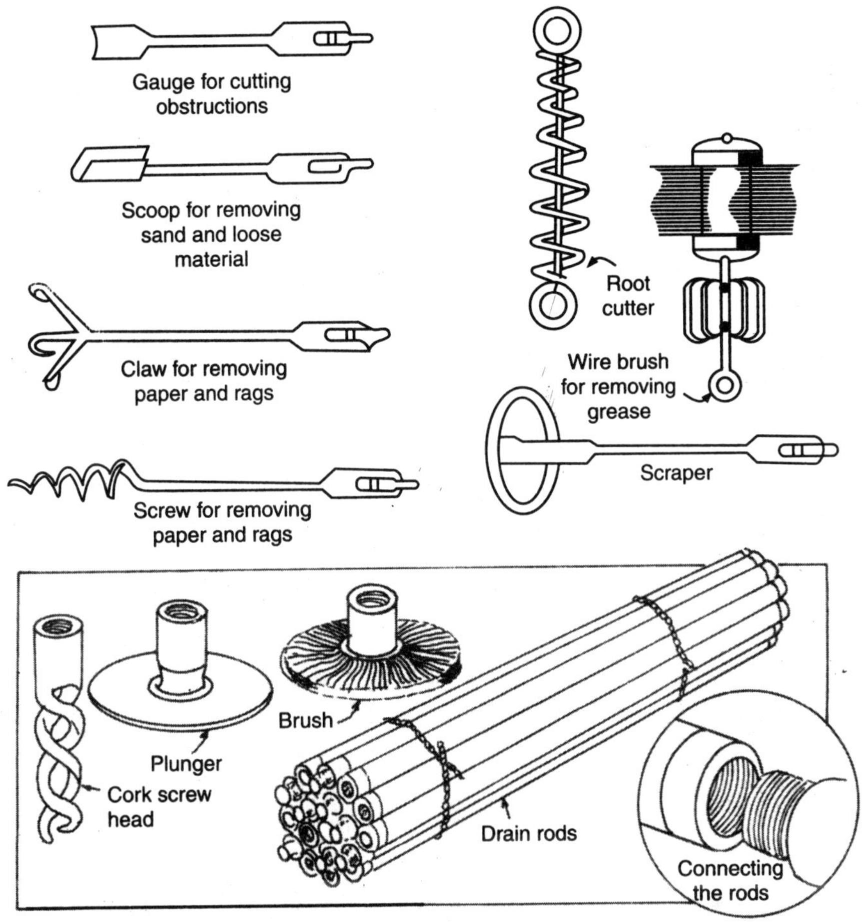

Fig. 11.4.

Turbine sewer cleaner (See Fig. 11.5) or scrapers are also used for obstruction removal in medium sized sewers. Turbine clear has revolving cutting blades which cut roots, fibrous material etc. The helical motion imparted to the water is useful in pushing material ahead of the machine and scrubbing the walls of the sewers.

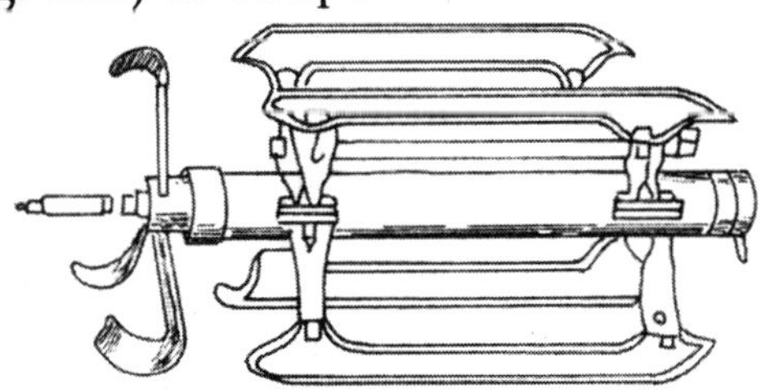

Fig. 11.5. Turbine sewer cleaner

For routine maintenance the city is divided into different units. Each unit is looked after by an inspector with some technical assistants and skilled labour. The equipment and material required by each group is given below:

(*i*) 150 to 200 m of sewer cleaning rods

(*ii*) Diaphragm pump of mud hog type, preferably power driven

(*iii*) Shovels

(*iv*) Picks

(*v*) Hammers
(*vi*) Root cutters and other tools
(*vii*) Buckets and ropes
(*viii*) Cement working tools
(*ix*) Rubber boots for workmen
(*x*) Dry cell torches and flash lights
(*xi*) Safety appliances e.g., gas masks and skull guards
(*xii*) First aid boxes etc.

11.11. CAUSES OF EXPLOSIONS IN SEWER

Explosions are not uncommon in the case of sewerage system that has not been properly ventilated. Moreover manhole covers may be blown into the air when the sewers are not properly ventilated. Sewer gases accumulate and when their concentration reaches a certain value, then little increase in pressure is enough to cause an explosion. Miscellaneous chemicals like calcium carbide may also cause explosions inside the sewers.

Gasoline enters the sewers from garages and cleaning establishment. I part of gasoline vapour +60 parts of air by volume is explosive at normal atmospheric temperature.

Sewer explosions may result not only serious damage to sewers and masonary work, but may prove a danger to workmen and traffic. These can be prevented by use of suitable traps which will intercept gas-producing liquids or dangerous gases.

11.12. SEWER REPAIRS

The next responsibility of a Public Health Maintenance Engineer is to repair the sewers that have been found damaged during inspection. The repair work is quite simple and is normally carried out with the aid of the normal workers gang. The usual repairs may be:

(*i*) Replacing broken manhole covers.
(*ii*) Tightening loose manhole covers.
(*iii*) Replacing broken inlet covers.
(*iv*) Raising or lowering of manhole heads to compensate change of pavement slope or sinking of foundations.
(*v*) Replacing broken-pipe.
(*vi*) Replacing loose bricks.
(*vii*) Reconstructing damaged house connections.
(*viii*) Repairing leaking joints.
(*ix*) Repairing or replacing cracked pipes.

MODEL QUESTIONS

1. (*i*) What are the causes for clogging of the sewer line?
   (*ii*) What are the cause for damaged or broken sewers?
   (*iii*) What are the causes of the nuisance due to odour in a building ?
2. Discuss the hazards of entering into sewers and the precautions you will take.
3. (*i*) What is the method of locating the troubles of blockages in sewers breakage of sewers and odour in houses ?
   (*ii*) How will you proceed to remove blockages of sewers ?
4. What are the different tool/equipment required for the maintenance of sewer lines?
5. Write short notes on the maintenance of sewers. How the sewers are cleaned?
6. (*a*) How will you lower the Man-hole ?
   (*b*) How will you get replaced the broken pipes ?
7. (*a*) What are the causes of explosions in sewer, how these can be prevented?
   (*b*) List the precaution to be taken while entering sewers.
8. Explain briefly the following:
   (*i*) Causes of sewer breakage or damage.
   (*ii*) Inspection of mains.
   (*iii*) Tests for presence of gases in sewer.
9. What is the object of a Catch pit and is it necessary? What precautions must be taken so that it serves its purpose well and does not become a nuisance?
10. Fill in the blanks:
    (*i*) Sewer of a small sizes are cleaned by ________ (Hand, Thrusting Rod)
    (*ii*) The presence of oxygen ________ be tested by lowering a flame before entering the sewer ________ (must not/must)
    (*iii*) Sewer become clogged due to deposition of ________
    (*iv*) If the size of sewer is large, then inspection is carried out by ________.
    (*v*) The capacity of sewer will ________ if there are more obstructions. (More/less)

    [**Ans.** (*i*) Thrusting rod, (*ii*) must, (*iii*) sand and other organic substance, (*iv*) floating in a boat, (*v*) less]
11. Draw the sketches and explain the function of various tools required for mainiaining sewers.
12. Explain briefly how sewer are flushed.
13. What are the precautions required to be taken while carrying out maintenance of sewers?
14. Enlist the causes of sewer Breakage or Damage.
15. Give at least Ten-precautions to be kept in inind while entering sewers.
16. Write short notes on :
    (*i*) Cleaning of catch pits.
    (*ii*) Methane gas test in sewer.
    (*iii*) Causes of Explosions in sewer.

OBJECTIVE TYPE QUESTIONS

WITH ANSWERS

IN

WASTE WATER ENGINEERING

MULTIPLE CHOICE OBJECTIVE TYPE QUESTIONS

1. The waste water from bath rooms, kitchen etc., is called
   (*a*) refuse (*b*) sullage
   (*c*) sewage (*d*) garbage
2. The liquid waste conveyed by a sewer is know as
   (*a*) sewer (*b*) sewerage
   (*c*) sewerage (*d*) all the above are correct
3. The Underground conduit constructed for removal of liquid waste of a Community is known as
   (*a*) canals (*b*) tunnels
   (*c*) sewer (*d*) sewerage
4. Fresh sewage is usually
   (*a*) alkaline (*b*) neutral
   (*c*) acidic (*d*) has pH value of 7
5. Stale sewage is usually
   (*a*) neutral (*b*) acidic
   (*c*) alkaline (*d*) of pH value 7
6. The solid content of sewage is usually
   (*a*) 99% (*b*) 80-85%
   (*c*) 10-15% (*d*) 0.1%
7. The minimum diameter of manhole opening is
   (*a*) 50 cm (*b*) 100 cm
   (*c*) 150 cm (*d*) 200 cm
8. The self-cleaning velocity normally adopted for sewers is
   (*a*) 0.1 m/sec (*b*) 0.2 m/sec
   (*c*) 0.4 m/sec (*d*) 0.85 m/sec
9. When a sewer gets discharged from two or more main sewers, it is called
   (*a*) leading sewer (*b*) trunk sewer
   (*c*) combining sewer (*d*) intercepting sewer
10. The shape of a sewer suitable to carry combined flow is
    (*a*) circular (*b*) elliptical
    (*c*) horse-shoe shaped (*d*) intercepting sewer
11. The colour of fresh sewage is
    (*a*) blue (*b*) green
    (*c*) pink (*d*) grey
12. The colour of septic sewage is
    (*a*) black (*b*) green
    (*c*) grey (*d*) blue

13. Stoneware pipes are used for the house drainage system mainly because

(*a*) they are strengest
(*b*) they are cheapest
(*c*) they interior surface is smooth and impervious
(*d*) their life is 100 years

14. The minimum diameter of sewer pipe is

(*a*) 1 cm (*b*) 5 cm
(*c*) 10 cm (*d*) 15 cm

15. Water carriage system is preferred to dry system because

(*a*) sewage is carried in covered conduits without having any nuisance
(*b*) treatment is given to sewage prior to its final disposal thus reducing chances of outbreak of epidemic
(*c*) there is no dearth of water
(*d*) (*a*) and (*b*) above are correct

16. In a combined sewer when only the industrial or sanitary sewage are flowing, then it is termed äs

(*a*) storm water flow (*b*) industrial flow
(*c*) ordinary flow (*d*) dry weather flow

17. In a combined system, the same sewer is intended to carry both

(*a*) domestic sewage and industrial wastes
(*b*) domestic sewage and storm water flow
(*c*) industrial waste and storm water flow
(*d*) dry weather flow and storm water flow

18. The wash water from stables which contains animal wastes is called

(*a*) barn sludge (*b*) animal sludge
(*c*) live sludge (*d*) horse sludge

19. Party or completely treated sewage flowing out of a sewage treatment tank or plain, is called

(*a*) treated sewage (*b*) sewage water
(*c*) discharged sewage (*d*) effluent

20. A fresh sewage becomes stale in

(*a*) 1 hour (*b*) 3 hour
(*c*) 6-8 hours (*d*) 24-36 hours

21. Waste water is said to be septic when

(*a*) it is known to contain disease producing bacteria
(*b*) it is undergoing decomposition
(*c*) it kills fish and plant life in streams receiving it
(*d*) the chief form of bacterial action in it is of the anaerobic type

22. The process of removing suspended and colloidal matter from sewage is called
(*a*) purification (*b*) clarification
(*c*) suspension (*d*) dewatering

23. Sewage that has received no purification treatment is called
(*a*) raw sewage (*b*) untreated sewage
(*c*) crude sewage (*d*) fresh sewage

24. The minimum infilteration of water in one km length of sewer pipe per day is
(*a*) 0 litre (*b*) 500 litres
(*c*) 2800 litres (*d*) 1,40,000 litres

25. A branch drain of which the last length of piping of the incoming drain before connection to the sewer is vertical, is called
(*a*) branch pipe (*b*) drain pipe
(*c*) drop connection (*d*) Y-connection

26. In combined sewers the velocity of flow of sewage should not be less than
(*a*) 0.6 m/sec (*b*) 0.75 m/sec
(*c*) 1 m/sec (*d*) 6 m/sec

27. The non-scouring velocity for cement concrete sewers is usually taken as
(*a*) 0.02 m/sec (*b*) 0.10 m/sec
(*c*) 0.12 m/sec (*d*) 0.25 m/sec

28. In separate sewers the minimum velocity of flow should not be less than
(*a*) 0.6 m/sec (*b*) 0.75 m/sec
(*c*) 1 m/sec (*d*) 6 m/sec

29. The maximum velocity of flow is limited to about 3 m/sec, through in practice it should preferably not exceed
(*a*) 2 m/sec (*b*) 2.25 m/sec
(*c*) 2.5 m/sec (*d*) 3 m/sec

30. The circular section of a sewer is best suitable when diameter is
(*a*) 0.75m (*b*) 1.25m
(*c*) 1.5m (*d*) 3m

31. While designing a sewerage system, the span of design period is generally taken äs
(*a*) one year (*b*) 5 years
(*c*) 10 years (*d*) 20 years

32. Where it is not possible to obtain self-cleaning velocities for sewers, flush tanks are installed with minimum available head of
(*a*) 1-2 m (*b*) 2-3 m
(*c*) 5-7 m (*d*) 10-15 m

33. pH value of fresh sewage is expected to be
(*d*) 1 (*b*) 3.5
(*c*) 7 (*d*) 8 to 11

34. The quantity of storm water from an area depends upon
(*a*) shape of the area (*b*) slope of the area
(*c*) nature of soil (*d*) all the above are correct

35. The ratio in between length of the sewer and velocity of flow when running full is known as
(*a*) inlet time (*b*) time of flow
(*c*) time of concentration (*d*) time intensity

36. The average annual at any station is the mean average of annual rainfall over a period of
(*a*) 10 years (*b*) 25 years
(*c*) 35 years (*d*) 50 years
(e) 100 years

37. Industrial sewage is generally expressed in
(*a*) litres per capita per day
(*b*) litres per day per sq. metre
(*c*) litres per hour per sq. metre
(*d*) litres per working (working in the industry) per day

38. The minimum infiltration of water in one km. length of water pipe per day is
(*a*) zero litres (*b*) 500 litres
(*b*) 2800 litres (*d*) 1,40,000 litres

39. A street sewer into which sewer from house connections is poured is called
(*a*) street sewer (*b*) main sewer
(*b*) lateral sewer (*d*) head sewer

40. A sewer which receives its flow from a number of transverse sewers or outlets is called
(*a*) head sewer (*b*) sewer reservoir
(*c*) connecting sewer (*d*) interceptor

41. The gas that is mainly responsible for explosion in sewer is
(*a*) ammonia (*b*) methane
(*c*) oxygen (*d*) carbon monoxide

42. The velocity of flow in sewers should be
(*a*) at least 30 cm/sec
(*b*) not more than 50 cm/sec
(*c*) less than cleansing velocity
(*d*) more than cleansing velocity

43. The lowest point of the interior of a sewer or drain at any cross-section is called
(*a*) bottom point (*b*) negative head point
(*c*) revetral point (*d*) invert

44. A street sewer, into which sewer from house connections pours, is called
(*a*) main sewer (*b*) street sewer
(*c*) head sewer (*d*) lateral sewer

45. Ventillation shafts are provided at the upper end of every branch sewer and they are generally spaced at a distance of
(*a*) 500 m (*b*) 300 m
(*c*) 100 m (*d*) 50 m

46. Generally salt glazed stone were pipes are manufactured in size 600 to 750 mm in diameter and their length is upto
(*a*) 60 to 90 cm (*b*) 3 m
(*c*) 5 m (*d*) 6 m

47. For large sewers, the maximum distance between sewers is usually limited to
(*a*) 50 metres (*b*) 300 metres
(*c*) 1000 metres (*d*) 1500 metres

48. The quantity of storm water from an area depends upon
(*a*) shape of the area (*b*) slope of the area
(*c*) nature of the soil (*d*) all the above are correct

49. The time that would be required for a drop of water to flow from the upper limit of the drainage area to the point where concentration or the maximum effect of flood is considered is called
(*a*) inlet time (*b*) compass
(*c*) time of concentration (*d*) time intensity

50. Sewer lines are usually laid with the help of a
(*a*) plain table (*b*) compass
(*c*) theodolite (*d*) sight rails
(e) time intensity

51. Alignment of sewer lines is started from
(*a*) the outfall (*b*) the high point
(*c*) the tail end (*d*) the last point

52. The proper gradient of sewer lines is ensured by
(*a*) compass (*b*) plane labelling
(*c*) dumpy level (*d*) telescope

53. Hydraulically the economical section of drains for large flow is
(*a*) circular (*b*) V-shaped
(*c*) rectangular (*d*) oval shaped

54. Hydraulically for dry weather flow, the best section is
(*a*) circular (*b*) V-shaped with circular invert
(*c*) rectangular shaped (*d*) oval shaped

55. The opening constructed on sewers or drains in order to enable men enter or leave the sewers is called
(*a*) lamp hole (*b*) manhole
(*c*) inspection chamber (*d*) street inlet

56. Manholes on sewer lines are provided for
(*a*) periodic cleaning (*b*) providing air for oxidation
(*c*) removal of part of sewerage (*d*) all of the above

57. A manhole is generally provided at each
(*a*) bend (*b*) junction and change of sewer dia
(*c*) change of gradient (*d*) all of the above

58. The minimum diameter of working chamber of a circular man hole should be
(*a*) 0.5 m (*b*) 0.9 m (*c*) 1.2 m (*d*) 2 m

59. The inlet with a basin which makes grit, sand or debris to settle and thus prevents it from entering into the sewer is called
(*a*) manhole (*b*) catch basin
(*c*) street inlet (*d*) inspection hole

60. A manhole of such depth that an access shaft is required in addition to the working chamber is called
(*a*) tube manhole (*b*) deep manhole
(*c*) well manhole (*d*) earth manhole

61. A manhole incorporating a vertical shaft or pipe in which sewage falls from a sewer at a higher level to a sewer at a lower level is called
(*a*) deep manhole (*b*) bottom manhole
(*c*) pit manhole (*d*) drop manhole

62. An access or opening having a removable cover for the obstruction to be cleared by means of drain rod is called
(*a*) Rodding-eye (*b*) rod-hole
(*c*) pit manhole (*d*) drop manhole

63. Manhole covers are usually made of
(*a*) cement (*b*) wood
(*c*) cast iron (*d*) steel

64. The opening provided in the gutter or street kerb in order to intercept the storm water or the surface wash and convey it into a storm or combined sever is known as
(*a*) manhole (*b*) street inlet
(*c*) catch basin (*d*) inlet hole

65. The location of flushing tank in a sewer in order to flush the sewer should be at its
(*a*) head (*b*) middle
(*c*) bottom (*d*) at any place

66. Bacteria in sewage are mostly
(*a*) anaerobic (*b*) pathogenic
(*c*) saprophytic (*d*) parasitic

67. The most frequent method of bacterial reproduction is by
(*a*) binding (*b*) cell division
(*c*) sexual process (*d*) colony formation

68. Single cell microorganisms in which organic matter diffuses into the cell and is consumed as food is known as
(*a*) bacteria (*b*) algae
(*c*) fungi (*d*) rotifers

69. The cells of bacteria are usually in the range of
(*a*) 0.1 to 0.2 microns (*b*) 0.5 to 5 microns
(*c*) 5 to 10 microns (*d*) 10 to 20 microns

70. Aerobes which feed on bacteria and algae and have a rigid shell structure are called
(*a*) algae (*b*) protozoa
(*c*) crust oceans (*d*) fungi

71. The cycle of life, death and decay involving organic nitrogenous matter, is known as
(*a*) the nitrogen cycle (*b*) the life cycle
(*c*) the hydrological cycle (*d*) the sulphur cycle

72. The temperature affects the
(*a*) viscosity of sewage
(*b*) solubility of gases in sewage
(*c*) biological activity of bacteria in sewage
(*d*) all of the above

73. Fungi
(*a*) grows in presence of sunlight
(*b*) grows in absence of sunlight
(*c*) is in organic mass
(*d*) grows in lime

74. Killing of pathogenic organisms is called
(*a*) endogeneous oxidation (*b*) liquidation
(*c*) pasteurization (*d*) disinfection

75. The bacteria which live on free oxygen of air or on the oxygen dissolved in water are called
(*a*) aerobic bacteria (*b*) anaerobic bacteria
(*c*) facultative bacteria (*d*) all the above are correct

76. The quantity of free oxygen in septic sewer (4 to 5 hours old) is approximately
(*a*) 10 to 15 mg/litre (*b*) 5 to 10 mg/litre
(*c*) 2 to 5 mg/litre (*d*) zero mg/litre

77. The process of spliting up of complex organic compounds into oxygen and other gases like carbon dioxide, methane, ammonia and hydrogen sulphide by anaerobic bacterias is known äs
(*a*) oxidation (*b*) putrefaction
(*c*) decomposition of sewage (*d*) all the above are correct

78. The process of transferring of liberated elements, viz, nitrogen, carbon and sulphur into stable and unobjectionable compounds of nitrates, carbohydrates and sulphates is known as
(*a*) oxidation
(*b*) putrefaction
(*c*) decomposition of sewage
(*d*) none of the above are correct

79. Test of BOD are usually made at
(*a*) 0°C-1 day (*b*) 10°C-2 days
(*d*) 20°C-5 days (*d*) 20°C-3 days

80. During the treatment of sewage it is necessary to kill the
(*a*) bacterias (*b*) algae
(*c*) fungi (*d*) pathogens

81. During putrefaction process the gas given off is
(*a*) carbon dioxide (*b*) methane
(*c*) ammonia (*d*) all of the above

82. In the nitrogen cycle, ammonia is produced from
(*a*) carbohydrates (*b*) cellulose
(*c*) proteins (*d*) sugar

83. The turbidity of sewage depends on the quantity of solid matters present in
(*a*) suspension state (*b*) soluble state
(*c*) collidal (*d*) all the above are correct

84. Direct destruction to pathogenic bacteria is caused by
(*a*) direct exposure to sunlight
(*b*) oxidation
(*c*) chlorination
(*d*) filtration

85. The process of mixing the sewage with large volume of water to keep in aerobic condition is known as
(*a*) dilution (*b*) current
(*c*) Sedimentation (*d*) oxidation

86. The major effect of bacteria in sewage treatment is to help in
(*a*) disinfection of sewage
(*b*) removal of objectionable odours
(*c*) the process of breaking the complex organic compounds into simple and stable compounds
(*d*) changing the colour and composition including pH value of the contents

87. "The pathogens in river flows are killed and helps in self-purification" It is due to
(*a*) sunlight (*b*) oxidation
(*c*) reduction (*d*) sedimentation

88. The depletion of dissolved oxygen due to BOD reaction if sewage is called
(*a*) deoxygenation (*b*) reaeration
(*c*) aeration (*d*) oxidation

89. The mass of sewage solids, both mineral and organic, which float at the surface of sewage treatment tank, is called
(*a*) solids (*b*) waste mass
(*c*) surface solids (*d*) scum

90. In biological treatment there is formation of a biological film of
(*a*) algel (*b*) anaerobic bacteria
(*c*) aerobic bacteria (*d*) protozoa

91. The maximurn efficiency of BOD removal is achieved in
(*a*) oxidation ditch (*b*) trickling filters
(*c*) aerated lagoons (*d*) digestion tanks

92. The septic tank should be cleaned at regular interval but ahould not exceed
(*a*) 5 years (*b*) 1 year
(*c*) 6 months (*d*) 3 years

93. Bacteria which use carbon dioxide as a source of carbon are known as
(*a*) autotrophic (*b*) hetrotropic
(*c*) aerobic (*d*) anaerobic

94. Which system is used for the treatment of sewage of small estates and colonies?
(*a*) Aerated lagoons (*b*) Oxidation pond
(*c*) Oxidation ditch (*d*) Any of the above

95. BOD means
(*a*) biochemical oxygen demand
(*b*) biological oxygen demand
(*c*) bacteriological oxygen demand
(*d*) none of these are correct

96. More the temperature, more will be biochemical activity, but the bacteria die if the temperature is more than

(*a*) 50°C (*b*) 40°C
(*c*) 30°C (*d*) 25°C

97. If organic matter increases in sewage, the demand of oxygen will

(*a*) increase (*b*) decrease
(*c*) remain constant (*d*) vary from organic-to-organic

98. The polluted water can be used for "Fish culture" if

(*a*) dissolved oxygen > 3 to 5 ppm
(*b*) CO_2 < 40 ppm
(*c*) both (*a*) and (*b*) are correct
(*d*) both (*a*) and (*b*)are not correct

99. The rate of de-oxygenation depends upon

(*a*) volume and BOD of sewage
(*b*) time available for decomposition
(*c*) temperature of the diluting water
(*d*) all the above are correct

100. The operation of destruction of organic matter by having it served as food for bacteria, is carried out in

(*a*) sludge setting tanks (*b*) sludge digestion tanks
(*c*) sludge beds (*d*) vacuum filters

101. In oxidation ponds, sewage is mainly treated by

(*a*) oxidation (*b*) reduction
(*c*) putrefaction (*d*) aerobic bacteria action

102. The BOD of waste water determines the number of milligrams per litre of oxygen required

(*a*) during stabilization of decomposable organic matter by aerobic bacterial action
(*b*) for the oxidation of sulphites and thiosulphates to sulphates
(*c*) to produce an equilibrium between the oxygen of the waste water and atmospheric oxygen
(*d*) to unite chemically with the inorganic matter present in the sample

103. A lest used to measure the strength of waste water is called

(*a*) detention period (*b*) pH
(*c*) BOD (*d*) surface setting rate

104. A detritus tank is provided in the primary treatment of sewage to remove

(*a*) suspended solids (*b*) grit
(*c*) stones (*d*) oils and greases

105. Before discharging the foul sewage into revers, it is treated by
(*a*) screening
(*b*) Sedimentation
(*c*) oxidation
(*d*) sludge digestion and disinfection

106. Operation generally done during preliminary treatment of sewage is
(*a*) chlorination of sewage
(*b*) oil and grease removal from skimming
(*c*) removal of grit and also floating material through screens
(*d*) (*b*) and (*c*) above

107. The spacing of steel bars in coarse screens used for the treatment of sewage is
(*a*) 10 mm (*b*) 20 mm
(*c*) 30 mm (*d*) 50 mm

108. The chlorination in sewage treatment plant is done to
(*a*) accelerate the sludge digestion process
(*b*) disinfect the sewage
(*c*) remove objectionable odour
(*d*) accelerate settlement of suspended matter

109. Grit is composed mostly of
(*a*) grease (*b*) paper
(*c*) wood (*d*) sand

110. The spacing of bars of perforation of fine screens used for the treatment of sewage is
(*a*) 2-3 mm (*b*) 8-10 mm
(*c*) 15-20 mm (*d*) 25-30 mm
(*e*) 50 mm

111. The bulking of sludge in activated sludge process can be remedied to some extent by
(*a*) 2.56 (*b*) 2.65
(*c*) 2.67 (*d*) 1.00

112. The bulking of sludge in activated sludge process can be remedied to some extent by
(*a*) chlorination
(*b*) reducing the aeration period
(*c*) reducing the pH value of the sewage
(*d*) addition of fresh water

113. An activated sludge is
(*a*) the sludge that is obtained by settling of sewage in presence of abundant oxygen
(*b*) the sludge that has been subjected to 50% anaerobic action

(*c*) the sludge that is discharged from primary setting tanks
(*d*) the sludge that is obtained from the sludge digestion tank

114. In a sludge digestion tank the process of sludge digestion is rapid when the pH value is in the range of
(*a*) 4–7 (*b*) 10–14
(*c*) 6.8–7.2 (*d*) 7–8

115. The gas produced during sludge digestion is
(*a*) nitrogen (*b*) methane
(*c*) carbon monoxide (*d*) carbon dioxide

116. A tank in which raw or partly treated sewage is collected, allowed to stay and discharged at such a rate as may be necessary for subsequent treatment, is known as
(*a*) surge tank (*b*) dosing tank
(*c*) sludge tank (*d*) sedimentation tank

117. Elutrition is the process of
(*a*) adding oxygen to the sludge
(*b*) washing digested sludge
(*c*) sludge digestion
(*d*) disposing off the sludge

118. Oil and grease can be removed by
(*a*) floatation tanks (*b*) rocks and screens
(*c*) biological growth (*d*) disinfection

119. Screening or skimming cannot remove
(*a*) large suspended matter (*b*) oil and fat
(*c*) grease (*d*) fine suspended matter

120. Fine suspended matter removed by
(*a*) screening (*b*) skimming
(*c*) sedimentation (*d*) filteration

121. The settling velocity does not depend on
(*a*) specific gravity of particles (*b*) depth of tank
(*c*) size or particles (*d*) temperature of liquid

122. The setting velocity depends on
(*a*) length of the tank (*b*) width of the tank
(*c*) depth of tank (*d*) length and width of tank

123. The coagulant widely used for sewage treatment is
(*a*) ferric chloride (*b*) line
(*c*) alum (*d*) ferric sulphate

124. The allowable detention period in sedimentation tank is
(*a*) 1 to 3 hours (*b*) 3 to 5 hours
(*c*) 5 to 10 hours (*d*) 12 to 24 hours

125. In case the surface area of a Sedimentation tank is increased, it will remove more
(*a*) fine particles (*b*) large particles
(*c*) particles of all sizes (*d*) water

126. Humus tanks are provided for
(*a*) primary sedimentation (*b*) secondary sedimentation
(*c*) filteration (*d*) sludge disinfection

127. The reduction of BOD in a sedimentation tank for sewage treatment may be expected in the range
(*a*) 10% (*b*) 10% to 25%
(*c*) 30% to 40% (*d*) 60% to 76%

128. The BOD is satisfied in biological treatment by
(*a*) screening (*b*) sedimentation
(*c*) filtration (*d*) disinfection

129. The normal trickling filter removes
(*a*) 50% of BOD (*b*) 80-90% of BOD
(*c*) 95% of BOD (*d*) 95-99% of BOD

130. The sewage filter used in modern sewage treatment works is
(*a*) intermittent sand filter (*b*) the contact beds
(*c*) the trickling filter (*d*) none of (*a*), (*b*), (*c*)

131. Vacuum filters are used for
(*a*) slowing down bacterial activity
(*b*) dewatering of sludge
(*c*) filteration of sludge
(*d*) filteration of released water

132. In a high rate rickling filter, high rate of loading is achieved by
(*a*) better filter media (*b*) recirculation of sewage
(*c*) dunbar filters (*d*) any of the above

133. Which of the following processes takes place in trickling filters?
(*a*) Filteration (*b*) Oxidation
(*c*) Disinfection (*d*) Biological action

134. Organisms used by both trickling filters and activated sludge system for the bulk of their purification capacity are
(*a*) anaerobic (*b*) parasitic
(*c*) aerobic (*d*) none of the above

135. Solids that cannot be removed by filtering are
(*a*) suspended solids (*b*) settleable solids
(*c*) dissolved solids (*d*) coarse solids

136. The main disadvantage of intermittent sand filters is
(*a*) requirement of large area (*b*) less depth
(*c*) use of costly ehemicals
(*d*) their suitability for acidic sludges only

137. Lagooning is
(*a*) method of sludge disinfection
(*b*) method of sludge dilution
(*c*) method of rapid sludge digestion
(*d*) method of sludge disposal

138. The normal rate for the disposal of sewage by irrigation on a sandy soil, in litres per hectare per day, is of the order
(*a*) 2,500 (*b*) 25,000
(*c*) 250,000 (*d*) 2,500,000

139. Pumps commonly used for sewage disposal are
(*a*) reciprocating pumps (*b*) turbine pumps
(*c*) vane pumps (*d*) mud pumps

140. The impeller for sewage pump is
(*a*) shrouded on both sides (*b*) shrouded on one side only
(*c*) open type (*d*) vane type

141. In distribution pipes, drain valves are located
(*a*) at junctions (*b*) at summit
(*c*) at lower joint (*d*) any where

142. Priming of a centrifugal pump may not be necessary, in case the pump is located
(*a*) at less than 10 m height above the reservoir level
(*b*) at less than 5 m height above the reservoir level
(*c*) immediately above the reservoir level
(*d*) below the reservoir level

143. A device used to measure the flow of waste water is
(*a*) comparator (*b*) weir
(*c*) sluice gate (*d*) comminutor

144. Which of the following disease is not considered as water borne?
(*a*) Typhoid (*b*) Jaundice
(*c*) Bacillary dysentry (*d*) Malaria

145. Fish life in receiving streams is most seriously affected by
(*a*) high coliform counts
(*b*) drop in dissolved oxygen level
(*c*) increase in dissolved mineral concentrations
(*d*) all of the above

146. Carbon dioxide, one of the gases given off by the decomposition of sludge, is not poisonous, but it may cause
(*a*) combustion (*b*) corrosion
(*c*) burns (*d*) asphyxiation

147. Raw waste water sludge should not be used as a garden fertilizer because it contains
(*a*) injurious acids
(*b*) pathogenic bacteria
(*c*) thermophilic organisms
(*d*) high sodium concentrations

148. A water closet is commonly made of
(*a*) stone ware (*b*) terracota
(*c*) porcelain (*d*) glazed earthenware

149. Normally the capacity of flushing cisterns for water closets is
(*a*) 1 to 2 litres (*b*) 2 to 5 litres
(*c*) 10 to 15 litres (*d*) 25 to 40 litres

150. The minimum height of a flushing cistern is
(*a*) 1 m (*b*) 3 m
(*c*) 5 m (*d*) 7.5 m

151. The system of collecting night soil in pots or pits for periodical removal outside the town area for burial, is known as
(*a*) potting (*b*) night washing
(*c*) conservancy (*d*) cleansing

152. From septic tank the effluents are discharged into
(*a*) soak pit (*b*) drainage
(*c*) oxidation pond (*d*) sewer

153. Traps are provided to
(*a*) stop flow of sewage
(*b*) separate the flow of liquids and solids
(*c*) avoid back flow of sewage
(*d*) prevent the escape of foul gases inside and outside the house

154. A trap provided to prevent the foul gases from going in the house, is
(*a*) Greavak trap (*b*) Anti-D trap
(*c*) Guilty trap (*d*) Intercepting trap

155. For a boarding house the standard number of WCs required is
(*a*) 1% (*b*) 2%
(*c*) 5% (*d*) 10%

156. For a cinema hall the number of WCs required for females is
(*a*) 1% (*b*) 2%
(*c*) 2% upto 200 and 1% thereafter
(*d*) 1% upto 200 and 2% thereafter

157. The minimum depth of water seal in a good trap is
(*a*) 2.5 to 7.5 cm (*b*) 10 to 15 cm
(*c*) 15 to 25 cm (*d*) 30 to 35 cm

158. The average spacing of urinal stalls in public places in kept as
(*a*) 30 cm (*b*) 60 cm
(*c*) 100 cm (*d*) 125 cm

159. For a cinema hall with seating capacity of 1000, the number of urinals to be provided as
(*a*) 5 (*b*) 10 (*c*) 25 (*d*) 40

160. The height of a porcelain urinal stall bowl type, above the foot level is usually
(*a*) 30 cm (*b*) 50 cm (*c*) 75 cm (*d*) 100 cm

161. The capacity of flushing cisterns for urinal is generally
(*a*) 1 litre (*b*) 5 litres
(*c*) 15 litres (*d*) 40 litres

162. A trickling filter is primarily a
(*a*) straining process to remove suspended solids from sewage
(*b*) biological oxidation process to remove BOD from sewage
(*c*) straining process to remove turbidity from water
(*d*) straining process to remove bacteria from water

163. A combined sewage is one which transports
(*a*) domestic sewage and storm water
(*b*) domestic sewage and industrial wastes
(*c*) domestic sewage and overhead flow
(*d*) domestic sewage and industrial wastes and storm water

164. A septic tank is
(*a*) an aerobic method of on-site sewage treatment
(*b*) an anaerobic method of on-site treatment
(*c*) a physical method of water treatment
(*d*) a physicochemical method of water treatment

165. Biochemical oxygen demand (BOD) of waste water is a measure of
(*a*) total concentration of biochemicals
(*b*) total concentration of organic matter
(*c*) concentration of biodegradable organic matter
(*d*) concentration of chemically oxidisable matter

166. In an atmosphere under super-adiabatic lapse rate conditions, the emission from a chimney produces a plume describable as
(*a*) coning (*b*) loffing
(*c*) looping (*d*) fumigation

167. Chemical Oxygen Demand (COD) of a sample is always greater than Biochemical Oxygen Demand (BOD) since it represents
(*a*) biodegradable organic matter only
(*b*) biodegradable and non-biodegradable
(*c*) non-biodegradable organic matter
(*d*) inogranic matter

168. A waste water sample diluted to 100 times with aeration water had and initial dissolved oxygen (DO) of 7.0 mg/L and after 5 days of incubation at 20°C, the DO was zero. The BOD of waste water is

(*a*) 700 mg/L (*b*) 100 mg/L
(*c*) cannot be determined (*d*) 1 mg/L

169. High COD to BOD ratio of an organic pollutant represents

(*a*) high biodegradability of the pollutant
(*b*) low biodegradability of the pollutant
(*c*) presence of free oxygen for aerobic decomposition
(*d*) presence of toxic material in the pollutant

170. Activated sludge is the

(*a*) aerated sludge in the aeration unit
(*b*) sludge settled in the humus tank
(*c*) sludge in the secondary tank after aeration and rich
(*d*) sludge in the secondary tank after aeration and rich in nutrients

171. During temperature inversion in atmosphere, air pollutants tends to

(*a*) accumulate above inversion layer
(*b*) accumulate below inversion layer
(*c*) disperse laterally
(*d*) disperse vertically

172. Symbiosis, the beneficial association between algae and bacteria is used for treatment of waste water in the following unit

(*a*) Activated sludge (*b*) Rotating biological disc
(*c*) Anaerobic digestor (*d*) Oxidation pond

173. The ultimate BOD of the waste water whose 5 days BOD $(BOD)_5$ and rate constant (base e) are respectively 150 mg/1 and 0.23 dassis

(*a*) 80 mg/1 (*b*) 150 mg/1
(*c*) 180 mg/1 (*d*) 220 mg/1

174. The drop manholes are provided in a sewerage system when there is

(*a*) change in alignment of sewer line
(*b*) change in size of sewers
(*c*) change in the elevation of ground level
(*d*) change from gravity system to pressure system

175. The main constituents of gas generated during the anaerobic digestion of sewage sludge are

(*a*) carbon dioxide and methane
(*b*) methane and ethane
(*c*) carbon dioxide and carbon monoxide
(*d*) carbon monoxide and nitrogen

176. A single rapid test to determine the pollutional status of river water is
(*a*) biochemical oxygen demand
(*b*) chemical oxygen demand
(*c*) total organic solids
(*d*) dissolved oxygen

177. Presence of excess nitrates in river water indicates
(*a*) recent pollution of water with sewage
(*b*) past pollution of water with sewage
(*c*) immediate pollution of water with sewage
(*d*) no pollution of water with sewage

178. The clariflocculator is the unit in which the following things will occur
(*a*) floc formation and its subsequent removal by filtration
(*b*) floc formation and its subsequent removal by sedimentation
(*c*) floc formation and its subsequent removal by decantation
(*d*) removal of bacteria by filtration and chlorination

179. A rapid test to indicate the intensity of pollution in river water is
(*a*) biochemical oxygen demand
(*b*) dissolved oxygen
(*c*) MPN
(*d*) total dissolved solids

180. Tricking filters are used to remove
(*a*) suspended solids (*b*) collidal solids
(*c*) organic matter (*d*) pathogenic bacteria

181. Under Indian conditions, the average per capita contribution of BOD is
(*a*) 10 to 20 gm/d (*b*) 20 to 35 gm/d
(*c*) 35 to 50 gm/d (*d*) 50 to 70 gm/d

182. Sewage sickness relates to
(*a*) toxicity of sewage interfering with response to treatment
(*b*) destruction of aquatic flora and fauna due to gross pollution of receiving bodies of water by sewage
(*c*) reduction is the waste purifying capacity
(*d*) clogging of pores in soil due to excessive application of sewage of land, obstructing aeration and leading of septic conditions

183. The slope of a 1.0 m diameter concrete sewer laid at a slope of 1 in 1000, develops a velocity of 1 m/s when flowing full when it is flowing half full, the velocity of flow through the sewer will be
(*a*) 0.5 m/s (*b*) 1.0 m/s
(*c*) 72 m/s (*d*) 2.0 m/.s

184. One litre of sewage when allowed to settle for 30 minutes gives a sludge volume of 27 cm^3. If the dry weight of this sludge is 3.0 gms then its sludge volume index will be

(*a*) 9 (*b*) 24
(*c*) 30 (*d*) 81

185. An aeration basin with a volume of 400 m^3 contains mixed liquor with suspended solid concentration of 1000 mg/1. The amount of mixed liquor suspended solids in the tank is

(*a*) 500 kg (*b*) 250 kg (*c*) 6600 kg (*d*) 400 kg

186. Which one of the following statement is true of trickling filter sludge?

(*a*) It has a comparatively low sludge volume index
(*b*) It is more difficult to dewater than activated sludge
(*c*) It has a comparatively low concentration of sludge solids
(*d*) It is bulky

187. An industrial waste water enters a steam having a BOD concentration of 10 mg/l and a flow of 20 m^3/s and its BOD concentration is 250 mg/l then the BOD concentration in the stream at a point down stream of the point of confluence of waste water with the stream will be

(*a*) 2.67 mg/l (*b*) 12.09 mg/l (*c*) 13.00 mg/l (*d*) 2.74 mg/l

188. The following reactions take place during anaerobic digestion of organics

(1) Methane production (2) Alkaline fermentation
(3) Acid fermentation (4) Acid regression

The correct sequence of these reactions is

(*a*) 3, 4, 2, 1 (*b*) 4, 3, 2, 1 (*c*) 3, 4, 1, 2 (*d*) 4, 3, 1, 2

189. Coal based thermal power stations pollute the atmosphere by addition

(*a*) NO_x and SO_2 (*b*) NO_x, SO_2 and SPM
(*c*) NO_x, SO_2 SPM and CO (*d*) NO_x, SPM and CO

190. Traps are used in household drainage systems to

(*a*) prevent entry of fond gases in the houses
(*b*) restrict the flow of water
(*c*) provide partial vacuum
(*d*) trap the solid wastes

191. For the combined sewage system egg-shaped sewers are preferred because

(*a*) their construction is economical
(*b*) they are structurally more stable
(*c*) the maintenance is easier
(*d*) they offer good flow velocity during the dry-weather-flow condition

192. Corrosion of concrete sewers occur due to
(*a*) high velocity of flow of sewage
(*b*) aerobic decomposition of sewage solids
(*c*) Anaerobic decomposition of sewage solids
(*d*) high pH value of sewage

193. The growth of algae is useful in
(*a*) Sedimentation tank (*b*) slow sand filter
(*c*) oxidation pond (*d*) sludge digestion tank

194. A polluted stress undergoes self-purification is four distinct zones
1. Zone of clear water 2. Zone of active decomposition
3. Zone of degradation 4. Zone of recovery
The correct sequence of these zones is
(*a*) 4, 3, 2, 1 (*b*) 2, 3, 4, 1 (*c*) 2, 4, 3, 1 (*d*) 3, 2, 4, 1

195. Under natural condition of flow on polluted river would contain
(*a*) more dissolved oxygen in summer than in winter
(*b*) less dissolved oxygen in summer than in winter
(*c*) more or less the same amount of dissolved oxygen in winter and summer
(*d*) the least amount of dissolved oxygen during the floods

196. Area method of land fllling is most suitable when
(*a*) area is unsuitable for excavation of trenches
(*b*) adequate depth of cover material is available at site
(*c*) the water table is near the surface
(*d*) natural or artificial depressions exist in the vicinity

197. Consider the following statements
Excessive growth of water-weeds in a water body is attributed to the
1. increase in the benthic organisms including bacteria
2. imbalance in the aquatic eco system
3. excessive inflow of nutrients
Of these statements
(*a*) 1,2 and 3 are correct (*b*) 1 and 2 are correct
(*c*) 1 and 3 are correct (*d*) 2 and 3 are correct

198. Which one of the following solid waste disposal methods is ecologically most acceptable?
(*a*) sanitary land fill (*b*) incineration
(*c*) composition (*d*) pyrolysis

199. Non-disposal of solid waste may cause the spread of
(*a*) Malaria (*b*) rodents related plague
(*c*) Typhoid (*d*) dysentery

200. Which of the following is not a major indoor pollutant in a developing country like India, in the rural areas?
(*a*) cooking and heating activities
(*b*) burning of cow dung
(*c*) burning of crop waste (*d*) building material

201. Velocity of sound in air is around
(*a*) 144 m/s (*b*) 244 m/s (*c*) 344 m/s (*d*) 444 m/s

202. Which of the following can be used for disinfection of swimming pools ?
(*a*) chlorine (*b*) cyanurates
(*c*) bromine (*d*) any of the above

203. Algae development in a swimming pool causes
(*a*) rapid rise in pH value (*b*) slimy growth on the walls
(*c*) increased chlorine consumption
(*d*) all of the above

204. Which of the following is not a biological process for destroying organic compounds
(*a*) compositing (*b*) trickling filters
(*c*) calcination (*d*) activated sludge

205. The proportion of colouring contents of water is reduced by
(*a*) coagulation (*b*) setting
(*c*) rapid and filteration (*d*) all of the above

206. The colour of water is measured on
(*a*) wavelength basis (*b*) platinum-cobalt scale
(*c*) pH scale (*d*) rockwell scale

207. pH value of drinking water should be in the range
(*a*) 1 to 10 (*b*) 2 to 6 (*c*) 6.5 to 8.5 (*d*) 8.5 to 10

208. Sea water has a total dissolved solids concentration of about
(*a*) 360,000 mg/litre (*b*) 36,000 mg/litre
(*c*) 3,600 mg/litre (*d*) 360 mg/litre

209. Which of the following chemical can be used for disinfection of water?
(*a*) chlorine (*b*) ozone
(*c*) chlorine-ammonia (*d*) any of the above

210. Incineration is
(*a*) a controlled combustion process for burning solid, liquid or gase- ous combustible waste
(*b*) a process of destroying solid wastes by oxidation
(*c*) a process of burning solid and liquid waste using its calorific value
(*d*) a process of powdering solids for remelting

211. Pollutants in air can be in the form of
(*a*) particulates (*b*) microorgranisms
(*c*) gases (*d*) any of the above

212. Air pollutants common to urban environment are
(*a*) sulphur dioxide (*b*) particulate matter
(*c*) hydrocarbon (*d*) all of the above

213. Swimming pool water requires
(*a*) pre-chlorination (*b*) super-chlorination
(*c*) dual-chlorination (*d*) de-chlorination

214. For minimal eye irritation pK value of swimming pool water should be in the range
(*a*) 6.2 to 6.8 (*b*) 7.0 to 7.2
(*c*) 6.5 to 7.0 (*d*) 7.5 to 7.6

215. Which of the following layer is closest to earth?
(*a*) troposphere (*b*) stratosphere
(*c*) ionosphere D layer (*d*) ionosphere F_2 layer

Answers

1. (*b*)	**2.** (*c*)	**3.** (*c*)	**4.** (*a*)	**5.** (*b*)	**6.** (*d*)
7. (*a*)	**8.** (*d*)	**9.** (*b*)	**10.** (*d*)	**11.** (*d*)	**12.** (*d*)
13. (*c*)	**14.** (*d*)	**15.** (*d*)	**16.** (*d*)	**17.** (*d*)	**18.** (*a*)
19. (*d*)	**20.** (*c*)	**21.** (*d*)	**22.** (*b*)	**23.** (*c*)	**24.** (*c*)
25. (*c*)	**26.** (*b*)	**27.** (*d*)	**28.** (*b*)	**29.** (*b*)	**30.** (*c*)
31. (*d*)	**32.** (*c*)	**33.** (*d*)	**34.** (*d*)	**35.** (*b*)	**36.** (*c*)
37. (*b*)	**38.** (*c*)	**39.** (*c*)	**40.** (*d*)	**41.** (*b*)	**42.** (*d*)
43. (*d*)	**44.** (*d*)	**45.** (*b*)	**46.** (*a*)	**47.** (*b*)	**48.** (*d*)
49. (*c*)	**50.** (*d*)	**51.** (*b*)	**52.** (*c*)	**53.** (*a*)	**54.** (*b*)
55. (*b*)	**56.** (*a*)	**57.** (*d*)	**58.** (*c*)	**59.** (*b*)	**60.** (*b*)
61. (*b*)	**62.** (*a*)	**63.** (*c*)	**64.** (*b*)	**65.** (*a*)	**66.** (*c*)
67. (*b*)	**68.** (*a*)	**69.** (*b*)	**70.** (*c*)	**71.** (*a*)	**72.** (*d*)
73. (*a*)	**74.** (*d*)	**75.** (*a*)	**76.** (*d*)	**77.** (*b*)	**78.** (*a*)
79. (*c*)	**80.** (*d*)	**81.** (*d*)	**82.** (*b*)	**83.** (*a*)	**84.** (*c*)
85. (*a*)	**86.** (*c*)	**87.** (*a*)	**88.** (*a*)	**89.** (*d*)	**90.** (*c*)
91. (*a*)	**92.** (*d*)	**93.** (*a*)	**94.** (*c*)	**95.** (*a*)	**96.** (*b*)
97. (*a*)	**98.** (*c*)	**99.** (*d*)	**100.** (*b*)	**101.** (*d*)	**102.** (*a*)
103. (*c*)	**104.** (*b*)	**105.** (*d*)	**106.** (*d*)	**107.** (*d*)	**108.** (*b*)
109. (*d*)	**110.** (*a*)	**111.** (*b*)	**112.** (*a*)	**113.** (*a*)	**114.** (*c*)
115. (*d*)	**116.** (*b*)	**117.** (*b*)	**118.** (*a*)	**119.** (*d*)	**120.** (*c*)
121. (*b*)	**122.** (*d*)	**123.** (*a*)	**124.** (*a*)	**125.** (*c*)	**126.** (*b*)
127. (*c*)	**128.** (*c*)	**129.** (*b*)	**130.** (*c*)	**131.** (*b*)	**132.** (*b*)
133. (*a*)	**134.** (*c*)	**135.** (*c*)	**136.** (*a*)	**137.** (*d*)	**138.** (*c*)
139. (*d*)	**140.** (*c*)	**141.** (*c*)	**142.** (*d*)	**143.** (*b*)	**144.** (*d*)

145. (*b*)	146. (*d*)	147. (*a*)	148. (*c*)	149. (*c*)	150. (*b*)
151. (*c*)	152. (*a*)	153. (*d*)	154. (*d*)	155. (*d*)	156. (*c*)
157. (*a*)	158. (*b*)	159. (*c*)	160. (*c*)	161. (*b*)	162. (*b*)
163. (*a*)	164. (*b*)	165. (*c*)	166. (*c*)	167. (*b*)	168. (*a*)
169. (*d*)	170. (*d*)	171. (*d*)	172. (*d*)	173. (*c*)	174. (*c*)
175. (*a*)	176. (*a*)	177. (*b*)	178. (*b*)	179. (*c*)	180. (*c*)
181. (*b*)	182. (*d*)	183. (*b*)	184. (*a*)	185. (*d*)	186. (*b*)
187. (*d*)	188. (*a*)	189. (*c*)	190. (*a*)	191. (*d*)	192. (*c*)
193. (*c*)	194. (*d*)	195. (*c*)	196. (*d*)	197. (*a*)	198. (*d*)
199. (*b*)	200. (*d*)	201. (*c*)	202. (*d*)	203. (*d*)	204. (*c*)
205. (*d*)	206. (*b*)	207. (*c*)	208. (*b*)	209. (*d*)	210. (*a*)
211. (*d*)	212. (*d*)	213. (*d*)	214. (*d*)	215. (*a*)	

TRUE OR FALSE TYPE QUESTIONS

1. Sewage consists of water only and solid matter is hardly present.
2. When rainfall is heavy in a region then it is better to provide separate system, because in combined system self-cleansing velocity will not be available for most of the period of year.
3. The flow which occurs during the day at about noon time is usually about two times the average flow.
4. While designing a sewerage system, domestic sewage is generally expressed as litres per capita per day.
5. As a thumb rule quantity of sewage may be considered to be equal to the quantity of water supplied neglecting the losses due to evaporation etc.
6. When sewage is conveyed through covered pipes then the sewer must not run full, otherwise the pressure might rise above the atmospheric pressure.
7. Sewers are normally designed to flow 1/2 or 2/3 running full when discharging the peak or maximum flow.
8. As compared to dry weather flow, the amount of storm water in a sewerage system is very high.
9. The size of manhole opening should be such as to allow a man go into and come out with ease, so 50 cm size is a satisfactory size.
10. The lamp hole is constructed of stoneware or cast iron pipe about 30 cm in diameter, based in concrete and built off a tee-junction on the sewer.
11. Junction chamber is provided where sewers intersect with horizontal angles between their axes less than 30°, so that ordinary type of junction is difficult to be eonstructed.
12. The organic matter accounts for about 45% of the total solids and consists of animal and vegetable wastes and proteins etc.
13. The normal temperature of sewage is higher than that of water supply.
14. "Streams get self-purified in less time in summer as compared to winter, because of temperature effect.

15. Due to algae, the dissolved oxygen increases in a BOD bottle, in the presence of light.
16. During purification process gases given off are carbon dioxide, methane and amonia.
17. For removal of very fine suspended matter by adding chemical coagulants in sewage, chemical precipitation tanks are used.
18. Alum is generally not used as coagulant for treating the sewage.
19. A trickling filter consists of four parts. Three parts are a tank, a filter medium and a distribution system. The fourth part is under drain.
20. The hydraulic mean depth of egg shaped sewer of equivalent section is the same as that of the circular sewers when running full.
21. Egg shaped sewers are not suitable for combined sewers where the flows are to vary considerably.
22. Ordinary sewages are normally slightly acidic when fresh.
23. Aluminium slats are suitable in treating sewage.
24. Recirculation reduces the thickness of the biological film on contact media by forced film sloughing.
25. The efficiency of BOD removal of Biofilters can be obtained by using formula $C = \dfrac{C}{2R+5}$
26. The aeration tanks in the activated sludge treatment plants are similar to trickling filters.
27. Imhoff tanks are three storeyed tanks.
28. The settled sludge from settling ranks is not required to be pumped.
29. The efficiency of BOD removal depends upon the temperature.
30. Environment information System was set up in India in the year 1960.
31. Domestic wastes and sewages are degradable pollutants.
32. For sewer upto 300 mm diameter, self-cleansing velocity of 1 m/sec should be adopted.
33. Degree of turbulence based on lapse rate and wind profile is expressed by Richardson number.
34. Concentration of pollutant may be kept within allowable limit by venting through stack'.
35. One gallon (UK) equals 5.446 liters.
36. Water pressure in distribution domestic pipe for 3 storyed building should be not more than 1 kgf/cm^2.
37. Minimum velocity of flow in a pipe should not be less than 0.5 m/sec.
38. Non-scouring velocity is less in stoneware sewer then brick lined sewer.
39. Globe valve and gate valve are the types of water pipe valves.
40. Friction factor in 1000 hours old pipe is same as is 50 hours old pipe, for identical case.
41. Delivery main has 100 mm diameter. 5 number of 50 mm diameter branch pipes, from main may be taken.
42. Hardness of water in ppm is measured by weight of calcium carbonate.

43. Hardness of 50 ppm indicates excessively hard water.
44. Nominal diameter of salt glazed stonewater pipe is limited to 300 mm.
45. Design of domestic sewer is done for about 6 times of average rate of annual flow.
46. Sewage production in (litres/person/day) in small city equals the water supplied (litres/person/day).

Answers

1. True	**2.** True	**3.** True	**4.** True	**5.** True
6. True	**7.** True	**8.** True	**9.** True	**10.** True
11. True	**12.** True	**13.** True	**14.** True	**15.** True
16. True	**17.** True	**18.** True	**19.** True	**20.** True
21. False	**22.** False	**23.** False	**24.** True	**25.** False
26. True	**27.** False	**28.** True	**29.** True	**30.** False
31. True	**32.** True	**33.** True	**34.** True	**35.** False
36. False	**37.** True	**38.** False	**39.** True	**40.** False
41. True	**42.** True	**43.** False	**44.** False	**45.** True
46. False				

FILL IN THE BLANKS TYPE QUESTIONS

1. A township is to treat 5,00,000 litres of sewage per day which has a 5 day BOD of 150 ppm. An oxidation pond is used for purpose. The effluent can have a BOD of 15 ppm. The loading is to 40 kg of 5 day BOD per hactare per day. The required area of the pond is ________
2. A 20 cm diameter sewer is laid at a slope of 0.004 and is designed to carry a discharge at a depth of 10 cm with Meaning's $n = 0.014$, the design discharge is ________
3. The reoxygenation coefficient K of a stream is 0.30 at 20°C. Its K value at 32°C is likely to be ________
4. In India committee on Environmental Coordination was set up in the year ________
5. Pollution Control Board in India was created in the year ________
6. National Waste Lands Board in India was created in the year ________
7. ________ sewage comprises of the surface runoff developed during and immediately after a reinstrom fallen over the area concerned.
8. While designing sewers an attempt is made to developed an average velocity of ________ cm/sec.
9. ________ ejectors are primarily used for lifting of sewage from the cellars of large building into the municipal decomposition of organic contents under ________ conditions.
10. BOD is the oxygen requirement of sewage for the biological decomposition of organic contents under ________ conditions.

11. Normal domestic sewage in India town and cities usually contains < 250 ppm of ________ day BOD at 20°C.
12. When sewage is finally disposed on land, disposal is called ________ irrigation.
13. When sewage is finally disposed in large bodies of water it is called the ________
14. Biological agencies in nature are utilised in the units of ________ treatment.
15. A ________ filter is a bed of coarse rough imperious material.
16. The ratio of recirculated flow to the flow of raw sewage is called the ________ ratio.
17. Foam in water is caused by ________
18. Flushing storage capacity for factories workshops, are per N.B.C. is ________
19. One brick thick wall with cement plaster provides good insulation for ________
20. Effect of air pollutant on paints is to ________
21. ________ valve is also known as reflux valve.
22. Population forecasting is done by ________
23. Dose of coagulant be added in water depends on ________
24. Normál dose of alluminium sulphate as coagulant is ________
25. Depth of sand bed for filtration should be about ________.

Answers

1. 1.875 ha	**2.** 476.6 lps	**3.** 0.52	**4.** 1972
5. 1980	**6.** 1981	**7.** storm	**8.** 90
9. pneumatic	**10.** aerobic	**11.** 5	**12.** broad
13. dilution	**14.** secondary	**15.** trickling	**16.** recirculation
17. trade wastes	**18.** 900 litres per W.C.		**19.** noise
20. discolour	**21.** check	**22.** apportionment	
23. pH value	**24.** 6 mg/litre	**25.** 75 cm	

❑❑❑

APPENDICES

APPENDIX-I

Conversion Tables

S.No. (1)	Quantity (2)	Conversion of (3)	Conversion to (4)	Multiplying factor (5)
1.	Length	Millimetres	Microns	1000
		Inches	Centimetres	2.54
		Feet	Centimetres	30.48
		Feet	Metres	0.3048
		Metres	Feet	3.281
		Miles	Kilometres	.1.6093
		Kilometres	Miles	0.6215
2.	Area	Square inches	Square centimetres	6.452
		Square feet	Square centimetres	929.0
		Square feet	Square inches	0.0929
		Square metres	Square feet	10.76
		Acres	Hectares	0.4047
		Square mile	Square kilometres	2.590
		Square mile	Hectare	258.99
		Square yard	Square metre	0.8361
3.	Volume or capacity	Cubic inches	Cubic centimetres	16.39
		Cubic feet	Cubic metres	0.0283
		Cubic yard	Cubic metre	0.7646
		Cubic metre	Cubic feet	35.815
		Litre	Cubic feet	0.0353
		UK (Imperial) gallons	Litres	4.546
		US gallon	Litres	3.785
		UK (Imperial) gallons	US gallons	1.20
4.	Weight	Grain	Gram	0.0648
		Ounce (oz)	Gram	28.3495
		Pounds	Gram	435.6
		Pounds	Kilograms	0.4536
		Ton	Tonnes	1.01605
		Kilogram	Pounds	2.2046
5.	Linear Density or, Density	Pounds per cubic feet	Kilogams per cubic meter	16.0185
		Pounds per cubic feet	Grams per litre	16.0185
		1 kilogram per cubic metre	Pound per cubic feet	0.0624
6.	Speed or Velocity	Feet per second	Metres per second	0.3042
		Miles per hour	Kilometres per hour	1.609
		Metres per hour	Metres per second	0.447
		Kilometers per hour	Centimetres per second	27.778

(Contd...)

7.	Moment of Inertia	in^4	cm^4	41.623
	Modulus Section	in^3	cm^3	16.387
	Radius of Gyration	in^2	cm^2	6.452
8.	Rate of Flow or Discharge	Cubic feet per second	Gallons per minute	374.4
		Cubic feet per second	Millions of gallon/day	0.538
		Cubic feet per second	Litres per second	28.32
		Cubic feet per second	Cubic metres per second	0.083
		Gallons per minute	Litres per second	0.0757
		Million gallons per day	Litres per second	52.00
		Million gallons per day	Cubic meters per sec	0.052
9.	Pressure and Head	Pounds per sq. inch	Kilograms/sq. centimetres	0.0701
		Pounds per sq. ft.	Kilograms/sq. metre	4.882
		Tons per sq. inch	Kg. per sq. millimetre	1.575
		kg. per sq. cm	Pounds per sq. inch	14.22
10.	Force	Pound force (lbf)	Newton	4.4482
		Pound force	Kilogram force (kgf)	0.4536
		Newton	Kilogram force (kgf)	0.10197
		Kilogram force (kgf)	Pound force (lbf)	2.2046
11.	Energy or Power	Foot-pounds	Kilograms-metres	0.1382
		Foot-pound/second	Kilowatts	0.00135
		Horse-power (550 ft, lb./sec)	Metric horse-power (75 kg/sec)	1.014
		Horse-power	Kilowatts	0.746
		Metric horse power	Kilowatts	0.736
12.	Temperature	Degree F	Degree C (°F-32) × 5/9	
13.	Viscosity	Foot units !b.sec/sq. ft. sec.	Poise-dynes sec. sq.cm	479
14.	Kinematic viscosity	Foot unit sq. ft.	Stokes sq.cm/sec	929
15.	Chezy's C	In F.P.S.	In C.G.S.	0.5525
16.	Illumination	Foot-candle	Lux (ix)	10.764
		Foot-lambert	Lambert	0.001076
17.	Young's modulus	Pounds/sq.inch	kg./sq.cm	0.07031
		E for steel(30×10^6 lb./in^2)	(2.1×10^6 Kg/cm^2)	
		E for concrete(2×10^6 lb/in^2)	(1.4×10^2 kg/cm^2)	
18.	Dosage	Parts per million	Milligram/litre	1.0
		Grains per gallon	Milligram/litre	14.25
		Pounds/million gallons	Milligram/litre	0.10
19.	Surfaceloading	Gallons/sq. ft./day	Litre/sq.metre/day	48.80
20.	Weirloading	Gallons/feet/day	Litre/metre/day	14.50

APPENDIX-II

List of I.S. Codes for Water Supply and Sanitary Engineering

IS : 4111(PartI)-1967	:	Part I Manholes
IS : 4111 (Part II)-1967	:	Part II Flushing tanks
IS : 4111 (Part III)-1967	:	Part III Inverted suphon
IS : 4111 (Part IV)-1968	:	Part IV Pumping stations and pumping mains (rising mains)
IS : 1172 -1971	:	Basic requirements for water supply drainage and sanitation (second revision)
IS : 1742-1972	:	Building drainage, code of pratice for (first revision)
IS : 6924-1973	:	Constuction of refuse chutes in multi-storeyed building, code of practice for
IS : 2470 (Part I)-1968	:	Design and construction of septic tanks code of practice for Part I small installation (first revision)
IS : 2470 (Part II) 1971	:	Part II Large installations (first revision)
IS : 6530-1972	:	Laying of asbestos cement pressure pipes, codes of practice for
IS : 3114-1965	:	Laying of cast iron pipes, code of practice for
IS : 782-1959	:	Laying of concrete pipes, code of practice for
IS : 4127-1967	:	Laying of glazed stoneware pipes code of practice for
IS : 5822-1970	:	Laying of welded steel pipes for water supply, code of practice for
IS : 5329-1969	:	Sanitary pipe work above ground for building
IS : 2401-1973	:	Selection, installation and maintenance of domestic water meters, code of practice for (First revision)
IS : 2064-1973	:	Selection, installations and maintenance of sanitary appliances, code of practice (first revision)
IS : 2685-1971	:	Selection, installation and maintenance of sluice valves, code of practice for (first revision)
IS : 5611-1970	:	Waste stabilization ponds facultative type, code of practice for

IS : 6295 -1971	:	Water supply drainage in high altitudes and/ or sub-zero temperature regions
IS : 2065-1963	:	Water supply in buildings, code of practice for
IS : 5434-1969	:	Alloy bottle traps, non-ferrous for marine use
IS : 1703-1968	:	Ball valves (horizontal plunger type) including floats for water supply perposes (first revision)
IS : 781-1967	:	Bib taps and stop taps for water services, sand cast brass screw-down (first revision)
IS : 18:775-1970	:	Brackets and supports, cast iron for wash basins and skims (second revision)
IS : 5219 (Part)-1969	:	Cast copper alloys tarps Partl 'P' and 'S' traps.
IS : 712-1966	:	Caulking lead (second revision)
IS : 1700-1973	:	Drinking foundations (first revision)
IS : 782-1973	:	Enamelled cast iron sanitary appliances, general requirements for (second revision)
IS : 3489-1966	:	Enamelled steel bath tubes
IS : 2692-1964	:	Ferrules for water services
IS : 2326 -1970	:	Flushing cisterns for uninals, automatic (first revision)
IS : 774-1971	:	Flushing cisterns for water closets and urinals, (valveless siphonic type) third revision
IS : 6249-1971	:	Flush valves and fittings for marine use
IS : 4038-1967	:	Foot valves for water works purposes
IS : 6411-1972	:	Gel-coated glass fibre reinforced polyster resin bath tubs
IS : 771-1963	:	Glazed earthenware, sanitary appliance (revised)
IS : 5061-1970	:	Gratings, cast iron, for drainage purposes
IS : 778-1971	:	Gunmetal gate, globe and check valves for general purposes (second revision)
IS : 1726-1967	:	Manhole covers and frames intended for use in drainage works, cast iron (first revision)
IS : 1701-1960	:	Mixing valves for ablutionary and domestic purposes
IS : 1795-1961	:	Pillar taps
IS : 5869-1970	:	Pillar taps for marine use
IS : 2548-1967	:	Plastic water-closed seats for covers (second revision)
IS : 3004-1964	:	Plug cocks for water supply purposes
IS : 776-1962	:	Seats and covers, wooden water closet (revised)

IS : 1711-1970	:	Self-closing taps (first revision)
IS : 6251-1971	:	Shower and shower fittings for marine use
IS : 3042-1965	:	Single faced sluice gates (200 to 1200 mm size)
IS : 2906-1969	:	Sluice valves for water works purposes (350 to 1200 mm size) first revision)
IS : 780-1969	:	Sluich valves ipr water works purposes (50 to 300 mm size) (fourth revision)
IS : 5455-1969	:	Steps for manholes, cast iron
IS : 3950-1966	:	Surface boxes for sluice valves
IS : 5312-(Part I)-1960	:	Swing check type reflux (non-return valves) Part I single door pattern
IS : 2556-(Part I)-1967	:	Vitreous sanitary appliances (vitreous chine) Part I General requirements (first revision)
IS : 2556-(Part II)-1973	:	Part II specific requirements of wash-down water closets (second revision)
IS : 2556-(Part III)-1973	:	Part III specific requirements of squatting pans (second revision)
IS : 2556-(Part IV)-1973	:	Part IV specific requirements of wash basins (second revision)
IS : 2556-(Part V)-1967	:	Part V specific requirements of laboratory sinks (first revision)
IS : 2556-(Part VI)-1967	:	Part VI specific requirements of urinals (first revision)
IS : 2556-(Part VII)-1967	:	Part VII specific requirements of half round channel (first revision)
IS : 2556-(Part VIII)-1973	:	Part VIII specific requirements of siphonic wash down water closets (second revision)
IS : 2556-(Part (IX)-1967	:	Part IX specific requirements of bidets (first revision)
IS : 2556-(Part X)-1967	:	Part X specific requirements of foot rests (first revision)
IS: 2556 (Part XI)-1792	:	Part XI specific requirements for shower rose
IS:2556 (Part-XIII)-1973	:	Part XIII specific requirments of traps for squatting pans
IS : 5917-1970	:	Vitreous (Vitreous china) wash basin for marine use
IS: 4326-1967	:	Washers for water taps for cold water services
IS: 29636-1964	:	Waste fittings for wash basins and sinks non-ferrous
IS : 3311 - 1965	:	Waste plug and its accessories for sinks and wash basins

IS : 773 – 1964	:	Water closets, railway coaching stock type enamelled cast iron (second revision)
IS : 2104 – 1962	:	Water meter boxes (domestic type)
IS : 2723 – 1973	:	Water meter (bulk type) (second revision)
IS : 779 – 1968	:	Water meter (domestic type) (fourth revision)
IS : 6784 – 1973	:	Water meter (domestic type) method for testing of
IS : 9279 – 1971	:	Equipment for gift removal devices
IS : 7090 – 1973	:	Guidelines for rapid mixing devices
IS : 6280 – 1971	:	Sewage screens
IS : 936 – 1965	:	Fire hydrant, double valve type underground
IS : 908 – 1965	:	Fire hydrant stand post type
IS : 5714 – 1970	:	Hydrant stand pipe for fire fighting
IS : 4410 (Part XI)	:	Hydrology

❑❑❑

APPENDIX-III

Table 1. Minimum sanitary conveniences, required for schools (as per BIS)

S.No.	Fitments	Nursery school	Schools other than nursery schools		Boarding schools	
			For boys	For girls	For boys	For girls
1.	Water closets	1 per 15 pupils	1 per 40 pupils	1 per 25 pupils	1 per 8 pupils	1 per 6 pupils
2.	Urinals	–	1 per 20 pupils	–	1 per 25 pupils	–
3.	Wash basin	1 per 15 pupils	1 per 40 pupils	1 per 40 pupils	1 per 8 pupils	1 per 6 pupils
4.	Baths	1 bath sink per 40 pupils	–	–	1 per 8 pupils	1 per 6 pupils
5.	Drinking water fountains	1 per 50 pupils	1 per 50 pupils	1 per 50 pupils	1 per 50 pupils	1 per 50 pupils

Table 2. Minimum sanitary conveniences require for Hotels (as per BIS)

S.No.	Fitments	For residents and residential staff		For non-residential staff		Room wherein outsiders are received	
		For males	For females	For males	For females	For males	For females
1.	Water closet	1 for 8 persons	1 for 6 persons	1 for 1–15 persons 2 for 16–35 persons 3 for 36–65 person 4 for 66–100 persons	1 for 1–12 persons 2 for 13–25 persons 3 for 26–40 persons 4 for 41–57 persons 5 for 58–77 persons 6 for 78–100 persons	1 per 100 persons Up to 400 persons For over 400 persons add @ 1 for 250 persons	2 per 100 person upto 200 persons For over 200 persons add @ 1 for 100 persons
2.	Urinals	1 for 25 persons	–	Nil upto 6 persons 1 for 7–20 persons 2 for 21–45 persons 3 for 46–70 persons 4 for 71–100 persons	–	1 for 50 persons	
3.	Wash basins	1 for 8 persons	1 for 6 persons	1 for 1–15 persons 2 for 16–35 persons 3 for 36–65 persons 4 for 66–100 persons	1 for 1–12 persons 2 for 13–25 persons 3 for 26–40 persons 4 for 41–57 persons 5 for 58–77 persons 6 for 78–100 persons	1 per each WC and urinals provided	1 per each WC and urinals provided
4.	Baths	1 for 8 persons	1 for 6 persons	–	–	–	–

Table 3. Minimum sanitary convenience required for hospitals

S.No.	Fitment	For indoor patients	Outdoor patients	
			For male	For female
1.	Water closet	One for 8 bed	One to 100 persons	2 for 100 persons
2.	Wash basins	2 upto 30 beds. Add one for every additional 30 beds	1 for 100 persons	1 for 100 persons
3.	Baths	1 bath with sower for evey 8 beds		
4.	Urinals	–	1 for 50 persons	1 for 50 persons
5.	Bed pans	1 for each ward	–	–

EXPERIMENTS

1 EXPERIMENT

Object: To determine turbidity in a given water sample.

Theory: The turbidity is a measure of the interference caused by the suspended matter present in water to the passage of light. The turbidity is caused mainly by colloidal suspension of clay, silt, microscopic organisms and organic matter. The turbidity is measured on silica scale in mg/l. It means if one mg silica is dissolved in one litre distilled water then it would give one unit of turbidity. The turbidity is measured by two techniques.

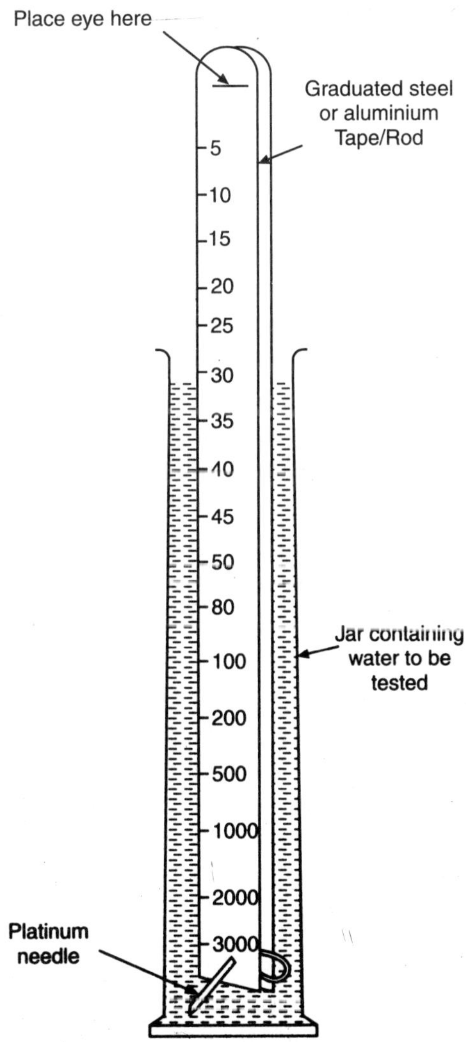

Fig. 1.1. Turbidity rod

(*i*) Based upon interference to transmission of light. The former is termed turbidimetry.

(*ii*) The instrumental method known as Nephetometry.

(A) Turbidity determination using turbidity rod.

Apparatus

Measuring cylinder, turbidity rod.

Procedure

(*i*) Pour the sample into cylinder.

(*ii*) Lower the rod vertically into water until the needle is not visible due to turbidity.

(*iii*) Read the turbidity on graduated scale.

(B) Nephetometric Method:

Apparatus

Turbidity meter flask.

Regents

(I) Stock turbidity suspension:

(*a*) **Solution I-**Dissolve one gm hydrazine sulphate $(NH_2)_2.H_2SO_4$ in distilled water and dilute to 100 ml in volumetric flask.

(*b*) **Solution II-** Dissolve 10 gms hexamethylene-tetramine, $(CH_2)_6N_4$ in distilled water and dilute to 100 ml in volumetric flask.

(*c*) In a 100 ml volumetric flask mix 5 ml solution -I and 5 ml solution II. Let stand for 24 hours at 25 ± 3°C. Dilute to mark and mix. The turbidity of this suspension is 400 NTU (Normal Turbidity Units). Prepare this solution daily for tests.

(2) Standard Turbidity suspensions: Dilute 10 ml stock turbidity suspension to 100 ml with turbidity free water. Prepare daily. The Turbidity of this suspension is defined as 40 NTU.

Procedure

(*i*) **Turbidimeter calibration**—Follow the manufacturer's operational instructions. Take standard suspension in sample cell and read the turbidity indicated on the scale if it deviates from 40 NTU then adjust it. The scale may be check in all ranges using different standard suspensions.

(*ii*) Take original sample in sample cell and measure its turbidity. If turbidity of given sample is more than 40 NTU then dilute it with one or more volumes of turbidity free water until turbidity falls between 30-40 NTU.

(*iii*) Compute turbidity of original sample from turbidity of dilute sample and dilution factor.

Observation

Turbidity of dilute sample = A NTU

Volume of sample taken for dilution = B ml

Volume of dilution water added = C ml

Calculation

Turbidity (NTU) = $A \times (B + C)/B$

Significance

Turbidity is not accepted in drinking water as well as in many manufacturing processes. The treatment given to raw water depends on amount of turbidity and use. It also affects disinfection process as well as coagulation.

(*c*) Determination of Turbidity by Jackson turbidity meter: It has already been explained in sanitary engineering

(*d*) Determination of Turbidity by Paterson turbidity meter.

Paterson turbidity meter: The Paterson turbidity meter is an optical instrument designed to enable the clarify of water or other liquids with ease and rapidly, and at the same time to ensure that such determination is accurate and comparable to recognised standard.

Briefly described, the apparatus consists of a vertical ebonite body which has been specially designed to hold one litre of water (although the quantity is not important provided the mirrors are completely covered). On the back and front faces of the body, protected by aluminium back plates, are mounted two mirrors (the reflecting surfaces of which face each other at a slight pre-determined angle). In each mirror an aperture is formed, one of these permiting the passage of a light ray from a lamp of standard strength, whilst the other allows the opposing mirror to be viewed. The ebonite head contains the mechanism by which a slider can be raised or lowered, and is removable for filling and/or cleaning purposes. This slider is actuated by the ebonite knob on the right hand side of the head.

On filling the instrument and switching on the attached electric light, multiple reflections are obtained of the original light spot. These reflections being visible as a row of light spots. Visibility is being a measure of the turbidity of the fluid under examination.

2 EXPERIMENT

Object

To determine dissolved oxygen (DO) in a given sample of water.

Theory

All living organisms depend upon oxygen in one form or another to maintain the Metabolic processes that produce energy for growth and are present in water/waste water namely, as aerobes and anaerobes. The dissolved oxygen is of concern for aerobic microorganisms and it is always desirable to maintain aerobic conditions in natural water bodies to avoid nuisance and foul odour. The oxygen is poorly soluble in water and solubility depends on temperature and partial pressure of oxygen. The solubility of oxygen in fresh water ranges from 14.6 mg per litre at 0°C to 7 mg/l at 35°C under one (1) Atmospheric pressure. The low solubility of oxygen is the major factor that limits the purification capacity of natural waters and necessitates treatment of waste to remove polutional matter before discharge to receiving streams. The DO is critical in summer, during summer months when temperature is high solubility decreases thus self-purification capacity is greatly affected.

Apparatus

DO bottles, burettes, pipette, conical flask.

Reagents

(*i*) Magnous sulphate $(MnSO_4)_2$
(*ii*) Alkali Azide/Sodium Azide
(*iii*) Concentrated sulphuric acid (H_2SO_4)
(*iv*) Starch Inidcator
(*v*) N/40 sodium thiosulphate.

Principle

The magnous sulphate reacts with KOH and thus forms brown precipitates of MnO_2. The H_2SO_4 reacts with $Mn(SO_4)_2$ which inturn react with KI and liberates free iodine. The free iodine is determined by titration with sodium thiosulphate. The NO_2 interfers and it is eliminated by adding sodium azide.

Reactions

$$MnSO_4 + KOH \rightarrow Mn(OH)_2 + K_2SO_4$$

$$Mn(OH)_2 + O_2 \rightarrow MnO_2 \downarrow + H_2O$$

Brown ppt (DO is present)

MnO_2	+	$2H_2,SO_4$	$\rightarrow$	$Mn(SO_4)_2$ (Mangamite sulphate)	+	$2H_2O$
$MnSO_4$	+	2KI	$\rightarrow$	$MnSO_4$	+	$K_2SO_4 + I_2$
$2Mn_2S_2O_3$	+	I_2	$\rightarrow$	2NaI (Sodium tetrathiomate)	+	Na_2SO_4

Interference Removal

$$2NaNO_2 + H_2SO_4 \rightarrow 2HNO_2 + Na_2SO_4$$

$$NH_3 + HNO_2 \rightarrow H_2O + N_2 + N_2O$$

Procedure

(*i*) Find out the exact capacity of DO bottle.

(*ii*) Fill the DO bottle completely with care to avoid contact of sample with air.

(*iii*) Add 1 ml manganous sulphate, 1 ml alkali azide, then insert stopper and shake well. During addition of reagents the pipette tip should be within water. It is referred as DO fixation.

(*iv*) The brown precipitates formed are allowed to settle. Now add 2 ml concentrated sulphuric acid and shake well so that precipitates are dissolved.

(*v*) Now calculate amount needed for titeration which is equivalent to 200 ml of original sample.

$$\frac{\text{Exact capacity of bottle} + 200}{\text{Exact capacity of bottle} - 2} = \text{ml sample}$$

(*vi*) Take the necessary amount in conical flask and titrate it against 0.025 N sodium thiosulphate till colour changes from blue to colourless using starch as indicator.

Result

DO in mg/l.

Significance

DO indicates freshness of water. It shows whether biological changes are brought about by aerobic or anaerobic microorganism in waste waters. The DO is very much important to maintain aerobic condition especially in natural waters to avoid nuisance conditions. It is also required to maintain aquatic life. It is significant in corrosion control of iron and steel pipes and in steam boilers.

3 EXPERIMENT

Object

To determine pH value of given sample of water.

Apparatus

Burette, pipette, conical flask, glazed tile, pH meter.

Reagents

Universal indicator, buffer solution of known pH.

Theory

pH is a term used rather, universally to express the intensities of the acid or alkaline conditions of a solution. More exactly, it is a way of expressing the hydrogen ion-concentration as a number which indicates negative logarithim of hydrogen ion concentration.

$$\text{Thus pH} = -\log H^+ \text{ or pH} = \log \frac{1}{(H)^+}$$

where H^+ is the moles per litre.

Chemistry of pH

The hydrogen electrode was found to be a suitable device for measuring H^+ concentration. It was found that pure water dissociate to yield a H^+ concentration equal to 10^{-7} mole/litre.

$$H_2O \Leftrightarrow H^1 + OH^{-1}$$

According to law of mass action

$$\frac{H^+(OH^{-1})^-}{H_2O} = K \qquad ...(1)$$

But since the concentration of water is so extremely large and is diminished so very little by the slight degree of ionization, it may be considered as constant and equation (*i*) may be written as

$(H^+) + (OH^{-1}) = Kw$ – (Ionization constant for water)

For pure water at about 25°C

$$(H^+)\ (OH^{-1}) = 10^{-14} = Kw$$

This is known as the ion product or ionization constant for water. The pH scale is usually represented as ranging from 0 – 14 with pH 7 absolute neutral. Pure water contains 0.0000001 gm of H^+ ions per litre and this is neutral solution neither acid or alkaline.

Method of pH Determination

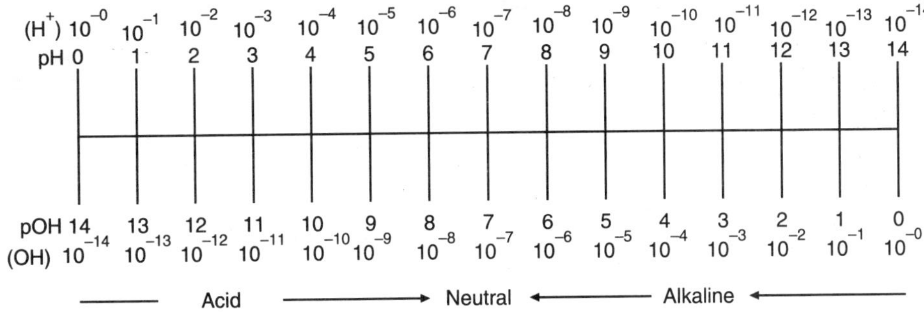

Fig. 3.1. pH scale-relationship between H, pH (OH-) and pOH in acid and alkaline-solution

Calorimetric Method: Calorimetric method depend upon the use of an indicator whose colour in solution is characteristics of pH value of this solution. A wide variety of indicators has been used for pH measurement. The indicator are salt of week acid and bases. Each indicator has a colour in ionized from different from that of natural molecules. An acid indicator will disiociant as

$$\text{H In} \Leftrightarrow \text{H}_+ + \text{In}^{-1}$$

$$\frac{\text{H}^+ + (\text{ln})^{-1}}{\text{H In}} = K$$

Taking log on both sides

$$\log \frac{(\text{ln})^{-1}}{\text{H In}} = \text{pH} - \text{PK}$$

where pH of solution is greater than –pK the indicator exists inionic form and solution assumes the colour of the ion, when pH is close to pK, mixed colour are provided at pH = pK the natural colour of indicator is provided. Some commonly used indicator are following:

Indicator	Useful pH range
Methyl orange	3.1 to 4.6
Phenol red	6.8 to 8.4
Phenophthalene	8.2 to 9.8
Universal indicator	4 to 11.0

This method is not suitable for coloured water.

Comments

The pH value of drinking water should be 6.5 to 8.5. The given water has pH value 7.2. Hence given water is good for drinking purpose.

❑❑❑

4 EXPERIMENT

Object

To determine the dose of coagulant or to perform jar test for coagulation.

Theory

As the water vary widely in quality. The optimum dosage is determined in practice by trial. Coagulants are used in nearly all treatment plants, in order to:

1. Remove natural suspended colloidal material.
2. Remove oils which do not settle in plain sedimentation.
3. Assist infiltration by farming mats on the same bed. Alum is the most widely used coagulant, commercial alum contains combination of $Al_2(SO_4)_3$,$18H_2O$ and $Al_2(SO_4)_3$, $9H_2O$, Alum reacts with natural alkalinity and produces a precipitation of Aluminium Hydroxide.

   $Al_2SO_4)_3 \cdot 18H_2O + 3\ Ca\ (HCO_3) \rightarrow Al\ (OH)_3 \downarrow + 2CaSO_4 + 18H_2O + 6CO_2\uparrow$

The precipitate grown by agglomeration of suspended impurities becomes heavier and settledown.

Reagents

1% alum solution 1 gm alum dissolved in 100 ml of distilled water.

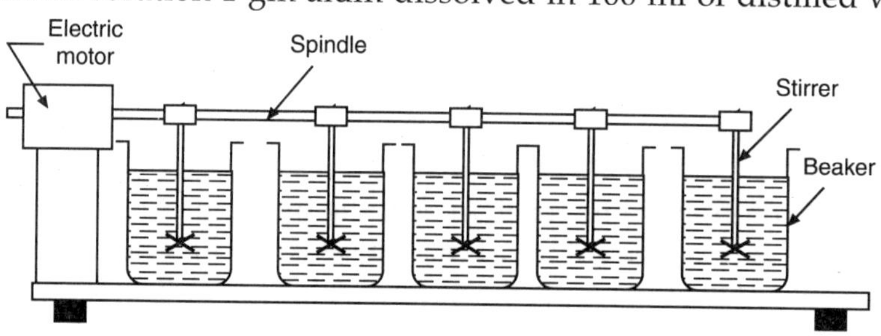

Fig. 4.1. Multiple stirrer procedure

(*i*) Measure 1.0 litre of sample in each beaker.

(*ii*) Place the beakers in multiple stirrer.

(*iii*) Add varying dosage of alum in increasing order, *i.e.*, 1 ml, 1.5 ml, 2.00 ml, 2.5 ml, 3.0 ml, 3.5 ml corresponding with 10 mg, 15 mg, 20 mg, 25 mg, 30 mg and 35 mg to the beaker.

(*iv*) Mix. the sample at highest speed for 2.0 minutes.

(*v*) Reduce the speed to 2 rpm and flocculate for 20 minutes.

(*vi*) Allow the floc to settle down for 30 minutes minimum.

(*vii*) Determine the turbidity of the sample without disturbing the sediments.

(*viii*) Note the dose (economical) which gives maximum turbidity removal.

5 EXPERIMENT

Object

The determine 5 day BOD at 20°C of given waste water sample.

Theory

The Biochemical oxygen demand (BOD) is usually defined as the amount of the oxygen required by bacteria while stability decomorable organic matter under aerobic conditions. The BOD parameter indicates pollutional strength of domestic and industrial waste in terms of oxygen that they will require if discharged into natural water courses in which aerobic conditions exist. The parameter is one of most important in stream pollution control activities. The BOD reaction follows first order kinetic reaction. It depends on temperature and amount of organic matter. The BOD at any time t is given by relation.

$$Lt = Le^{-Kt}$$

The BOD exerted at any time 't' is given by

$$y = L(1 - e^{-Kt})$$

where Lt = BOD remaining at time t

K = Reaction rate constant

t = time

y = BOD exerted in t days.

It has been observed that BOD exertion takes place in two steps.

(*a*) Carbonaceous demand (*b*) Nitrofication. The large portion of total BOD is exerted in 5 days at 20°C. The standard conditions for India is BOD at 37°C for 5 days.

Apparatus

BOD bottles, Burette, pipette and indicator.

Reagent

1. Magnesium, sulphate solution
2. Calcium chloride solution
3. Ferric chloride solution
4. Phosphate buffer solution
5. Mangnous sulphate solution
6. Alkali azide solution
7. Conc. sulphuric acid
8. Starch indicator
9. 0.025 N sodium thiosulphate solution

Principle

The BOD test is based upon determination of dissolved oxygen. If sample has low BOD i.e., below 7 mg/l it is directly used for BOD determination. If waste has high BOD then it is diluted using dilution water. The dilution water is prepared by distilled or demineralized water. It contains various salts to give buffering capacity and Nutrient requirement of Microorganisms. It has also read with domestic waste to provide adequate population of mixed organisms.

Procedures

(A) Preparation of dilution water

1. Placed desired volume of distilled water in a bottle and saturate the water by shaking the partially filled bottle or by aeration.
2. Add 1 ml each of phosphate buffer, magnesium sulphate, calcium chloride and ferric chloride per litre of water.
3. Add 2 ml per litre of domestic settled sewage.

(B) Dilution of waste water: The waste water is required to be diluted using dilution water depending upon type of waste. A general guidelines is given below:

Type of waste	Dilution mixture
Strong trade mix	0.1–1
Raw settled sewage	1-5
Oxidized effluent	5-25
Polluted river water	25-100

(C)

(*i*) Take 3 BOD bottles and determine their exact capacity.

(*ii*) Fill up two bottles using dilution water and mixture of waste water.

(*iii*) Incubate at 20°C for 5 days.

(*iv*) Find out DO in both bottles.

(*v*) Find of undiluted sewage sample.

Significance

The BOD parameter have wide application in environmental engineering practice. It includes strength of waste in terms of oxygen required for stabilization. It indicates rate at which oxidation will occur in receiving bodies of water. Therefore it is very important parameter for storm pollution control. It is used to check the ineffluent quality and to deside textent of treatment necessary before discharging waste.

6 EXPERIMENT

Object

To determine chlorine dose of a given water sample.

Theory

The chlorine demand of water is the difference between the amount of residual chlorine remaining at the end of the contact period. The chlorine demand is different with different water and even with a given water will vary with amount of chlorine applied desired residual time of contact pH and temperature. The chlorine combines with water to form hypochlorious acid.

$$Cl_2 + H_2O = HOCl + H^+ + Cl^-$$

The HOCl is very weak acid and it dissociates to form hypochlorites ions which has less disinfection capacities.

$$HOCl \rightarrow H^+ + OCl^-$$

The chlorine hypochloreous acid and hypochloriterous are referred as free residual chlorine. The chlorine reacts with ammonia, if it present in water and form chloromines. It is refered as combined residual chlorine.

The chlorine demand can determined by treating as free residual chlorine of sample with known but varying dosages of chlorine. After desired contact period total residual chlorine is determined by OTA method. A graph can be plotted between total residual chlorine and dose applied. And dose required corresponding to desired free residual chlorine can be determined.

Apparatus

Beakers, comparator cells, pipette etc.

Reagents

(*i*) Chlorine water 0.5 gm/litr

(*ii*) Sodium Arsenite (SA)

(*iii*) Orehotaliding reagent

Procedure

(*i*) Take 200 ml sample in each of sample bottle.

(*ii*) Add 0.2, 0.5, 0.8, 1.00, 2.00, 3.00 ml or chlorine water which corresponding to 0.5, 0.9, 1.00, 1.5, 2.00 mg/litr. chlorine dose.

(*iii*) Allow a contact time 30 minutes.

(*iv*) Lable the comparison cell as A, B, C.

(*v*) Add 10 ml sample and 0.5 ml OT solution to cell *A* and compare the colour after 5 minutes. It will give total residual chlorine. The interference due to NO_2. Mn and Fe. Let residual Cl_2, *i.e.*, K_3

(*vi*) Repeat step No. (*v*) for each sample.

(*vii*) Add 5 ml SA solution and 10 ml sample in each cell B mix quickly and 0.5 ml to 0.7 ml solution.

(*viii*) After 5 minutes compare the colour with standard and determined residual value chlorine which is equivalent to interfering substance. Let it give residual chlorine as k_1.

(*ix*) Calculate the total chlorine *i.e.*, $k_3 - k_1$.

(*x*) Plot a graph between residual chlorine as ordinate and chlorine dose applied.

Significance

The chlorination is essential in water purification to distroy pathogens, determine the chlorine demand of water or waste water is an important consideration in design.

7 EXPERIMENT

Objects:

To determine temporary and permanent hardness in a given sample of water.

Theory:

Hardness is caused by divalent metallic cations such as Ca^{++}, Mg^{++}, Fe^{++}, Sr^{++}, Mn^{++}, etc. The important anions associated are HCO_3^-, Sou^-, Cl^-, NO_3^- etc. The hardness in water is derived largely from contact with the soil and rock formations. The rainwater percolating through soil contains considerable amount of CO_2 and under low pH condition that develop, basic materials, particularly lime stone formations are dissolved. Since lime stone is not pure carbonate but includes impurities such as sulfates, chlorides, silicates etc., these materials too dissolved in water. The hardness is classified in two ways:

1. With respect to metallic ion as calcium, magnesium hardness.
2. With respect to anions associated as carbonate of temporary hardness and non-carbonate or permanent hardness. The temporary hardness is caused by HCO_3^- and permanent hardness is caused by other anions.

Apparatus

Conical flask, burette, pipette and Hot bath.

Reagent

1. Eriochrome black 'T' indicator.
2. EDTA (Ethylene diaminetetraacitic acid) Salt—solution
3. Ammonia buffer.

Procedure

1. Take 100 ml of sample in conically flask. If sample is turbid, filter and proceed. Add 1 ml ammonia buffer solution and 3 drops of ariochrome black T indicator.
2. Now titrate it against EDTA solution till colour changes from urine red to blue. Record the ml of EDTA used.
3. Take 100 ml sample in another flask. Boil it for 15 minutes, cool it and filter.

Observation

Sample volume ml.	EDTA			
	Intial burette reading	Final burette reading	Volume used ml.	Hardness mg/l. as $CaCO_3$
Unboiled	–	–	–	–
Boiled	31.8	–	–	–

Calculation

Total hardness mg/l as $CaCO_3$

$$= \frac{\text{ml. of EDTA used} \times 1000 \text{ ml (unboiled sample)}}{\text{ml. of sample}}$$

Permanent hardness mg/l as $CaCO_3$

$$= \frac{\text{ml. of EDTA used} \times 1000 \text{ ml (boiled sample)}}{\text{ml. of sample}}$$

Temporary hardness = Total hardness – Permanent hardness

4. Repeat the above procedure and record EDTA ml used.

Result

Total hardness = 207 mg/l as $CaCO_3$

Permanent hardness = 67 mg/l as $CaCO_3$

Temporary hardness = 140 mg/l as $CaCO_3$

Significance

Hard water is as satisfactory for human consumption as soft water. The hardness is important consideration in determining suitability of water for industrial and domestic use. It creates problems like scale formation and excessive soap consumption. It serve as a basis for control of software process.

Comment

As the total hardness 207 mg as $CaCO_3$ in the given water sample which is less than desirable limit of 300 mg/l as $CaCO_3$. Hence water is good for drinking purposes.

8 EXPERIMENT

Object

To determine chloride content in a given water sample.

Theory

Chloride occurs in all natural water in widely varying concentration. The chloride content normally increase as the mineral content increases. The upland and mountain water are low in chlorides and ocean water content very high chloride. The source of chlorides are solvent powder in water during percolation from top soil and disappear formation and human waste particularly urine which contain chlorides. The often source are industrial waste which contain chloride.

Principle

The chlorides are measured readily by Mohar method. The sample is titrated against N/35.5 silver nitrate using potassium dicromate as indicator. The following reactions takes place.

$$Ag^{+} + Cl^{-} \Leftrightarrow AgCl \text{ (white PPT)}$$

$$2Ag^{+} + Crom^{-} \Leftrightarrow Ag_2\,CrO_4$$

Radish brown PPT (End point). The pH should be maintained between 7 to 8.

Apparatus

Burrette, pipette and conical flask.

Reagents

Potassium dicromate indicator.

N/3 5 Silver nitrate

Procedure

1. Take 25 ml sample in conical flask.
2. Determine pH value of sample if pH is not between 7 to 8 adjust it by using $NaCl/H_2SO_4$.
3. Add two drops of potassium dicromate indicator and titrate with N/35.5 silver nitrate till colour changes from yellow reddish brown.
4. Record the ml of silver nitrate use.
5. Take 25 ml of distilled water in a conical flask and repeat the above observation.
6. Note the ml of silver nitrate as flask.

Observation

Result

The chlorine content is found as 20 mg/l.

Significance

In many areas loads of chlorides in water is an important consideration in the relation of supplies for human use. It is used in determining selection of dosolting apparatus. The excess of chloride have no effect in human body but it effect taste which is objectionable. It indicator pollution if there is sudden increase in its level.

Comment

The chloride content is within the permissible limit which is upto 250 mg/litr. Hence the water is good for drinking purposes.

9 EXPERIMENT

Object

To determine the most probable number of caliform bacteria in given water sample by British technique.

Theory

Water and sewage contains a variety of bacterial flora. There are some which are capable of causing diseases called pathogenic bacteria and other indicate the presence of pathogens by their presence and exclude the possibility of their presence by their absence. The MPN caliform shows the number of those bacteria which are called as bacteriological indicators. The method involves incubating those bacteria in liquid food at a suitable temperature.

Apparatus and Reagents

1. Test tube (Sterlised)
2. Darhami's tubes
3. Incubator
4. Maconkey's Broths.

Procedure:

1. Collect the sample in sterlize bottle intended for bacteriological sampling.
2. Incubate the sample in an exponential order 10.0 ml, 1.0 ml 0.1 ml in three sets of tubes each of macconkey's Broth under complete aseptic conditions.
3. Incubate for 24 ±2 hrs at 37°C ± 0.5°C
4. Note the number and order of positive tubes. Positively is determined by the change in colour from purple to yellow and accumulation of gas in Darham's tube.
5. Further incubate the tubes for 24 hrs and note positive tubes record the MPN of probability tube.
6. Record the MPN from probability table.

Results:

S.No.	Description of sample	No. of tubes inoculated	24 hrs. presumtive			48 hrs. confirmed		
			10 ml	1 ml	0.1 ml	10 ml	1 ml	0.1 ml

Discussion

In this test we detect the *E-coli,* these are aerobic gram negative non-sporing bacilli which ferment lactose (sugar) at 37°C with gas and acid formation and which can resist bile salt. They are a harmless variety of a bacteria residing in large number in the instantine of all worm blooded animals including the human being and are exerted out by them in large numbers. There presence in water therefore indicates contiamination by human excerta.

Those bacteria are chosen as indicator of pollution due the following:

1. They are excreted out in very large number by all human being eq. 2 to 3 million per capita/day the number of pathogens in sewage will be obviously less depending upon the prevalence of sickness in the town.
2. They are non-resistant external environment than the pathogens.
3. They can be more easily detected by carrying out simple laboratory tests.

EXAMINATION PAPER

WATER & WASTE WATER ENGINEERING

Time: 3 hrs. **[M.M. : 100]**

Note:

1. *Part 'A' may be attempted in first 6 pages of answer sheet.*
2. *Part 'B' in rest of the sheets of answer sheet.*
3. *Answer may be given in English or Hindi.*

PART-A

1. Attempt any 10 questions: **(10×2=20)**

1. What is M.P.N.?
2. What is meant by E-coli?
3. Permisible level of suspended solids in drinking water as per BIS is ....
4. Coagulation is done to remove .... from water.
5. Most common method of disinfection is ....
6. What are the salts responsible for permanent hardness?
7. Rate of filtration of rapid sand filter is ....
8. What is meant by sullage?
9. What is B.O.D. ?
10. Name the treatment techniques which removes organic dissolved solids from waste water.
11. What is meant by sludge?
12. Biogas is produced from Aerobic decomposition. (True/False)
13. Name the two major methods of disposal of treated sewage.
14. UV radiation is used in the water treatment for ....

2. Attempt any 5 questions: **(5×4=20)**

1. What is meant by break point chlorination?
2. What is Reverse Osmosis (RO) and when it can be done?
3. What are the chemical treatments generally adopted for sewage?
4. What is Grit chamber? What are it's uses?
5. Differentiate manhole and lamp hole.

6. Differentiate Rapid sand filter and Pressure filter.
7. Write a note on sludge digestion.
8. Differentiate BODs and Ultimate BOD.

PART-B

Attempt any 3 questions: **(3×20=60)**

3. (*a*) What are the various methods of forecasting population? Explain any one of it in detail.

(*b*) What is an Intake? What are the feature to be considered to locate an Intake?

4. (*a*) Draw the sectional plan and elevation of a rapid sand filter.

(*b*) Explain the various layout of water distribution system.

5. (*a*) What is Dry Weather Flow? Explain the procedure of design of sanitary sewer?

(*b*) Explain physical and chemical characteristics of sewage.

6. (*a*) What is skimming tank? Compare it with sedimentation tank?

(*b*) Explain the theory and working of Imhoff tank.

7. Write short notes on:

(*a*) Oxidation pond

(*b*) Sewage sickness

(*c*) Self cleansing velocity

(*d*) Various sewar appurtenances.

Index

L...

M...

N...

O...

P...

Q...

R...

S...

T...

U...

V...

W...

Y...

Win Prizes

ATTENTION READERS

Win Prizes

We request for your frank assessment regarding various aspects of the book as given below :

WATER SUPPLY & WASTE WATER ENGINEERING

By: A K Upadhyay & D. Lal

Please Fill up the Sheet in Neat and Clean Writing

(i) What topics of your syllabus, important for your examination point of view, are not covered in this book ?

....

....

....

....

(ii) What are the chapters wherein the treatment of the subject matter is not systematic or organised or updated ?

....

....

....

....

(iii) Have you come across misprints/mistakes/factual inaccuracies in the book? Please specify the chapters and the page numbers.

....

....

....

....

(iv) What other books on the same subject have you found/heard better than the present book? Please specify in terms of price and quality.

....

....

....

....

(v) Further suggestions and comments for the improvement of this book.

....

....

....

....

(PS: If need be please attach a separate sheet to write your views)

OTHER DETAILS

(i) Who recommended you this book ? (Please tick (✓) in the box near the option relevant to you.

☐ Teacher ☐ Friends ☐ Bookseller

....

....

....

....

(ii) Name of the recommending teacher, his designation and address :

....

....

....

....

(iii) Name and address of the bookseller you purchased the book from :

....

....

....

....

(iv) Name and address of your institution (Please mention the University or Board as the case may be)

....

....

....

....

(v) Your name and complete postal address :

....

....

....

....

(vi) Course you are enrolled in :

....

....

....

....

The best assessment will be awarded each month: The award will be in the form of our publications as decided by the Editorial Board.

Please mail the filled up coupon at your earliest to :

S.K. Kataria & Sons®
4885/109, Prakash Mahal, Dr. Subhash Bhargav Lane,
Opposite Delhi Medical Association, Daryaganj, New Delhi-110002
Phones: +91-11-23243489, 43551243; Mobile: +91-9871775858
e-mail: katariabook@gmail.com; katariabooks@yahoo.com
Website: www.skkatariaandsons.com